P9-EDD-204

GROSS ANATOMY
2nd edition

Board Review Series

GROSS ANATOMY
2nd edition

Kyung Won Chung, Ph.D.

Professor and Vice Chairman
Department of Anatomical Sciences
College of Medicine
University of Oklahoma Health Sciences Center
Oklahoma City, Oklahoma

**Williams
& Wilkins**

Board Review Series from Williams & Wilkins
Baltimore, Hong Kong, London, Sydney

Harwal Publishing Company, Malvern, Pennsylvania

Williams & Wilkins

Managing editor: Susan Kelly
Editorial assistant: Joan Coper
Production coordinator: Laurie Forsyth

Figures 2.8, 2.11, and 3.8 have been reproduced with permission from Anderson, JE: *Grant's Atlas of Anatomy*, 8th edition, Baltimore, Williams & Wilkins, 1983.

Library of Congress Cataloging-in-Publication Data

Chung, Kyung Won.
 Gross anatomy/Kyung Won Chung.—2nd ed.
   p.   cm.—(Board review series)
 Includes bibliographical references and index.
 ISBN 0-683-01565-6
 1. Human anatomy—Outlines, syllabi, etc. 2. Human anatomy—
Examinations, questions, etc. I. Title. II. Series.
 [DNLM: 1. Anatomy—examination questions. 2. Anatomy—programmed
instruction. QS 18 C559g]
 QM31.C54R6413 1991
 611—dc20
 DNLM/DLC
 for Library of Congress   91-15340
                                                                 CIP

10 9 8 7 6 5

Dedication

To My Wife, Young Hee,

and

My Sons, Harold and John

Contents

Preface to the Second Edition

This concise review of human anatomy is designed for medical and dental students, and it is intended primarily to help these students prepare for the United States Medical Licensing Examination Step 1 (USMLE) exam as well as other examinations. It presents the essentials of human anatomy in the form of condensed descriptions and simple illustrations. The text is tightly outlined with related Board-type questions following each section. I have attempted to include all Board-relevant information without introducing a vast amount of material or entangling students in a web of details. However, because this book is in summary form, students are encouraged to consult a standard textbook for a comprehensive study of difficult concepts and fundamentals.

Organization

As with the previous edition, the second edition begins with a brief introduction to the skeletal, muscular, nervous, and circulatory systems. Following are chapters on regional anatomy, which include the upper limb, lower limb, thorax, abdomen, perineum and pelvis, back, and head and neck.

I believe that anatomy is a visual science of the configuration of the body, and thus the success of learning and understanding it depends largely on the quality of dissection and the illustrations of the human structure. Many of the illustrations are simple schematic drawings used to enhance the student's understanding of the descriptive text. A few of the illustrations are more complex, attempting to exhibit important anatomical relationships. The considerable number of tables of muscles will prove particularly useful as a summary and review. In addition, several summary charts for muscle innervation and action, cranial nerves, autonomic ganglia, and foramina of the skull are included in order to highlight and summarize pertinent aspects of the system.

Test questions at the end of each chapter are designed to emphasize important information and hence lead to a better understanding of the material. These questions also serve as a self-evaluation to help the student uncover areas of weakness. Answers and explanations are provided after the questions.

Features of the new edition

• Questions have been revised to reflect the guidelines set forth beginning in 1991 by the National Board of Medical Examiners. A few K-type questions have been retained because of their educational value.

• A Comprehensive Examination has been added. It serves as a practice exam and self-assessment tool to help the student diagnose weaknesses prior to beginning a review of the subject. It also serves as a self-examination upon completion of the review book prior to examination.

• Illustrations have been improved and more have been added.

• Roentgenograms and computer tomograms are used in the test questions to aid in the study of anatomical structures and their relationships.

• The practical application of anatomical knowledge is included throughout the book as clinical considerations.

Kyung Won Chung

Acknowledgments

I wish to express my sincere thanks to many medical students, colleagues, and friends who have made valuable suggestions regarding the preparation of this new edition. My appreciation is extended to Ms. Diane Abeloff, Medical Illustrator, for her book, *Medical Art, Graphics for Use,* which was used for illustrations with a little modification in some cases; Shawn C. Schlinke, M.D., and Ms. Laura Barton for their excellent illustrations and full cooperation; and Mr. Ben Han for his quality photographic services. I am deeply indebted to Mr. Alexander Kugushev for his critical advice and constructive suggestions for improvement of the original text; and to Ms. Yvonne Strong and Ms. Jung W. Yoon for their copyediting during the preparation of this manuscript. Finally, I greatly appreciate and enjoy the privilege of working with Mr. John Gardner, Vice President and Editor-in-Chief of Williams & Wilkins Company, the talented and dedicated Managing Editor, Ms. Susan Kelly, and other members of Harwal Publishing Company, a division of Williams & Wilkins Company, for their constant guidance, enthusiasm, and unfailing support throughout the production of this book.

1

Introduction

Skeleton and Joints

I. Bones

—are calcified connective tissue consisting of cells (**osteocytes**) in a matrix of ground substance and collagen fibers.

—act as levers on which muscles act to produce the movements permitted by joints.

—serve as a reservoir for calcium and phosphorus.

—contain internal soft tissue, the **marrow,** where blood cells are formed.

—are classified, according to shape, into long, short, flat, irregular, and sesamoid bones; they also are classified according to their developmental history (i.e., endochondral and membranous bones).

A. Long Bones

—are longer than they are wide; they include the clavicle, humerus, radius, ulna, femur, tibia, fibula, metacarpals, and phalanges.

—develop by replacement of hyaline cartilage plate.

—have a shaft (**diaphysis**) and two ends (**epiphyses**).

1. Diaphysis

—is the central region and is composed of a thick collar of compact bone surrounded by the periosteum.

—contains the marrow cavity and the metaphysis, which is the more recently developed part adjacent to an epiphyseal disk.

2. Epiphyses

—are composed of a trabecular bony meshwork surrounded by a thin layer of compact bone.

—have articular surfaces that are covered by hyaline cartilage.

B. Short Bones

—are found only in the wrist and ankle and are approximately cuboidal-shaped.

—are composed of spongy bone and marrow surrounded by a thin outer layer of dense compact bone.

C. Flat Bones

—include the ribs, sternum, scapulae, and bones in the vault of the skull.

—consist of two layers of compact bone separated by spongy bone and marrow space (diploë).

1

—have articular surfaces that are covered with cartilage of fibrous tissues.
—grow by replacement of connective tissue.

D. Irregular Bones

—include bones of mixed shapes, such as bones of the skull, vertebrae, and coxal bones.
—contain mostly spongy bone enveloped by a thin outer layer of dense compact bone.

E. Sesamoid Bones

—develop in certain tendons and serve to reduce friction on the tendon, thus protecting it from excessive wear.
—are commonly found where tendons cross the ends of long bones in the limbs and the wrist.

II. Joints

—are places of union between two or more bones of the skeleton.
—are innervated as follows: The nerve supplying a joint also supplies the muscles that move the joint and the skin covering the insertion of such muscles (Hilton's law).
—are classified on the basis of their structural features into **fibrous, cartilaginous,** and **synovial** types.

A. Fibrous Joints (Synarthroses)

1. Sutures

—are connected by fibrous connective tissue.
—are found between the flat bones of the skull.

2. Syndesmoses

—are connected by fibrous connective tissue.
—occur as the inferior tibiofibular and tympanostapedial syndesmoses.

B. Cartilaginous Joints

1. Synchondroses (Primary Cartilaginous Joints)

—are united by hyaline cartilage.
—permit no movement but growth in the length of the bone.
—include epiphyseal cartilage plates (the union between the epiphysis and the diaphysis of a growing bone) and spheno-occipital and manubriosternal synchondroses.

2. Symphyses (Secondary Cartilaginous Joints)

—are joined by a plate of fibrocartilage and are slightly movable joints.
—include the pubic symphysis (symphysis means "grown together") and the intervertebral disks (unions between the bodies of the vertebrae).

C. Synovial Joints (Diarthrodial Joints)

—permit a great degree of free movement.
—are characterized by four distinguishing features, including a joint cavity, an articular cartilage, a synovial membrane (which produces synovial fluid), and an articular capsule.
—are classified according to axes of movement into plane, hinge, pivot, ellipsoidal, saddle, and ball-and-socket joints.

1. **Plane Joints**
   —are limited in movement by the articular capsule.
   —allow simple gliding or sliding movement between two flat surfaces.
   —occur in the proximal tibiofibular, intercarpal, intermetacarpal, carpo-metacarpal, and acromioclavicular joints.

2. **Hinge (Ginglymus) Joints**
   —allow movement around one axis (uniaxial) at right angles to the bones (have 1° of freedom).
   —allow movements of flexion and extension only.
   —occur in the elbow, knee, ankle, and interphalangeal joints.

3. **Pivot (Trochoid) Joints**
   —allow movement around one axis (uniaxial), which is a longitudinal axis. A central bony pivot rotates within a bony ring.
   —allow rotation only (have 1° of freedom).
   —occur in the superior and inferior radioulnar joints and in the joint between the first and second cervical vertebrae.

4. **Ellipsoidal (Condyloid) Joints**
   —allow movement in two directions (biaxial) at right angles to each other (have 2° of freedom).
   —allow flexion/extension and abduction/adduction but no rotation.
   —occur in the wrist (radiocarpal) and metacarpophalangeal joints.

5. **Saddle (Sellar) Joints**
   —allow movement around two horizontal axes (biaxial) at right angles to each other (have 2° of freedom).
   —allow movement in several directions but allow less free movements of flexion/extension, abduction/adduction, and rotation than do the ball-and-socket type joints.
   —occur in the carpometacarpal joint of the thumb.

6. **Ball-and-Socket (Spheroidal) Joints**
   —allow movement in many directions (multiaxial; have 3° of freedom, the greatest freedom of motion).
   —allow flexion and extension, abduction and adduction, medial and lateral rotations, and circumduction.
   —occur in the shoulder and hip joints.

Muscular System

I. Muscle
—consists predominantly of contractile cells.
—produces the movements of various parts of the body by contraction.
—has three types: **skeletal, cardiac,** and **smooth** muscles.

A. Skeletal Muscle
—is under voluntary control and makes up about 40% of the total body mass.
—has two attachments, an origin and an insertion. The origin is usually the more fixed and proximal attachment, the insertion the more movable and distal attachment.

—is enclosed by **epimysium**, a thin layer of connective tissue. Smaller bundles of muscle fibers are surrounded by **perimysium**. Each muscle fiber is enclosed by **endomysium**.

B. Cardiac Muscle

—is known as **myocardium** and forms the middle layer of the heart.

—is innervated by the autonomic nervous system but contracts spontaneously without any nerve supply.

—responds to increased demands by increasing the size of its fiber; this is known as compensatory hypertrophy.

C. Smooth Muscle

—is generally arranged in two layers, circular and longitudinal, in the walls of many visceral organs.

—is innervated by the autonomic nervous system, regulating the size of the lumen of a tubular structure.

—undergoes rhythmic contractions called **peristaltic waves** in the walls of the gastrointestinal tract, uterine tubes, ureters, and other organs.

II. Structures Associated with Muscles

A. Tendons

—are fibrous bands of dense connective tissue that always have one end attached to muscle and the other end blending with the fibrous connective tissue of the structure to which they attach (usually bone).

—are supplied by sensory fibers extending from muscle nerves.

B. Bursa

—is a flattened sac of synovial membrane containing a viscid fluid to moisten its wall in order to facilitate movement by minimizing friction.

—is found where a tendon rubs against a bone, ligament, or other tendon.

—is prone to fill with fluid when infected or injured.

C. Synovial Tendon Sheaths

—are tubular sacs wrapped around the tendons; they are similar to bursae in their fundamental structure and are filled with synovial fluid.

—occur where tendons pass under ligaments, retinacula, and through osseofibrous tunnels, thus facilitating movement by reducing friction.

—its lining, like synovial membrane, responds to infection by forming more fluid and by proliferating more cells, causing adhesions and thus restriction of movement of the tendon.

D. Aponeurosis

—is a flat fibrous sheet or expanded broad tendon that attaches to muscles and serves as the means of origin or insertion of a flat muscle.

E. Ligament

—is a fibrous band or sheet connecting bones or cartilages, or a fold of peritoneum serving to support and strengthen joints, muscles, and visceral structures.

F. Fascia

—is a fibrous sheet that envelops the body under the skin and invests the muscles.

—may limit the spread of pus and extravasated fluids such as urine and blood.

1. Superficial Fascia
—is a loose connective tissue between the dermis and the deep (investing) fascia.
—contains fat, cutaneous vessels, and nerves.
—has a fatty superficial layer and a membranous deep layer.

2. Deep Fascia
—is a sheet of fibrous tissue that invests the muscles and helps to support them by serving as an elastic sheath or stocking, providing origins and insertions for muscles and forming retinacula and fibrous sheaths for tendons.
—forms potential pathways for infection or extravasation of fluids.
—has no sharp distinction from epimysium.

Nervous System

I. Nervous System
—is divided anatomically into the central nervous system (CNS), consisting of the brain and spinal cord, and the peripheral nervous system (PNS), consisting of 12 pairs of cranial nerves and 31 pairs of spinal nerves.
—is divided functionally into the somatic nervous system, which controls primarily voluntary activities, and the visceral (autonomic) nervous system, which controls primarily involuntary activities.
—is composed of nerve cells (**neurons**), which typically have two types of processes (**dendrites** and **axons**), and **neuroglia,** which are of three types (**astrocytes, oligodendrocytes,** and **microglia**).
—controls and integrates the activity of various parts of the body.

II. Neurons
—are the structural and functional units of the nervous system (neuron doctrine).
—consist of cell bodies and their processes (dendrites and axons).
—their **dendrites** (dendron means "tree") are usually short and highly branched and carry impulses to the cell body.
—their **axons** are usually single and long, have fewer branches (collaterals), and carry impulses away from the cell body.
—are unipolar, bipolar, or multipolar in shape.
—are specialized for the reception, integration, transformation, and transmission of information.

A. Classification

1. Unipolar Neurons
—have only one process, which divides into a central branch that functions as an axon, and a peripheral branch that serves as a dendrite.
—are sensory neurons of the peripheral nerve (i.e., cerebrospinal ganglion cells).

2. Bipolar Neurons
—have two processes, one dendrite and one axon, and are found in the olfactory epithelium, in the retina of the eye, and in the inner ear.

3. Multipolar Neurons

—have several dendrites and one axon and are most common in the CNS (i.e., motor cells in anterior and lateral horns of the spinal cord, autonomic ganglion cells).

B. Cells That Support Neurons

—include Schwann cells and satellite cells in the peripheral nervous system.
—are called neuroglia in the CNS and are composed mainly of three types: oligodendrocytes, astrocytes, and microglia.

C. Myelin

—is the fat-like substance forming a sheath around certain nerve fibers.
—is formed by Schwann cells in the peripheral nervous system and oligodendrocytes in the CNS.

III. Central Nervous System

A. Brain

—is enclosed within the cranium, or brain case.
—has a cortex, which is the outer part of the cerebral hemispheres, and is composed of **gray matter,** which consists largely of the nerve cell bodies.
—has an interior part composed of **white matter,** which consists largely of axons forming tracts or pathways, and ventricles, which are filled with cerebrospinal fluid.

B. Spinal Cord

—is cylindrical, occupies approximately the upper two-thirds of the vertebral canal, and is enveloped by the meninges.
—has cervical and lumbar enlargements for the nerve supply of upper and lower limbs, respectively.
—has centrally located **gray matter,** in contrast to the cerebral hemispheres, and peripherally located **white matter.**
—has a conical end known as the **conus medullaris.**
—grows more slowly than the vertebral column during fetal development, and hence its terminal end gradually shifts to a higher level.
—ends at the level of L2 (or between L1 and L2) in the adult and at the level of L3 in the newborn.

C. Meninges (see p 283)

—consist of three layers of connective tissue membranes (**pia, arachnoid,** and **dura mater**) that surround and protect the brain and spinal cord.
—contain the subarachnoid space, which is the interval between the arachnoid and pia mater, filled with cerebrospinal fluid.

IV. Peripheral Nervous System

A. Cranial Nerves

—consist of 12 pairs, which, along with 31 pairs of spinal nerves, make up most of the peripheral nervous system.
—are connected to the brain rather than to the spinal cord.
—have motor fibers with cell bodies located within the CNS, and sensory fibers with cell bodies that form sensory ganglia located in the nerve outside the CNS.
—emerge from the ventral aspect of the brain (except for the trochlear nerve [fourth cranial nerve]).

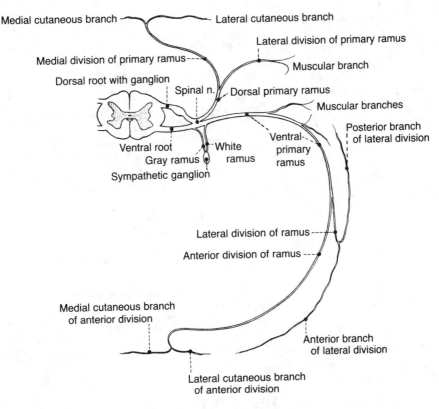

Figure 1.1. Typical spinal nerve.

—contain all four functional components of the spinal nerves (GSA, GSE, GVA, GVE) and an additional three components (SSA, SVA, SVE). [See sections B and C below and Chapter 8.]

B. Spinal Nerves (Figure 1.1)
—have 31 pairs: 8 cervical, 12 thoracic, 5 lumbar, 5 sacral, and 1 coccygeal.
—are formed from dorsal and ventral roots, and each dorsal root has a ganglion that is within the intervertebral foramen.
—are connected with the sympathetic chain ganglia by rami communicantes.
—contain sensory fibers with cell bodies in the dorsal root ganglion (GSA and GVA); motor fibers with cell bodies in the anterior horn of the spinal cord (GSE); and motor fibers with cell bodies in the lateral horn of the spinal cord (only segments between T1 and L2) [GVE].
—are divided into the ventral and dorsal primary rami. The ventral primary rami enter into the formation of plexuses (i.e., cervical, brachial, and lumbosacral), whereas the dorsal primary rami innervate the skin and deep muscles of the back.

C. Nerve Components in Peripheral Nerves (Figures 1.2 and 1.3)

1. General Somatic Afferent (GSA) Nerves
—transmit pain, temperature, touch, and proprioception from the body to the CNS.

2. General Somatic Efferent (GSE) Nerves
—carry motor impulses to skeletal muscles of the body.

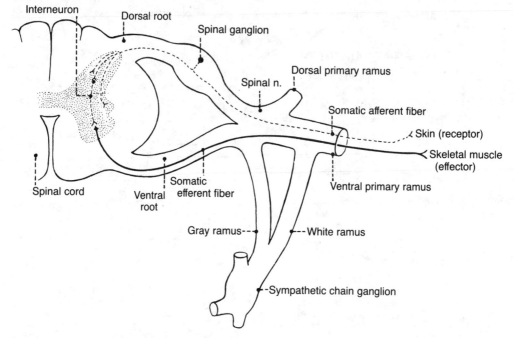

Figure 1.2. General somatic afferent and efferent nerves.

3. **General Visceral Afferent (GVA) Nerves**

   —convey sensory impulses from the visceral organs to the CNS.

4. **General Visceral Efferent (GVE) Nerves (Autonomic Nerves)**

   —transmit motor impulses to smooth muscle, cardiac muscle, and glandular tissues.

5. **Special Somatic Afferent (SSA) Nerves**

   —convey special sensory impulses of vision, hearing, and equilibration to the CNS.

6. **Special Visceral Afferent (SVA) Nerves**

   —transmit smell and taste sensations to the CNS.

7. **Special Visceral Efferent (SVE) Nerves**

   —conduct motor impulses to the muscles of the head and neck; they arise from branchiomeric structures, such as muscles for mastication, muscles for facial expression, and muscles for elevation of the pharynx and movement of the larynx.

V. Autonomic Nervous System

—is divided into the **sympathetic** (thoracolumbar outflow) and **parasympathetic** (craniosacral outflow) systems and is composed of two neurons, **preganglionic** and **postganglionic**, which are GVE neurons.

A. Sympathetic Nerve Fibers (see Figure 1.3)

—have preganglionic nerve cell bodies that are located in the lateral horn of the thoracic and upper lumbar levels (L2 or L1–L3) of the spinal cord.

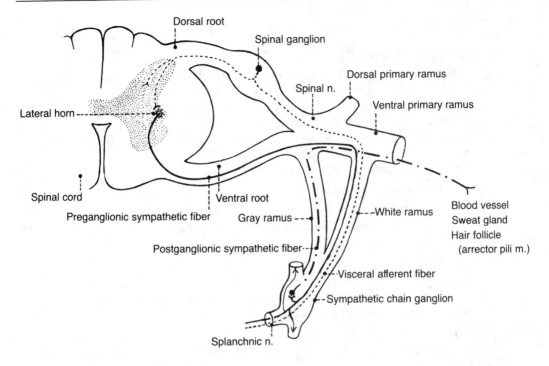

Figure 1.3. General visceral efferent (autonomic) and afferent nerves.

—have preganglionic fibers that pass through ventral roots, spinal nerves, and white rami communicantes. These fibers enter adjacent sympathetic chain ganglia, where they synapse or travel up or down the chain to synapse in remote ganglia or run further through the splanchnic nerves to synapse in collateral ganglia, located along the major abdominal blood vessels.

—have postganglionic fibers from the chain ganglia that return to spinal nerves by way of gray rami communicantes and supply the skin with secretory fibers to sweat glands, motor fibers to smooth muscles (arrectores pilorum muscles) of the hair follicles, and vasomotor fibers to the blood vessels.

—function primarily in emergencies, preparing individuals for fight or flight and thus increase heart rate, inhibit gastrointestinal motility and secretion, and dilate pupils and bronchial lumen.

B. Parasympathetic Nerve Fibers

—comprise the preganglionic fibers that arise from the brainstem (cranial nerves III, VII, IX, and X) and sacral part of the spinal cord (second, third, and fourth sacral segments).

—are, with few exceptions, characterized by long preganglionic fibers and short postganglionic fibers.

—are distributed to the walls of the visceral organs and glands of the digestive system but not to the skin or to the periphery.

—decrease heart rate, increase gastrointestinal peristalsis, and stimulate secretory activity.

—function primarily in homeostasis, tending to promote quiet and orderly processes of the body.

Circulatory System

I. Vascular System

—functions to transport vital materials between the external environment and the internal fluid environment of the body.

—consists of the heart and vessels (arteries, capillaries, veins) that transport blood through all parts of the body.

—includes the lymphatic vessels, a set of channels that begin in the tissue spaces and return excess tissue fluid to the bloodstream.

—has two circulatory loops for blood:

1. Pulmonary Circulation

—pumps blood from the right ventricle to the lungs through the pulmonary arteries and returns it to the left atrium of the heart through the pulmonary veins.

2. Systemic Circulation

—pumps blood from the left ventricle through the aorta to all parts of the body and returns it to the right atrium through the superior and inferior vena cavae and the cardiac veins.

A. Heart

—is a hollow, muscular, four-chambered organ that pumps blood through the pulmonary and systemic circulations.

—receives the venous blood from the body in the right atrium and then passes it into the right ventricle, which pumps it to the lungs for oxygenation.

—receives the oxygenated blood from the lungs in the left atrium and passes it to the left ventricle, which pumps it through the arteries to supply the tissues of the body.

—is regulated in its pumping rate and strength by the autonomic nervous system, which controls a pacemaker (the sinoatrial node).

B. Fetal Circulation

1. In the fetus, blood is oxygenated in the placenta rather than in the lungs.

2. Before birth, three shunts exist to partially bypass the lungs and liver:

a. Foramen Ovale

—shunts blood from the right atrium to the left atrium, partially bypassing the lungs (pulmonary circulation).

b. Ductus Arteriosus

—shunts blood from the pulmonary trunk to the aorta, partially bypassing the lungs (pulmonary circulation).

c. Ductus Venosus

—shunts oxygenated blood from the umbilical vein (returning from the placenta) to the inferior vena cava, without passing through the liver (portal circulation).

C. Blood Vessels

—carry blood to the lungs, where carbon dioxide is exchanged for oxygen.

—carry blood to the intestines, where nutritive materials in fluid form are absorbed, and to the endocrine glands, where hormones pass through their walls and are distributed to target cells.

—transport the waste products of tissue fluid to the kidneys, intestines, lungs, and skin, where they are excreted.

—are of four types: arteries, veins, capillaries and sinusoids.

1. Arteries

—carry blood away from the heart and distribute it to all parts of the body.

—have thicker and stronger walls than those of veins.

—have three main types: elastic arteries, muscular arteries, and arterioles.

2. Veins

—return the deoxygenated venous blood to the heart (except the pulmonary veins).

—closely accompany an artery as a pair or more, known as **venae comitantes,** which are found especially in the extremities.

3. Capillaries

—are composed of endothelium and its basement membrane and connect the arterioles to the venules.

—are the exchange sites where oxygen and nutritive materials from oxygenated blood diffuse across the endothelial wall of the arteriolar end of the capillary into the tissue spaces, whereas metabolic waste products and carbon dioxide diffuse from the tissue spaces into the blood through the wall of the venous end.

—are absent in the cornea, epidermis, and hyaline cartilage.

—may not be present in some areas where the arterioles and venules have direct connections. These arteriovenous anastomoses (AV shunts) bypass the capillaries and are especially numerous in the skin of the nose, lips, fingers, and ears, where they conserve body heat.

4. Sinusoids

—are wider and more irregular than capillaries.

—substitute for capillaries in the liver, spleen, red bone marrow, carotid body, adenohypophysis, suprarenal cortex, and parathyroid glands.

—have walls that consist largely of phagocytic cells.

—form a part of the reticuloendothelial system, which is concerned chiefly with phagocytosis and antibody formation.

II. Lymphatic System

—provides an important immune mechanism for the body. (When foreign proteins are drained from an infected area by the lymphatic capillaries, immunologically competent cells produce an antibody specific to the foreign protein, or lymphocytes, which have an important role in the development of antibodies and immune reactions, are dispatched to the infected area.)

—is involved in the spread (metastasis) of cancer cells.

A. Lymphatic Vessels

—serve as a one-way drainage toward the heart and return lymph to the bloodstream through the thoracic duct, the largest lymphatic vessel.

—are not generally visible in dissections but are the major route by which carcinoma metastasizes.

—function to absorb the large protein molecules and transport them to the bloodstream because they cannot pass through the walls of the blood capillaries back into the blood.

—carry lymphocytes from lymphatic tissues to the bloodstream.

—have valves to ensure the flow of lymph away from the tissues and toward the venous system.

—are constricted at the sites of valves, showing a knotted or beaded appearance.

—are absent in the brain, spinal cord, eyeballs, bone marrow, splenic pulp, hyaline cartilage, nails, and hair.

B. Lymphatic Capillaries

—begin blindly in most tissues, collect tissue fluid, and join to form large collecting vessels that pass to regional lymph nodes.

—are wider than blood capillaries.

—absorb lymph from tissue spaces and transport it back to the venous system.

—are called **lacteals** in the villi of the small intestine, which absorb emulsified fat.

C. Lymph Nodes

—are organized collections of lymphatic tissue permeated by lymphatic channels.

—have fibroelastic capsules from which connective tissue trabeculae extend into the substance of the organ.

—have a **cortex** (outer part), which contains collections of lymphatic cells called **germinal centers,** and a **medulla** (inner part), which contains cords of lymphatic cells.

—produce lymphocytes and plasma cells.

—trap bacteria drained from an infected area and contain reticuloendothelial cells and phagocytic cells (macrophages) that ingest these bacteria.

—serve as filters. (Thus, the cancer cells in lymph vessels migrate or metastasize to lymph nodes and tend to remain within them, proliferating and gradually destroying them.)

—are hard and often palpable when there is a metastasis and are enlarged and tender during infection.

D. Lymph

—is a clear, watery fluid that is collected from the intercellular spaces. It contains no cells until lymphocytes are added in its passage through the lymph nodes.

—contains constituents similar to those of blood plasma (e.g., proteins, fats, lymphocytes).

—often contains fat droplets (called **chyle**) when it comes from intestinal organs.

—is filtered by passing through several lymph nodes before entering the venous system.

Review Test

DIRECTIONS: Each of the numbered items or incomplete statements in this section is followed by answers or by completions of the statement. Select the **one** lettered answer or completion that is **best** in each case.

1. Which of the following structures is a fibrous sheet or band that covers the body under the skin and invests the muscles?

(A) Tendon
(B) Fascia
(C) Synovial tendon sheath
(D) Aponeurosis
(E) Ligament

2. Which of the following statements concerning cranial nerves is true?

(A) They consist of 31 pairs.
(B) They have motor fibers with cell bodies located in the nerve outside the CNS.
(C) They are connected to the brain.
(D) All but the trochlear nerve emerge from the dorsal aspect of the brain.
(E) They are connected to the sympathetic chain ganglia by gray rami communicantes.

3. Each statement below concerning neurons is correct EXCEPT

(A) they are the basic structural and functional units of the nervous system.
(B) they consist of cell bodies, axons, and dendrites.
(C) their dendrites carry impulses to the cell bodies.
(D) they are specialized for the reception, integration, and transmission of information.
(E) their axons carry impulses toward the cell bodies.

4. Which blood vessel or group of vessels below carries richly oxygenated blood to the heart?

(A) Superior vena cava
(B) Pulmonary arteries
(C) Pulmonary veins
(D) Ascending aorta
(E) Cardiac veins

5. Each statement below concerning fetal circulation is true EXCEPT

(A) the right atrium receives oxygenated blood from the placenta.
(B) right atrial pressure is higher than left atrial pressure.
(C) blood flows from the pulmonary artery to the aorta through the ductus arteriosus.
(D) the foramen ovale shunts blood from the left atrium to the right atrium, partially bypassing the lungs.
(E) the ductus venosus shunts oxygenated blood from the umbilical vein to the inferior vena cava, bypassing the liver.

DIRECTIONS: Each group of items in this section consists of lettered options followed by a set of numbered items. For each item, select the **one** lettered option that is most closely associated with it. Each lettered option may be selected once, more than once, or not at all.

Questions 6–10

Match each example or description below with the appropriate type of joints.
(A) Syndesmoses
(B) Synchondroses
(C) Hinge joints
(D) Ellipsoidal joints
(E) Ball-and-socket joints

6. Radiocarpal and metacarpophalangeal joints

7. Allow flexion and extension only

8. Fibrous joints

9. Allow movement in many directions

10. Primary cartilaginous joints; contain epiphyseal cartilage plates

Answers and Explanations

1–B. The fascia is a fibrous sheet or band that covers the body under the skin and invests the muscles.

2–C. The cranial nerves consist of 12 pairs; some have sensory fibers with cell bodies located in the nerve outside the CNS. All cranial nerves except the trochlear nerve emerge from the ventral aspect of the brain. The trochlear nerve emerges from the posterior aspect of the midbrain.

3–E. The dendrites of neurons carry impulses to the cell bodies; axons carry impulses away from the cell bodies.

4–C. Pulmonary arteries carry deoxygenated blood from the heart to the lungs for oxygen renewal. Pulmonary veins return the oxygenated blood to the heart. The aorta carries oxygenated blood from the left ventricle to all parts of the body.

5–D. The foramen ovale shunts blood from the right atrium to the left atrium, partially bypassing the lungs.

6–D. The wrist (radiocarpal) and metacarpophalangeal joints are ellipsoidal (condyloid) joints, which allow flexion/extension and abduction/adduction.

7–C. Hinge (ginglymus) joints allow flexion and extension only. The elbow and interphalangeal joints are hinge joints.

8–A. Syndesmoses are fibrous joints. The tympanostapedial and inferior tibiofibular joints are syndesmoses.

9–E. Ball-and-socket (spheroidal) joints allow movement in many directions (multiaxial). The shoulder and hip joints are ball-and-socket joints.

10–B. Synchondroses are united by hyaline cartilage and contain epiphyseal cartilage plates.

2
Upper Limb

Cutaneous Nerves, Superficial Veins, and Lymphatics

I. **Cutaneous Nerves**

A. **Supraclavicular Nerve**
—arises from the cervical plexus (C3, C4) and supplies the skin over the upper pectoral, deltoid, and outer trapezius areas.

B. **Medial Brachial Cutaneous Nerve**
—arises from the medial cord of the brachial plexus and supplies the medial side of the arm.

C. **Medial Antebrachial Cutaneous Nerve**
—arises from the medial cord of the brachial plexus and supplies the medial side of the forearm.

D. **Lateral Brachial Cutaneous Nerve**
—arises from the axillary nerve and supplies the lateral side of the arm.

E. **Lateral Antebrachial Cutaneous Nerve**
—arises from the musculocutaneous nerve and supplies the lateral side of the forearm.

F. **Posterior Brachial and Antebrachial Cutaneous Nerves**
—arise from the radial nerve and supply the posterior side of the arm and forearm, respectively.

G. **Intercostobrachial Nerve**
—is the lateral cutaneous branch of the second intercostal nerve and emerges from the second intercostal space by piercing the intercostal and serratus anterior muscles.
—may communicate with the medial brachial cutaneous nerve.

II. **Superficial Veins** (Figure 2.1)

A. **Cephalic Vein**
—begins as a radial continuation of the dorsal venous network and runs on the lateral side of the forearm and the front of the elbow.
—is often connected with the basilic vein by the median cubital vein in front of the elbow.
—ascends along the lateral surface of the biceps, pierces the brachial fascia, and lies in the deltopectoral triangle with the deltoid branch of the thoraco-acromial trunk.
—pierces the clavipectoral fascia and empties into the axillary vein.

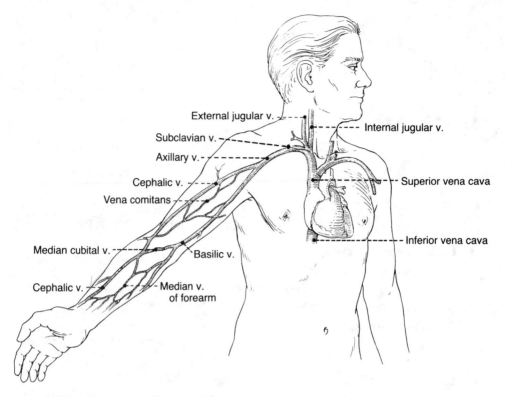

Figure 2.1. Venous drainage of the upper limb.

B. Basilic Vein

—arises from the dorsum of the hand and accompanies the medial antebrachial cutaneous nerve.

—ascends on the posteromedial surface of the forearm and passes anterior to the medial epicondyle.

—pierces the deep fascia of the arm and joins the two brachial veins, the venae comitantes of the brachial artery, to form the axillary vein at the lower border of the teres major muscle.

C. Median Cubital Vein

—connects the cephalic vein to the basilic vein.

—lies superficial to the bicipital aponeurosis, which separates it from the brachial artery.

—is frequently used for intravenous injections and blood transfusions.

D. Median Antebrachial Vein

—arises in the palmar venous network, ascends on the front of the forearm, and terminates in the median cubital or the basilic vein.

III. Superficial Lymphatics

A. Lymphatics of the Finger

—drain into the plexus on the dorsum and palm of the hand.

B. Medial Group of Lymphatic Vessels

—accompanies the basilic vein, passes through the cubital or supratrochlear nodes, and ascends to enter the lateral axillary nodes.

C. **Lateral Group of Lymphatic Vessels**
—accompanies the cephalic vein and drains into the lateral axillary nodes and also into the deltopectoral (infraclavicular) nodes. The deltopectoral nodes drain into the apical nodes.

Bones and Joints

I. Bones

A. **Clavicle (Collarbone) [Figure 2.2]**
—forms the girdle of the upper limb with the scapula.
—is the first bone to begin ossification during fetal development, but it is the last one to complete ossification, at about the twenty-first year.
—is the only long bone to be ossified intermembranously.
—has its medial two-thirds convex forward, whereas the lateral one-third is flattened with a marked concavity.
—articulates with the sternum at the sternoclavicular joint and with the scapula at the acromioclavicular joint.
—is a commonly fractured bone.

B. **Scapula (Shoulder Blade) [see Figure 2.2]**
—is a triangular-shaped flat bone.

1. **Spine**
—is a triangular-shaped process and continues laterally as the acromion.
—divides the dorsal surface into the upper supraspinous and lower infraspinous fossae.
—provides an origin for the deltoid muscle and an insertion for the trapezius muscle.

2. **Acromion**
—is the lateral end of the spine and articulates with the clavicle.
—provides an origin for the deltoid muscle and an insertion for the trapezius muscle.

3. **Coracoid Process**
—provides the origin of the coracobrachialis and biceps brachii muscles and the insertion of the pectoralis minor muscle.
—gives attachment to the coracoclavicular, coracohumeral, and coracoacromial ligaments and the costocoracoid membrane.

4. **Scapular Notch**
—is bridged by the superior **transverse scapular ligament** and converted into a foramen, which permits passage of the suprascapular nerve.

5. **Glenoid Cavity**
—articulates with the head of the humerus.
—is deepened by a fibrocartilaginous lip (glenoidal labrum).

6. **Supraglenoid and Infraglenoid Tubercles**
—provide origins for the tendons of the long heads of the biceps brachii and triceps brachii muscles, respectively.

C. **Humerus** (Figure 2.3; see Figure 2.2)

1. **Head**
—has a smooth, rounded, articular surface and articulates with the scapula at the glenohumeral joint.

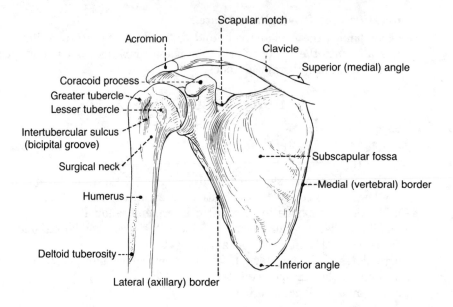

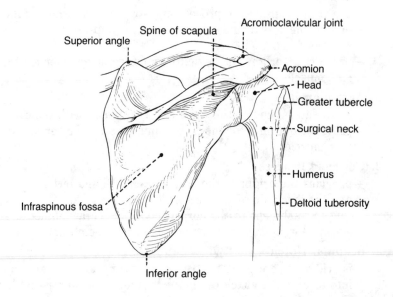

Figure 2.2. Pectoral girdle and humerus.

2. Anatomical Neck

—is an indentation distal to the head and provides for the attachment of the articular capsule.

3. Greater Tubercle

—lies on its lateral side, just lateral to the anatomical neck.

—provides attachments for the supraspinatus, infraspinatus, and teres minor muscles.

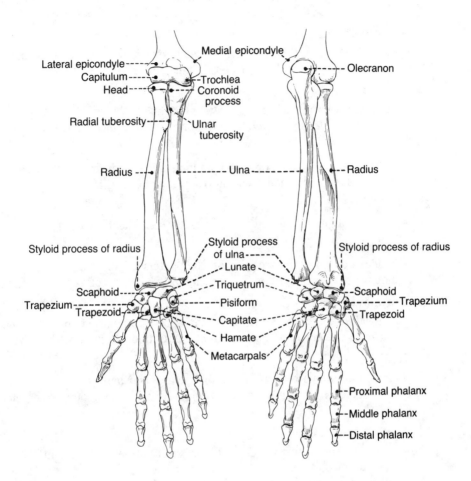

Figure 2.3. Bones of the forearm and hand.

4. Lesser Tubercle

—lies on its medial side of the front, just distal to the anatomical neck.

—provides an insertion for the subscapularis muscle.

5. Intertubercular (Bicipital) Groove

—lies between the greater and lesser tubercles.

—lodges the tendon of the long head of the biceps brachii muscle, which forces the humeral head medially into the joint.

—is spanned by the transverse humeral ligament, which restrains the tendon of the biceps long head.

—provides insertions for the pectoralis major on its lateral lip, the teres major on its medial lip, and the latissimus dorsi on its floor.

6. **Surgical Neck**

—is a narrow area distal to the tubercles and a common site of fracture.

—is in contact with the axillary nerve and the posterior humeral circumflex artery.

7. **Deltoid Tuberosity**

—is a V-shaped roughened area on the lateral aspect of the midshaft and marks the insertion of the deltoid muscle.

8. **Spiral Groove**

—is a groove for the radial nerve, separating the origin of the lateral head of the triceps above and the origin of the medial head below.

9. **Trochlea**

—is shaped like a spool or pulley and has a deep depression between two margins, which articulates the trochlear notch of the ulna.

10. **Capitulum**

—is globular in shape and articulates with the head of the radius.

11. **Olecranon Fossa**

—is located above the trochlea on the posterior aspect of the humerus.

—houses the olecranon of the ulna upon full extension of the forearm.

12. **Coronoid Fossa**

—is located above the trochlea on the anterior aspect of the humerus.

—receives the coronoid process of the ulna upon flexion of the elbow.

13. **Radial Fossa**

—is located above the capitulum on the anterior aspect.

—is occupied by the head of the radius during full flexion of the elbow joint.

14. **Medial Epicondyle**

—projects from the trochlea and is larger and more prominent than the lateral epicondyle.

—gives attachment to the ulnar collateral ligament, the pronator teres muscle, and the common tendon of the flexor muscles of the forearm.

15. **Lateral Epicondyle**

—projects from the capitulum and provides the origin of the supinator and extensor muscles of the forearm.

D. **Radius** (see Figure 2.3)

—its **head (proximal end)** articulates with the capitulum of the humerus and the radial notch of the ulna.

—its distal end articulates with the proximal row of carpal bones, including the scaphoid, lunate, and triquetral bones but excluding the pisiform bone.

—its **tuberosity** is an oval prominence just distal to the neck and gives attachment to the biceps brachii tendon.

—is shorter than the ulna and has, on its distal end, a **styloid process** (which is about 1 cm distal to that of the ulna). The styloid process is palpable in the anatomical snuff-box (which is located between the extensor pollicis longus and brevis tendons) and provides insertion of the brachioradialis muscle.

—is characterized by displacement of the hand dorsally and radially when fractured at its distal end (Colles' fracture).

E. **Ulna** (see Figure 2.3)
   —its **olecranon** is the curved projection on the back of the elbow and gives attachment to the triceps tendon.
   —its **coronoid process,** located below the trochlear notch, gives attachment to the brachialis.
   —its **trochlear notch** receives the trochlea of the humerus.
   —its **radial notch** accommodates the head of the radius.
   —its **head (distal end)** articulates with the articular disk of the distal radioulnar joint and contains a styloid process.

F. **Carpal Bones** (see Figure 2.3)
   —are arranged in two rows of four:
   1. **Proximal row** (lateral to medial): scaphoid, lunate, triquetral, and pisiform.
      —in the proximal row (except for the pisiform), they articulate with the radius and the articular disk (the ulna has no contact with the carpal bones).
   2. **Distal row** (lateral to medial): trapezium, trapezoid, capitate, and hamate.

G. **Metacarpals**
   —are miniature long bones consisting of bases (proximal ends), shafts (bodies), and heads (distal ends).
   —their heads form the knuckles of the fist.

H. **Phalanges**
   —are miniature long bones consisting of bases, shafts, and heads.
   —there are three in each finger but two in the thumb.
   —the heads of the proximal and middle phalanges form the knuckles.

II. Joints (see Figures 2.2 and 2.3)

A. **Shoulder (Glenohumeral) Joint**
   —is a multiaxial ball-and-socket (spheroidal) joint located between the glenoid cavity of the scapula and the head of the humerus.
   —both articular surfaces are covered with hyaline cartilage.
   —its capsule lies deep to the tendon of the musculotendinous cuff and is attached to the glenoid outside the labrum and to the anatomical neck of the humerus.
   —its cavity communicates with the subscapular bursa.
   —is supplied by the axillary, suprascapular, and lateral pectoral nerves.

B. **Elbow Joint**
   —forms a hinge (ginglymus) joint between the capitulum of the humerus and the head of the radius (humeroradial joint) and between the trochlea of the humerus and the trochlear notch of the ulna (humeroulnar joint).
   —also includes the proximal radioulnar joint within a common articular capsule.
   —is innervated by the musculocutaneous, median, radial, and ulnar nerves.

C. **Proximal Radioulnar Joint**
   —forms a pivot (trochoid) joint in which the head of the radius articulates with the radial notch of the ulna.

D. Distal Radioulnar Joint

—forms a pivot joint between the head of the ulna and the ulnar notch of the radius.

E. Wrist (Radiocarpal) Joint

—is an ellipsoid (condyloid) joint formed superiorly by the radius and the articular disk and inferiorly by the proximal row of carpal bones (scaphoid, lunate, and triquetral), exclusive of the pisiform. (The ellipsoidal surface of the carpal bones fits into the concave surface of the radius and articular disk.)

—allows flexion, extension, abduction, adduction, and circumduction.

—its articular capsule is strengthened by radial and ulnar collateral ligaments and dorsal and palmar radiocarpal ligaments.

F. Midcarpal Joint

—is an articulation between the proximal and distal rows of carpal bones.

—forms an ellipsoid synovial joint by fitting the hamate bone and the head of the capitate bone into the concavity of the scaphoid, lunate, and triquetral bones.

—also forms a plane joint by joining the scaphoid with the trapezium and trapezoid bones.

G. Carpometacarpal Joints

—form saddle (sellar) joints between the trapezium and the base of the first metacarpal bone.

—also form plane joints between the carpal bones and the medial four metacarpal bones.

H. Metacarpophalangeal Joints

—are ellipsoid joints.

—are supported by a palmar ligament and two collateral ligaments.

I. Interphalangeal Joints

—are hinge joints.

—have a strong palmar ligament and two collateral ligaments.

Pectoral Region and Axilla

I. Breast and Mammary Gland (Figure 2.4)

A. Breast

—consists of mammary gland tissue, fibrous and fatty tissue, blood and lymph vessels, and nerves.

—extends from the second to sixth ribs and from the sternum to the midaxillary line.

—has glandular tissue, which lies in the superficial fascia.

—is supplied by the anterior perforating branches of the internal thoracic artery and the lateral mammary branches of the lateral thoracic artery.

—is innervated by the anterior and lateral cutaneous branches of the second to sixth intercostal nerves.

—has suspensory ligaments (Cooper's), which are strong fibrous processes that support the breast and run from the dermis of the skin to the deep layer of the superficial fascia through the breast.

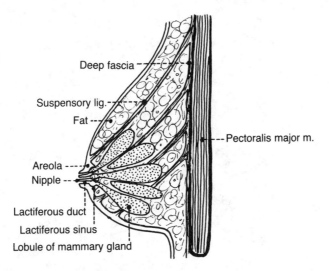

Figure 2.4. Breast.

—the nipple usually lies at the level of the fourth intercostal space.
—the areola is a ring of pigmented skin around the nipple.
—has an axillary tail, which is a normal extension of the mammary gland into the axilla.
—may present more than one pair of breasts (polymastia) and more than one pair of nipples (polythelia).

B. Mammary Gland
—is a modified sweat gland located in the superficial fascia.
—has 15 to 20 lobes of glandular tissue, each of which opens by a **lactiferous duct** onto the tip of the nipple. Each duct enlarges to form a **lactiferous sinus,** which serves as a reservoir for milk during lactation.

C. Lymphatic Drainage of the Breast or the Mammary Gland (Figure 2.5)
—drains mainly (75%) to the axillary nodes, more specifically to the pectoral nodes (including drainage of the nipple).
—follows the perforating vessels through the pectoralis major muscle and the thoracic wall to enter the parasternal (internal thoracic) nodes, which lie along the internal thoracic artery.
—also drains to the apical nodes and may connect to lymphatics draining the opposite breast and to lymphatics draining the anterior abdominal wall.
—is of great importance in view of the frequent development of cancer and subsequent dissemination of cancer cells through the lymphatic stream.

D. Breast Cancer
—forms a palpable mass in the advanced stage.
—enlarges, attaches to Cooper's suspensory ligaments, and produces shortening of the ligaments, causing depression or dimpling of the overlying skin.
—may also attach to and shorten the lactiferous ducts, causing the nipple to become retracted or inverted.
—may invade the deep fascia of the pectoralis major muscle, so that contraction of this muscle produces a sudden upward movement of the whole breast.
—may be detected and treated by the following procedures:

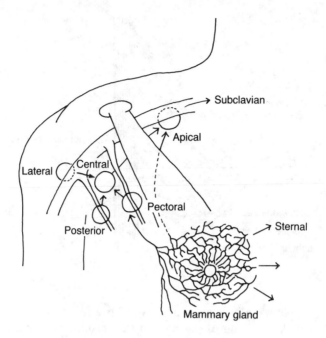

Figure 2.5. Lymphatic drainage of the breast and axillary lymph nodes.

1. **Mammography**
    —is a roentgenographic examination of the breast to detect breast cancer.

2. **Radical Mastectomy**
    —is an extensive surgical removal of the breast and its related structures, including the pectoralis major and minor muscles, axillary lymph nodes and fascia, and part of the thoracic wall.
    —may injure the long thoracic and thoracodorsal nerves.
    —may cause postoperative swelling (edema) of the upper limb as a result of lymphatic obstruction caused by removal of most of the lymphatic channels that drain the arm, or by venous obstruction caused by thrombosis of the axillary vein.

3. **Modified Radical Mastectomy**
    —involves excision of the entire breast and axillary lymph nodes, with preservation of the pectoralis major and minor muscles. (The pectoralis minor muscle often is retracted or severed near its insertion into the coracoid process.)

II. Fasciae of Pectoral and Axillary Regions

A. Clavipectoral Fascia
    —extends between the coracoid process, the clavicle, and the thoracic wall.
    —envelops the subclavius and pectoralis minor muscles.

B. Costocoracoid Membrane
    —is a part of the clavipectoral fascia between the first rib and the coracoid process and covers the deltopectoral triangle.
    —is pierced by the **cephalic vein**, the **thoracoacromial artery**, and the **lateral pectoral nerve**.

C. **Pectoral Fascia**

—covers the pectoralis major muscle, is attached to the sternum and clavicle, and is continuous with the axillary fascia.

D. **Axillary Fascia**

—is continuous anteriorly with the pectoral and clavipectoral fasciae, laterally with the brachial fascia, and posteromedially with the fascia of the latissimus dorsi and serratus anterior muscles.

E. **Axillary Sheath**

—is a fascial prolongation of the prevertebral layer of the cervical fascia into the axilla, enclosing the axillary vessels and the brachial plexus.

III. Boundaries of the Axilla

A. **Medial wall:** upper ribs and their intercostal muscles and serratus anterior muscle.

B. **Lateral wall:** humerus.

C. **Posterior wall:** subscapularis, teres major, and latissimus dorsi muscles.

D. **Anterior wall:** pectoralis major and pectoralis minor muscles.

E. **Base:** axillary fascia.

F. **Apex:** interval between the clavicle, scapula, and first rib.

IV. Muscles (Figure 2.6)

A. **Pectoralis Major**

—arises from the medial half of the clavicle, the manubrium and body of the sternum, and the upper six costal cartilages.
—inserts on the lateral lip of the intertubercular groove (the crest of the greater tubercle) of the humerus.
—is supplied by the medial and lateral pectoral nerves.
—adducts and medially rotates the arm; the clavicular part can rotate the arm medially and flex it, whereas the sternocostal part depresses the arm and shoulder. Its lower fibers can help extend the arm when it is flexed.
—forms the anterior wall of the axilla, and its lateral border forms the anterior axillary fold.

B. **Pectoralis Minor**

—arises from the external surfaces of the second to fifth ribs.
—inserts into the coracoid process and depresses the shoulder.
—is innervated chiefly by the medial pectoral nerve but also by the lateral pectoral nerve as a result of their communication.
—is invested by the clavipectoral fascia, divides the axillary artery into three parts, forms part of the anterior wall of the axilla, and crosses the cords of the brachial plexus.

C. **Subclavius**

—originates from the junction of the first rib and its cartilage.
—inserts on the lower surface of the clavicle.
—is supplied by the nerve to the subclavius.
—assists in depressing the lateral portion of the clavicle.

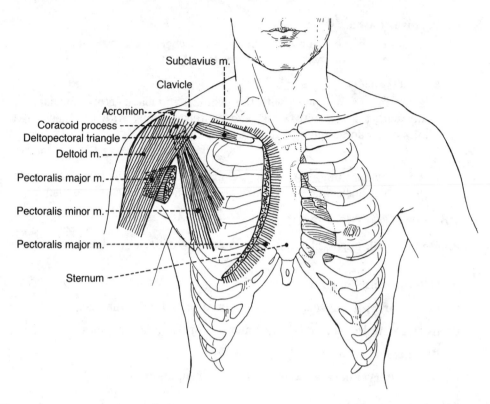

Figure 2.6. Muscles of the pectoral region.

D. Serratus Anterior
—arises from the external surfaces of the upper eight ribs.
—is inserted on the medial border of the scapula.
—is innervated by the long thoracic nerve.
—rotates the scapula upward so that the inferior angle swings laterally and abducts the arm and elevates it above a horizontal position.

V. Axillary Artery (Figures 2.7 and 2.8)
—extends from the outer border of the first rib to the inferior border of the teres major muscle, where it becomes the brachial artery.
—is divided into three parts by the pectoralis minor muscle.
—is considered to be the central structure of the axilla and is bordered on its medial side by the axillary vein.

A. Supreme Thoracic Artery
—supplies the first and second intercostal spaces.

B. Thoracoacromial Artery
—is a short trunk from the first or second part of the axillary artery and has pectoral, clavicular, acromial, and deltoid branches.
—pierces the costocoracoid membrane (or clavipectoral fascia).

C. Lateral Thoracic Artery
—runs along the lateral border of the pectoralis minor muscle.
—supplies the pectoralis major, pectoralis minor, and serratus anterior muscles and the axillary lymph nodes, and gives rise to lateral mammary branches.

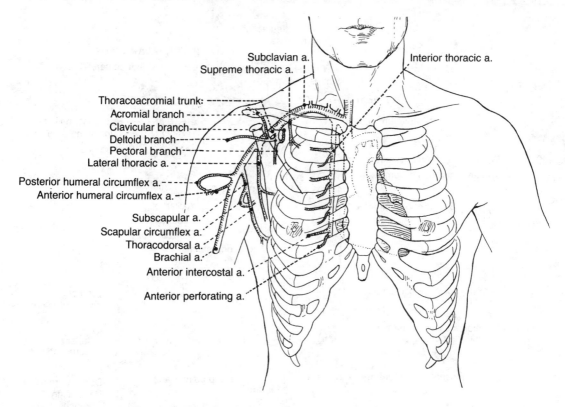

Figure 2.7. Blood supply to the pectoral and axillary regions.

D. Subscapular Artery

—is the largest branch of the axillary artery.

—arises at the lower border of the subscapularis muscle and descends along the axillary border of the scapula.

—divides into the thoracodorsal and circumflex scapular arteries.

1. Thoracodorsal Artery

—accompanies the thoracodorsal nerve and supplies the latissimus dorsi muscle and the lateral thoracic wall.

2. Circumflex Scapular Artery

—passes posteriorly into the triangular space bounded by the subscapularis muscle and the teres minor muscle above, the teres major muscle below, and the long head of the triceps brachii laterally.

—ramifies in the infraspinous fossa and anastomoses with branches of the dorsal scapular and suprascapular arteries.

E. Anterior Humeral Circumflex Artery

—passes anteriorly around the surgical neck of the humerus.

—anastomoses with the posterior humeral circumflex artery.

F. Posterior Humeral Circumflex Artery

—runs posteriorly with the axillary nerve through the quadrangular space bounded by the teres minor and teres major muscles, the long head of the triceps brachii, and the humerus.

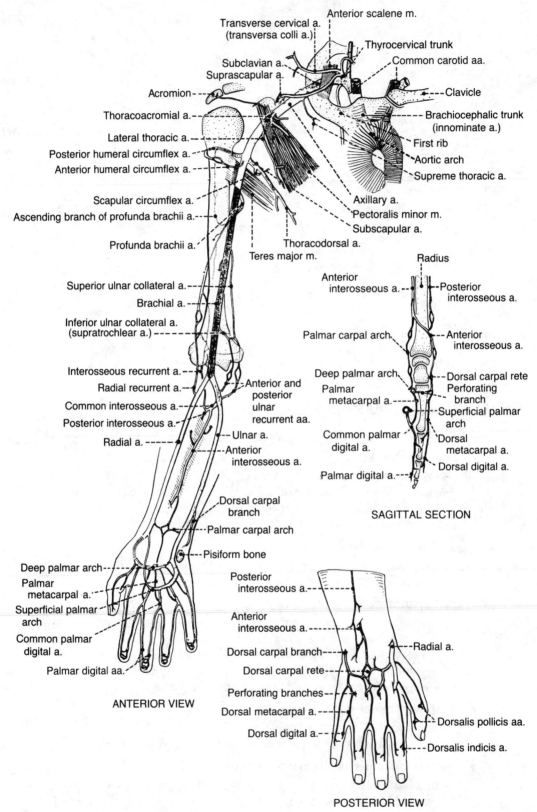

Figure 2.8. Blood supply to the upper limb.

—anastomoses with the anterior humeral circumflex artery and an ascending branch of the profunda brachii artery and also sends a branch to the acromial rete.

VI. Veins

A. Axillary Vein (see Figure 2.1)

—begins at the lower border of the teres major muscle as the continuation of the basilic vein and ascends along the medial side of the axillary artery.
—continues as the subclavian vein at the inferior margin of the first rib.
—commonly receives the thoracoepigastric veins directly or indirectly and thus provides a collateral circulation if the inferior vena cava becomes obstructed.
—its tributaries are the cephalic vein, brachial veins, and veins that correspond to the branches of the axillary artery, with the exception of the thoracoacromial vein.

B. Cephalic Vein (see Figure 2.1)

—runs in the deltopectoral triangle along with the deltoid branch of the thoracoacromial trunk.
—perforates the costocoracoid membrane and usually terminates in the axillary vein.

VII. Axillary Lymph Nodes (see Figure 2.5)

A. Central Nodes

—lie near the base of the axilla between the lateral thoracic and subscapular veins, receive lymph from the lateral, pectoral, and posterior groups of nodes, and drain into the apical nodes.

B. Lateral Nodes

—lie posteromedial to the axillary veins, receive lymph from the upper limb, and drain into the central nodes.

C. Subscapular Nodes

—lie along the subscapular vein and drain lymph from the posterior thoracic wall and the posterior aspect of the shoulder to the central nodes.

D. Pectoral Nodes

—lie along the inferolateral border of the pectoralis minor muscle and drain lymph from the anterior and lateral thoracic walls, including the breast.

E. Apical Nodes

—lie at the apex of the axilla medial to the axillary vein and above the upper border of the pectoralis minor muscle, receive lymph from all of the other axillary nodes (and occasionally from the breast), and drain into the subclavian trunks.

VIII. Brachial Plexus (Figure 2.9)

—is formed by the ventral primary rami of the lower four cervical nerves and the first thoracic nerve (C5–T1).
—has roots that pass between the scalenus anterior and medius muscles.

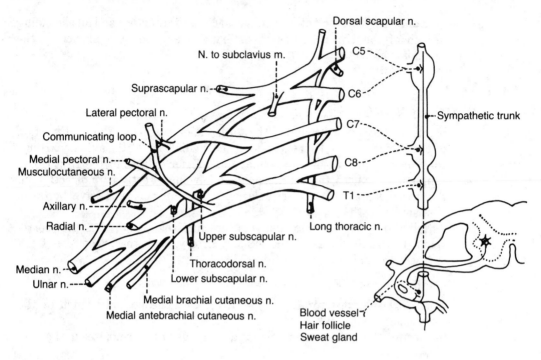

Figure 2.9. Brachial plexus.

—is enclosed with the axillary artery and vein in the axillary sheath.
—has the following subdivisions:

A. Roots

1. Dorsal Scapular Nerve (C5)

—pierces the scalenus medius muscle to reach the posterior cervical triangle and descends deep to the levator scapulae and the rhomboid minor and major muscles.
—supplies the rhomboids and frequently the levator scapulae muscles.

2. Long Thoracic Nerve (C5–C7)

—descends behind the brachial plexus and runs on the external surface of the serratus anterior muscle, which it supplies.
—when damaged, causes winging of the scapula and makes elevating the arm above a horizontal position impossible.

B. Upper Trunk

1. Suprascapular Nerve (C5–C6)

—runs laterally across the posterior cervical triangle.
—passes through the scapular notch under the superior transverse scapular ligament, whereas the suprascapular artery passes over the ligament. (Thus, it can be said that the army [artery] runs over the bridge [ligament], and the navy [nerve] runs under the bridge.)
—supplies the supraspinatus muscle and the shoulder joint and then descends through the notch of the scapular neck to innervate the infraspinatus muscle.

2. **Nerve to Subclavius (C5)**
   —descends in front of the brachial plexus and the subclavian artery and behind the clavicle to reach the subclavius muscle.
   —also supplies the sternoclavicular joint.
   —frequently branches to the accessory phrenic nerve (C5), which enters the thorax to join the phrenic nerve.

C. **Lateral Cord**

1. **Lateral Pectoral Nerve (C5–C7)**
   —supplies the pectoralis major muscle primarily, but by way of a nerve loop, it also supplies the pectoralis minor muscle.
   —sends a branch over the first part of the axillary artery to the medial pectoral nerve and forms a nerve loop through which the lateral pectoral nerve conveys motor fibers to the pectoralis minor muscle.
   —pierces the costocoracoid membrane of the clavipectoral fascia.
   —is accompanied by the pectoral branch of the thoracoacromial artery.

2. **Musculocutaneous Nerve (C5–C7)**
   —pierces the coracobrachialis muscle, descends between the biceps brachii and brachialis muscles, and innervates these three muscles.
   —continues into the forearm as the lateral antebrachial cutaneous nerve.

D. **Medial Cord**

1. **Medial Pectoral Nerve (C8–T1)**
   —passes forward between the axillary artery and vein and forms a loop in front of the axillary artery with the lateral pectoral nerve.
   —enters and supplies the pectoralis minor muscle and reaches the overlying pectoralis major muscle.

2. **Medial Brachial Cutaneous Nerve (C8–T1)**
   —runs along the medial side of the axillary vein.
   —supplies skin on the medial side of the arm.
   —may communicate with the intercostobrachial cutaneous nerve.

3. **Medial Antebrachial Cutaneous Nerve (C8–T1)**
   —runs between the axillary artery and vein and then runs medial to the brachial artery.
   —supplies skin on the medial side of the forearm.

4. **Ulnar Nerve (C7–T1)**
   —runs down the medial aspect of the arm but does not branch in the brachium (see p 46).

E. **Medial and Lateral Cords: Median Nerve (C5–T1)**
   —is formed by a head from both the medial and lateral cords.
   —runs down the anteromedial aspect of the arm but does not branch in the brachium (see p 43).

F. **Posterior Cord**

1. **Upper Subscapular Nerve (C5–C6)**
   —supplies the upper portion of the subscapularis muscle.

2. **Thoracodorsal Nerve (C7–C8)**
   —runs behind the axillary artery and accompanies the thoracodorsal artery to enter the latissimus dorsi muscle.

3. **Lower Subscapular Nerve (C5–C6)**
—innervates the lower part of the subscapularis and teres major muscles.
—runs downward behind the subscapular vessels to the teres major muscle.

4. **Axillary Nerve (C5–C6)**
—innervates the deltoid muscle by its anterior and posterior branches and the teres minor muscle by its posterior branch.
—gives rise to the **lateral brachial cutaneous nerve**.
—passes posteriorly through the quadrangular space accompanied by the posterior circumflex humeral artery.
—winds around the surgical neck of the humerus.

5. **Radial Nerve (C5–T1)**
—is the largest branch of the brachial plexus and occupies the musculospiral groove on the back of the humerus with the profunda brachii artery (see p 44).

IX. Functional Components of the Peripheral Nerves

A. **Somatic Motor Nerves** (radial, axillary, median, musculocutaneous, and ulnar nerves)
—contain nerve fibers with cell bodies that are located in the following areas:

1. **Dorsal root ganglia** (for GSA and GVA fibers).

2. **Anterior horn of the spinal cord** (for GSE fibers).

3. **Sympathetic chain ganglia** (for sympathetic postganglionic GVE fibers).

B. **Cutaneous Nerves** (medial brachial and medial antebrachial cutaneous nerves)
—contain nerve fibers with cell bodies that are located in the following areas:

1. **Dorsal root ganglia** (for SSA and GVA fibers).

2. **Sympathetic chain ganglia** (for sympathetic postganglionic GVE fibers).

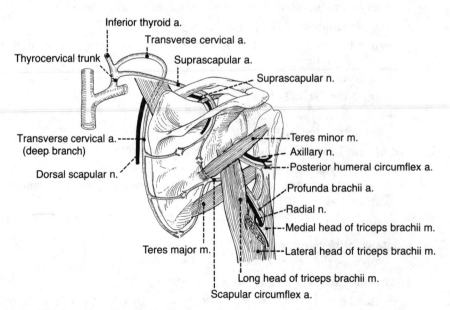

Inferior thyroid a.
Transverse cervical a.
Thyrocervical trunk
Suprascapular a.
Suprascapular n.
Transverse cervical a. (deep branch)
Dorsal scapular n.
Teres minor m.
Axillary n.
Posterior humeral circumflex a.
Profunda brachii a.
Radial n.
Medial head of triceps brachii m.
Lateral head of triceps brachii m.
Teres major m.
Long head of triceps brachii m.
Scapular circumflex a.

Figure 2.10. Blood supply to the dorsal scapular region.

Shoulder Region

I. Muscles

Muscle	Origin	Insertion	Nerve	Action
Deltoid	Lateral third of clavicle, acromion, and spine of scapula	Deltoid tuberosity of humerus	Axillary n.	Abducts, adducts, flexes, extends, and rotates arm medially
Supraspinatus	Supraspinous fossa of scapula	Superior facet of greater tubercle of humerus	Suprascapular n.	Abducts arm
Infraspinatus	Infraspinous fossa	Middle facet of greater tubercle of humerus	Suprascapular n.	Rotates arm laterally
Subscapularis	Subscapular fossa	Lesser tubercle of humerus	Upper and lower subscapular n.	Rotates arm medially
Teres major	Dorsal surface of inferior angle of scapula	Medial lip of intertubercular groove of humerus	Lower subscapular n.	Adducts and rotates arm medially
Teres minor	Upper portion of lateral border of scapula	Lower facet of greater tubercle of humerus	Axillary n.	Rotates arm laterally
Latissimus dorsi	Spines of T7–T12 thoracolumbar fascia, iliac crest, ribs 9–12	Floor of bicipital groove of humerus	Thoracodorsal n.	Adducts, extends, and rotates arm medially

II. Quadrangular and Triangular Spaces (Figure 2.10)

A. Quadrangular Space
—is bounded superiorly by the teres minor and subscapularis muscles, inferiorly by the teres major muscle, medially by the long head of the triceps, and laterally by the surgical neck of the humerus.
—transmits the axillary nerve and the posterior humeral circumflex vessels.

B. Triangular Space
—is bounded superiorly by the teres minor muscle, inferiorly by the teres major muscle, and laterally by the long head of the triceps.
—contains the circumflex scapular vessels.

C. Triangular Interval
—is formed superiorly by the teres major muscle, medially by the long head of the triceps, and laterally by the medial head of the triceps.
—contains the radial nerve and the profunda brachii (deep brachial) artery.

D. Triangle of Auscultation

—is bounded by the upper border of the latissimus dorsi muscle, the lateral border of the trapezius muscle, and the medial border of the scapula; its floor is formed by the rhomboid major muscle.

—is most prominent when the shoulders are drawn forward.

—is the site at which breathing sounds are heard most clearly.

III. Arteries (see Figure 2.10)

A. Suprascapular Artery

—is a branch of the thyrocervical trunk.

—passes over the superior transverse scapular ligament (whereas the suprascapular nerve passes under the ligament).

—anastomoses with the deep branch of the transverse cervical artery (dorsal scapular) and the circumflex scapular artery around the scapula, providing a collateral circulation.

—supplies the supraspinatus, infraspinatus, shoulder, and acromioclavicular joints.

B. Dorsal Scapular or Descending Scapular Artery

—arises from the subclavian artery but may be a deep branch of the transverse cervical artery.

—accompanies the dorsal scapular nerve and supplies the levator scapulae, rhomboids, and serratus anterior muscles.

C. Arterial Anastomoses Around the Scapula

—occur between:

1. Suprascapular and circumflex scapular arteries.

2. Descending scapular and circumflex scapular arteries.

3. Descending scapular and posterior intercostal arteries.

4. Suprascapular, acromial, and posterior humeral circumflex arteries.

IV. Nerves (see Figures 2.9 and 2.10)

A. Suprascapular Nerve (C5–C6)

—runs laterally across the posterior cervical triangle and passes under the superior transverse scapular ligament, whereas the suprascapular artery passes over that ligament.

—supplies the supraspinatus muscle and the shoulder joint and then enters the infraspinous fossa lateral to the spine of the scapula to end at the infraspinatus muscle.

B. Dorsal Scapular Nerve (C5)

—pierces the scalenus medius muscle and descends deep to the levator scapulae and the rhomboid minor and major to supply these muscles.

C. **Spinal Accessory Nerve**

—lies on the levator scapulae in the posterior cervical triangle.
—accompanies the superficial branch of the transverse cervical artery on the deep surface of the trapezius muscle.
—supplies the sternocleidomastoid and trapezius muscles.

V. Shoulder Joint and Associated Structures

A. **Shoulder (Glenohumeral) Joint**

—is a multiaxial ball-and-socket joint located between the glenoid cavity of the scapula and the head of the humerus.
—both articular surfaces are covered with hyaline cartilage.
—is supplied by the axillary, suprascapular, and lateral pectoral nerves.
—its capsule lies deep to the tendon of the **musculotendinous cuff** and is attached to the margin of the glenoid cavity and to the anatomic neck of the humerus.
—its cavity is deepened by the fibrocartilaginous glenoid lip and communicates with the subscapular bursa.
—its anterior dislocation stretches the fibrous capsule, avulses the glenoid labrum, and may injure the axillary nerve.

B. **Acromioclavicular Joint**

—is a plane joint between the acromion and the lateral border of the clavicle.
—its articular surfaces are covered by fibrous cartilages.
—its stability is provided by the coracoclavicular ligament, which consists of the conoid and trapezoid ligaments.

C. **Rotator (Musculotendinous) Cuff**

—contributes the stability of the shoulder joint by keeping the head of the humerus pressed into the glenoid fossa.
—is formed by the tendons of the subscapularis, supraspinatus, infraspinatus, and teres minor muscles.

D. **Bursae Around the Shoulder**

—form a lubricating mechanism between the rotator cuff and coracoacromial arch during movement of the shoulder joint.

1. **Subacromial Bursa**

—lies between the acromion and the coracoacromial ligament superiorly and the supraspinatus muscle inferiorly.
—frequently communicates with the subdeltoid bursa but normally does not communicate with the synovial cavity of the glenohumeral joint.
—facilitates the movement of the deltoid muscle over the joint capsule and the supraspinatus tendon.

2. **Subdeltoid Bursa**

—lies between the deltoid muscle and coracoacromial arch and the tendon of the supraspinatus muscle.
—frequently communicates with the subacromial bursa.

3. Subscapular Bursa

—lies between the tendon of the subscapularis and the neck of the scapula.
—communicates with the synovial cavity of the shoulder joint.

E. Ligaments

1. Coracohumeral Ligament

—extends from the coracoid process to the greater tubercle of the humerus.

2. Glenohumeral Ligaments

a. Superior Glenohumeral Ligament

—extends from the supraglenoid tubercle to the upper part of the lesser tubercle of the humerus.

b. Middle Glenohumeral Ligament

— extends from the supraglenoid tubercle to the lower anatomic neck.

c. Inferior Glenohumeral Ligament

—extends from the supraglenoid tubercle to the lower part of the lesser tubercle.

3. Transverse Humeral Ligament

—extends between the greater and lesser tubercles and holds the tendon of the long head of the biceps in the intertubercular groove.

4. Coracoacromial Ligament

—extends from the coracoid process to the acromion.

VI. Clinical Considerations

A. Referred Pain to the Shoulder

—most probably indicates involvement of the phrenic nerve (or diaphragm).
—has the same origin as the phrenic nerve (C3–C5), which supplies the diaphragm and the supraclavicular nerve (C3–C4), which supplies sensory fibers over the shoulder.

B. Inferior Dislocation of the Humerus

—is not uncommon because the inferior aspect of the shoulder joint is not supported by muscles.
—may damage the axillary nerve and the posterior humeral circumflex vessels.

Arm and Forearm

I. Muscles of the Arm

Muscle	Origin	Insertion	Nerve	Action
Coracobra-chialis	Coracoid process	Middle third of medial surface of humerus	Musculocutaneous n.	Flexes and adducts arm

(Continued on next page)

Muscle	Origin	Insertion	Nerve	Action
Biceps brachii	Long head, supraglenoid tubercle; short head, coracoid process	Radial tuberosity of radius	Musculocutaneous n.	Flexes arm and forearm, supinates forearm
Brachialis	Lower anterior surface of humerus	Coronoid process of ulna and ulnar tuberosity	Musculocutaneous n.	Flexes forearm
Triceps	Long head, infraglenoid tubercle; lateral head, superior to radial groove of humerus; medial head, inferior to radial groove	Posterior surface of olecranon process of ulna	Radial n.	Extends forearm
Anconeus	Lateral epicondyle of humerus	Olecranon and upper posterior surface of ulna	Radial n.	Extends forearm

II. Muscles of the Anterior Forearm

Muscle	Origin	Insertion	Nerve	Action
Pronator teres	Medial epicondyle and coronoid process of ulna	Middle of lateral side of radius	Median n.	Pronates forearm
Flexor carpi radialis	Medial epicondyle of humerus	Bases of second and third metacarpals	Median n.	Flexes forearm, flexes and abducts hand
Palmaris longus	Medial epicondyle of humerus	Flexor retinaculum, palmar aponeurosis	Median n.	Flexes hand and forearm
Flexor carpi ulnaris	Medial epicondyle, medial olecranon, and posterior border of ulna	Pisiform, hook of hamate, and base of fifth metacarpal	Ulnar n.	Flexes and adducts hand, flexes forearm
Flexor digitorum superficialis	Medial epicondyle, coronoid process, oblique line of radius	Middle phalanges of finger	Median n.	Flexes proximal interphalangeal joints, flexes hand and forearm
Flexor digitorum profundus	Anteromedial surface of ulna, interosseous membrane	Bases of distal phalanges of fingers	Ulnar and median nn.	Flexes distal interphalangeal joints and hand

(Continued on next page)

Muscle	Origin	Insertion	Nerve	Action
Flexor pollicis longus	Anterior surface of radius, interosseous membrane, and coronoid process	Base of distal phalanx of thumb	Median n.	Flexes thumb
Pronator quadratus	Anterior surface of distal ulna	Anterior surface of distal radius	Median n.	Pronates forearm

III. Muscles of the Posterior Forearm

Muscle	Origin	Insertion	Nerve	Action
Brachioradialis	Lateral supracondylar ridge of humerus	Base of radial styloid process	Radial n.	Flexes forearm
Extensor carpi radialis longus	Lateral supracondylar ridge of humerus	Dorsum of base of second metacarpal	Radial n.	Extends and abducts hand
Extensor carpi radialis brevis	Lateral epicondyle of humerus	Posterior base of third metacarpal	Radial n.	Extends fingers and abducts hands
Extensor digitorum	Lateral epicondyle of humerus	Extensor expansion, base of middle and digital phalanges	Radial n.	Extends fingers and hand
Extensor digiti minimi	Common extensor tendon and interosseous membrane	Extensor expansion, base of middle and distal phalanges	Radial n.	Extends little finger
Extensor carpi ulnaris	Lateral epicondyle and posterior surface of ulna	Base of fifth metacarpal	Radial n.	Extends and adducts hand
Supinator	Lateral epicondyle, radial collateral and annular ligaments	Lateral side of upper part of radius	Radial n.	Supinates forearm
Abductor pollicis longus	Interosseous membrane, middle third of posterior surfaces of radius and ulna	Lateral surface of base of first metacarpal	Radial n.	Abducts thumb and hand

(Continued on next page)

Muscle	Origin	Insertion	Nerve	Action
Extensor pollicis longus	Interosseous membrane and middle third of posterior surface of ulna	Base of distal phalanx of thumb	Radial n.	Extends distal phalanx of thumb and abducts hand
Extensor pollicis brevis	Interosseous membrane and posterior surface of middle third radius	Base of proximal phalanx of thumb	Radial n.	Extends proximal phalanx of thumb and abducts hand
Extensor indicis	Posterior surface of ulna and interosseous membrane	Extensor expansion of index finger	Radial n.	Extends index finger

IV. Arteries (see Figure 2.8)

A. Brachial Artery

—extends from the inferior border of the teres major muscle to its bifurcation in the cubital fossa.

—lies on the triceps brachii and brachialis muscles medial to the coracobrachialis and biceps brachii muscles and is accompanied by the basilic vein in the middle of the arm.

—lies in the center of the cubital fossa, medial to the biceps tendon, lateral to the median nerve, and deep to the bicipital aponeurosis.

—provides muscular branches and terminates by dividing into the radial and ulnar arteries at the level of the radial neck, about 1 cm below the bend of the elbow, in the cubital fossa.

1. Profunda Brachii (Deep Brachial) Artery

—descends posteriorly with the radial nerve and gives off an ascending branch, which anastomoses with the descending branch of the posterior humeral circumflex artery.

—divides into the middle collateral artery, which anastomoses with the interosseous recurrent artery, and the radial collateral artery, which follows the radial nerve through the lateral intermuscular septum and ends in front of the lateral epicondyle by anastomosing with the radial recurrent artery of the radial artery.

2. Superior Ulnar Collateral Artery

—pierces the medial intermuscular septum, accompanies the ulnar nerve behind the septum and medial epicondyle, and anastomoses with the posterior ulnar recurrent branch of the ulnar artery.

3. Inferior Ulnar Collateral Artery

—arises just above the elbow, descends in front of the medial epicondyle, and anastomoses with the anterior ulnar recurrent branch of the ulnar artery.

B. **Radial Artery**

—arises as the smaller lateral branch of the brachial artery in the cubital fossa and descends laterally under cover of the brachioradialis muscle, with the superficial radial nerve on its lateral side, on the supinator and flexor pollicis longus muscles.

—curves over the radial side of the carpal bones beneath the tendons of the abductor pollicis longus muscle, the extensor pollicis longus and brevis muscles, and over the surface of the scaphoid and trapezium bones.

—runs through the anatomical snuff-box, enters the palm by passing between the two heads of the first dorsal interosseous muscle and then between the heads of the adductor pollicis muscle, and divides into the princeps pollicis artery and the deep palmar arch.

—gives off the following branches:

1. **Radial Recurrent Artery**

—arises from the radial artery just below its origin and ascends on the supinator muscle and then between the brachioradialis and brachialis muscles.

—anastomoses with the radial collateral branch of the profunda brachii artery.

2. **Palmar Carpal Branch**

—joins the palmar carpal branch of the ulnar artery and forms the palmar carpal arch.

3. **Superficial Palmar Branch**

—passes through the thenar muscles and anastomoses with the superficial branch of the ulnar artery to complete the superficial palmar arterial arch.

4. **Dorsal Carpal Branch**

—joins the dorsal carpal branch of the ulnar artery and the dorsal terminal branch of the anterior interosseous artery to form the dorsal carpal rete (or arch).

5. **Deep Palmar Arch**

—is formed by the terminal part of the radial artery in conjunction with the deep palmar branch of the ulnar artery.

C. **Ulnar Artery**

—is the larger medial branch of the brachial artery in the cubital fossa.

—descends behind the ulnar head of the pronator teres muscle and lies between the flexor digitorum superficialis and profundus muscles.

—enters the hand anterior to the flexor retinaculum, lateral to the pisiform bone, and medial to the hook of the hamate bone.

—divides into the superficial palmar arch and the deep palmar branch, which passes between the abductor and flexor digiti minimi brevis muscles and runs medially to join the radial artery to complete the deep palmar arch.

—gives off the following branches:

1. **Anterior Ulnar Recurrent Artery**

—anastomoses with the inferior ulnar collateral artery.

2. **Posterior Ulnar Recurrent Artery**

—anastomoses with the superior ulnar collateral artery.

3. **Common Interosseous Artery**

—arises from the lateral side of the ulnar artery and divides into the anterior and posterior interosseous arteries.

a. **Anterior Interosseous Artery**

—descends with the anterior interosseous nerve in front of the interosseous membrane, located between the flexor digitorum profundus and the flexor pollicis longus muscles.

—perforates the interosseous membrane to anastomose with the posterior interosseous artery and join the dorsal carpal network.

b. **Posterior Interosseous Artery**

—gives off an interosseous recurrent artery, which anastomoses with a middle collateral branch of the profunda brachii artery.

—descends behind the interosseous membrane in company with the posterior interosseous nerve.

—anastomoses with the dorsal carpal branch of the anterior interosseous artery.

4. **Palmar Carpal Branch**

—joins the palmar carpal branch of the radial artery to form the palmar carpal arch.

5. **Dorsal Carpal Branch**

—passes around the ulnar side of the wrist and joins the dorsal carpal rete.

6. **Deep Palmar Branch**

—anastomoses with the radial artery and completes the **deep palmar arch**.

7. **Superficial Palmar Arch**

—is formed by the ulnar artery and usually is completed by the superficial palmar branch of the radial artery.

D. **Anastomoses Around the Elbow Joint**

—occur between the following arteries:

1. Radial collateral artery and the radial recurrent artery in front of the lateral epicondyle.

2. Middle collateral artery and the interosseous recurrent artery behind the lateral epicondyle.

3. Inferior ulnar collateral artery and the anterior ulnar recurrent artery in front of the medial epicondyle.

4. Superior ulnar collateral artery and the posterior ulnar recurrent artery behind the medial epicondyle.

V. Nerves (Figures 2.11–2.13; see Figure 2.9)

A. **Musculocutaneous Nerve (C5–C7)**

—pierces the coracobrachialis muscle and descends between the biceps and brachialis muscles.

—innervates all of the flexor muscles in the anterior compartment of the arm, such as the coracobrachialis, biceps, and brachialis muscles.

—continues into the forearm as the lateral antebrachial cutaneous nerve.

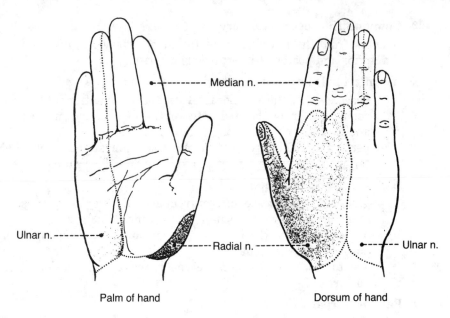

Median n.

Ulnar n.

Radial n.

Ulnar n.

Palm of hand

Dorsum of hand

Figure 2.11. Cutaneous innervation of the hand.

B. Median Nerve (C5–T1)

—runs down the anteromedial aspect of the arm (has no muscular branches in the arm).

—passes through the cubital fossa, deep to the bicipital aponeurosis and medial to the brachial artery.

—enters the forearm between the humeral and ulnar heads of the pronator teres muscle and then passes between the flexor digitorum superficialis and the flexor digitorum profundus muscles.

—gives off in the cubital fossa the **anterior interosseous nerve,** which descends on the interosseous membrane between the flexor digitorum profundus muscle and the flexor pollicis longus muscle, and then passes behind the pronator quadratus muscle, supplying these three muscles.

—innervates all of the muscles of the forearm except the flexor carpi ulnaris and the ulnar half of the flexor digitorum profundus.

—enters the hand through the carpal tunnel, deep to the flexor retinaculum (see p 51).

C. Axillary Nerve (C5–C6)

—passes posteriorly through the quadrangular space accompanied by the posterior humeral circumflex artery.

—winds around the surgical neck of the humerus (may be injured when this part of the bone is fractured).

—innervates the deltoid muscle (by its anterior and posterior branches) and the teres minor muscle (by its posterior branch).

—gives rise to the **lateral brachial cutaneous nerve.**

D. Radial Nerve (C5–T1)

—is the largest branch of the brachial plexus.

—runs down the posterior aspect of the arm and lies in the radial groove on the back of the humerus with the profunda brachii artery.

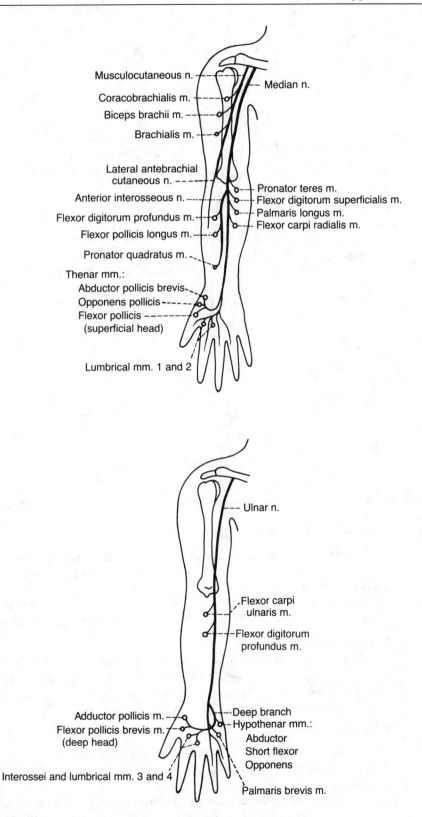

Musculocutaneous n.

Median n.

Coracobrachialis m.

Biceps brachii m.

Brachialis m.

Lateral antebrachial cutaneous n.

Pronator teres m.

Anterior interosseous n.

Flexor digitorum superficialis m.

Palmaris longus m.

Flexor digitorum profundus m.

Flexor carpi radialis m.

Flexor pollicis longus m.

Pronator quadratus m.

Thenar mm.:
Abductor pollicis brevis
Opponens pollicis
Flexor pollicis
(superficial head)

Lumbrical mm. 1 and 2

Ulnar n.

Flexor carpi ulnaris m.

Flexor digitorum profundus m.

Adductor pollicis m.

Deep branch

Flexor pollicis brevis m.
(deep head)

Hypothenar mm.:
Abductor
Short flexor
Opponens

Interossei and lumbrical mm. 3 and 4

Palmaris brevis m.

Figure 2.12. Distribution of the musculocutaneous, median, and ulnar nerves.

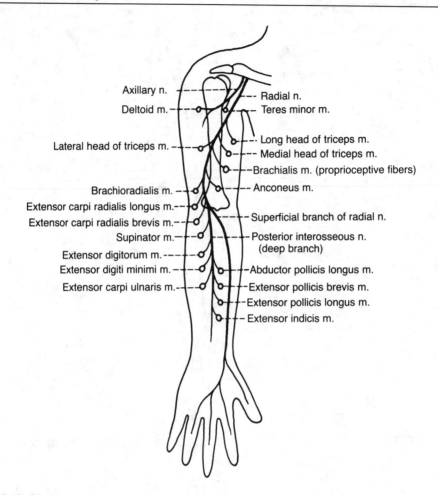

Axillary n.
Deltoid m.
Lateral head of triceps m.
Brachioradialis m.
Extensor carpi radialis longus m.
Extensor carpi radialis brevis m.
Supinator m.
Extensor digitorum m.
Extensor digiti minimi m.
Extensor carpi ulnaris m.

Radial n.
Teres minor m.
Long head of triceps m.
Medial head of triceps m.
Brachialis m. (proprioceptive fibers)
Anconeus m.
Superficial branch of radial n.
Posterior interosseous n. (deep branch)
Abductor pollicis longus m.
Extensor pollicis brevis m.
Extensor pollicis longus m.
Extensor indicis m.

Figure 2.13. Distribution of the axillary and radial nerves.

—gives off the **posterior brachial** and **antebrachial cutaneous nerves**.
—may be damaged by a midhumeral fracture, causing paralysis of the extensor muscles of the hand.
—passes anterior to the lateral epicondyle, between the brachialis and brachioradialis muscles, where it divides into superficial and deep branches.

1. **Deep Branch**

    —enters the supinator muscle, winds laterally around the radius in the substance of the muscle, and continues as the **posterior interosseous nerve** with the posterior interosseous artery.
    —supplies the muscles of the back of the forearm.

2. **Superficial Branch**

    —descends in the forearm under cover of the brachioradialis and then passes dorsally around the radius under the tendon of the brachioradialis.
    —runs distally to the dorsum of the hand to supply the radial side of the hand and the radial two and one-half digits over the proximal phalanx. This nerve does not supply the skin of the distal phalanges.

E. Ulnar Nerve (C7–T1)

—runs down the medial aspect of the arm and behind the medial epicondyle in a groove.

—descends between and supplies the flexor carpi ulnaris and flexor digitorum profundus muscles.

—enters the hand superficial to the flexor retinaculum and lateral to the pisiform bone and divides into superficial and deep branches (see pp 51–52).

—may be damaged by a fracture of the medial epicondyle and produce "funny-bone" symptoms.

VI. Cubital Fossa

—is a V-shaped interval on the anterior aspect of the elbow.

—is bounded laterally by the brachioradialis muscle and medially by the pronator teres muscle.

—its upper limit is an imaginary horizontal line connecting the epicondyles of the humerus.

—its floor is formed by the brachialis and supinator muscles.

—contains (from lateral to medial) the radial nerve, biceps tendon, brachial artery, and median nerve.

—its lower end is where the brachial artery divides into the radial and ulnar arteries, and its fascial roof is strengthened by the bicipital aponeurosis.

VII. Aponeurosis and Ligaments

A. Bicipital Aponeurosis

—originates from the medial border of the biceps tendon.

—lies on the brachial artery and the median nerve and passes downward and medially to blend with the deep fascia of the forearm.

B. Interosseous Membrane of the Forearm

—is a broad sheet of dense connective tissue extending between the radius and the ulna.

—its proximal border and the oblique cord form a gap through which the posterior interosseous vessels pass. (The oblique cord extends from the ulnar tuberosity to the radius, somewhat distal to its tuberosity.)

—provides extra surface area for attachment of the deep extrinsic flexor, extensor, and abductor muscles of the hand.

—its distal part is pierced by the anterior interosseous vessels.

C. Annular Ligament

—is a fibrous band that forms nearly four-fifths of a circle around the head of the radius; the remainder is formed by the radial notch.

—attaches to the anterior and posterior lips of the radial notch of the ulna.

—forms a collar around the head of the radius and thus serves as a restraining ligament, preventing withdrawal of the head of the radius from its socket.

—fuses with the radial collateral ligament and blends with the articular capsule of the elbow joint.

—attaches to the origin of the supinator muscle.

D. Radial Collateral Ligament

—is a fibrous thickening that extends from the lateral epicondyle to the anterior and posterior margins of the radial notch of the ulna and the annular ligament of the radius.

E. Ulnar Collateral Ligament

—is triangular-shaped and composed of anterior, posterior, and oblique bands.

—extends from the medial epicondyle to the coronoid process and the olecranon of the ulna.

VIII. Additional Characteristics of the Arm and Forearm

A. Carrying Angle

—is the angle formed by the axis of the arm and forearm when the forearm is extended, because the medial edge of the trochlea projects more inferiorly than its lateral edge; thus, the long axis of the humerus lies at an angle of about 170° to the long axis of the ulna.

—carries the hand away from the sides of the body in extension and pronation and is wider in women than in men.

—disappears when the forearm is flexed or pronated, because the trochlea runs in a spiral direction from anterior to posterior aspects.

B. Pronation and Supination

—occur at the superior and inferior radioulnar joints.

—in supination, the palm faces forward (lateral rotation); in pronation, the radius rotates over the ulna, and thus the palm faces backward (medial rotation about a longitudinal axis, in which case the shafts of the radius and ulna cross each other).

—are movements in which the upper end of the radius nearly rotates within the annular ligament.

—have unequal strengths, with supination being the stronger.

C. Radial Pulse

—can be felt proximal to the wrist between the tendons of the brachioradialis and flexor carpi radialis muscles.

—may also be palpated in the anatomical snuff-box between the tendons of the extensor pollicis longus and brevis muscles.

D. Ulnar Pulse

—is palpable just to the radial side of the insertion of the flexor carpi ulnaris into the pisiform bone.

Hand

I. Fascia, Aponeurosis, and Synovial Sheaths

A. Flexor Retinaculum

—serves as an attachment (origin) for muscles of the thenar eminence.

—forms a **carpal (osteofacial) tunnel** on the anterior aspect of the wrist.

—is attached medially to the triquetrum and pisiform bones and the hook of hamate and laterally to the tubercles of the scaphoid and trapezium bones.

B. Extensor Retinaculum

—is a thickening of the antebrachial fascia on the back of the wrist.
—extends from the lateral margin of the radius to the styloid process of the ulna, the pisiform, and the triquetrum.
—is crossed by the superficial branch of the radial nerve.

C. Palmar Aponeurosis

—is a triangular fibrous layer overlying the tendons in the palm.
—is continuous with the palmaris longus tendon, the thenar and hypothenar fasciae, the flexor retinaculum, and the palmar carpal ligament.
—its thickening, shortening, and fibrosis produce **Dupuytren's contracture**, a flexion deformity in which the fingers are pulled toward the palm.

D. Fascial Spaces of the Palm

—are large fascial spaces in the hand divided by a midpalmar (oblique) septum into the **thenar space** and the **midpalmar space**.

1. Thenar Space

—is a palmar space lying between the middle metacarpal bone and the tendon of the flexor pollicis longus muscle.

2. Midpalmar Space

—is a palmar space lying between the middle metacarpal bone and the radial side of the hypothenar eminence.

E. Synovial Flexor Sheaths

1. Common Synovial Flexor Sheath (Ulnar Bursa)

—envelops or contains both the tendons of the flexor digitorum superficialis and profundus muscles.

2. Synovial Sheath for Flexor Pollicis Longus (Radial Bursa)

—envelops the tendon of the flexor pollicis longus muscle.

F. Extensor Aponeurosis (Figure 2.14)

—is the expansion of the extensor tendon over the metacarpophalangeal joint and is referred to by clinicians as the **extensor hood**.
—provides the insertion of the lumbrical and interosseous muscles.

G. Carpal Tunnel

—is formed anteriorly by the flexor retinaculum and posteriorly by the carpal bones.
—transmits the flexor tendons and the median nerve, which may be compressed.

H. Anatomical Snuff-Box

—is a triangular-shaped interval bounded medially by the tendon of the extensor pollicis longus muscle and laterally by the tendons of the extensor pollicis brevis and abductor pollicis longus muscles.
—is limited proximally by the styloid process of the radius.
—its floor is formed by the scaphoid and trapezium bones and is crossed by the radial artery.

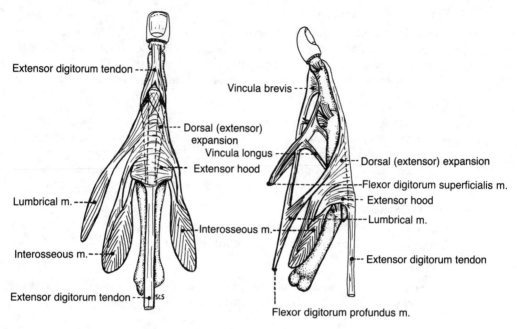

Figure 2.14. Dorsal (extensor) expansion of the middle finger.

II. Muscles of the Hand (Figures 2.15 and 2.16)

Muscle	Origin	Insertion	Nerve	Action
Abductor pollicis brevis	Flexor retinaculum, scaphoid, and trapezium	Lateral side of base of proximal phalanx of thumb	Median n.	Abducts thumb
Flexor pollicis brevis	Flexor retinaculum and trapezium	Base of proximal phalanx of thumb	Median n.	Flexes thumb
Opponens pollicis	Flexor retinaculum and trapezium	Lateral side of first metacarpal	Median n.	Opposes thumb to other digits
Adductor pollicis	Capitate and bases of second and third metacarpals (oblique head); palmar surface of third metacarpal (transverse head)	Medial side of base of proximal phalanx of the thumb	Ulnar n.	Adducts thumb
Palmaris brevis	Medial side of flexor retinaculum, palmar aponeurosis	Skin of medial side of palm	Ulnar n.	Wrinkles skin on medial side of palm

(Continued on next page)

Muscle	Origin	Insertion	Nerve	Action
Abductor digiti minimi	Pisiform and tendon of flexor carpi ulnaris	Medial side of base of proximal phalanx of little finger	Ulnar n.	Abducts little finger
Flexor digiti minimi brevis	Flexor retinaculum and hook of hamate	Medial side of base of proximal phalanx of little finger	Ulnar n.	Flexes proximal phalanx of little finger
Opponens digiti minimi	Flexor retinaculum and hook of hamate	Medial side of fifth metacarpal	Ulnar n.	Opposes little finger
Lumbricals (4)	Lateral side of tendons of flexor digitorum profundus	Lateral side of extensor expansion	Median (two lateral) and ulnar (two medial) nn.	Flex metacarpophalangeal joints and extend interphalangeal joints
Dorsal interossei (4)	Adjacent sides of metacarpal bones	Lateral sides of bases of proximal phalanges; extensor expansion	Ulnar n.	Abduct fingers; flex metacarpophalangeal joints; extend interphalangeal joints
Palmar interossei (3)	Medial side of second metacarpal; lateral sides of fourth and fifth metacarpals	Bases of proximal phalanges in same sides as their origins; extensor expansion	Ulnar n.	Adduct fingers; flex metacarpophalangeal joints; extend interphalangeal joints

III. Nerves (see Figures 2.11 and 2.12)

A. Median Nerve

—enters the palm of the hand through the carpal tunnel deep to the flexor retinaculum, gives off a muscular branch (the recurrent branch) to the thenar muscles, and terminates by dividing into three common palmar digital nerves, which then divide into the palmar digital branches.

—supplies the lateral two lumbricals, the skin of the lateral side of the palm, and the palmar side of the lateral three and one-half fingers, as well as the dorsal side of the index finger, middle finger, and one-half of the ring finger.

B. Ulnar Nerve

—enters the hand superficial to the flexor retinaculum and terminates by dividing into the superficial and deep branches at the root of the hypothenar eminence.

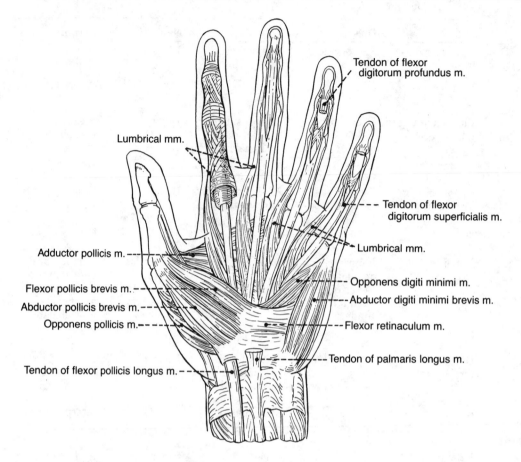

Tendon of flexor
digitorum profundus m.

Lumbrical mm.

Tendon of flexor
digitorum superficialis m.

Lumbrical mm.

Adductor pollicis m.

Opponens digiti minimi m.

Flexor pollicis brevis m.

Abductor digiti minimi brevis m.

Abductor pollicis brevis m.

Opponens pollicis m.

Flexor retinaculum m.

Tendon of palmaris longus m.

Tendon of flexor pollicis longus m.

Figure 2.15. Superficial muscles of the hand.

1. Superficial Branch

—supplies the palmaris brevis and the skin of the hypothenar eminence
and terminates in the palm by dividing into three palmar digital
branches, which supply the skin of the little finger and the medial side
of the ring finger.

2. Deep Branch

—arises at about the level of the pisiform bone and enters the palm under
cover of the flexor digiti minimi brevis muscle.

—turns around the hook of hamate and then turns laterally with the deep
palmar arterial arch under cover of the flexor tendons.

—supplies all three of the hypothenar muscles, the third and fourth lumbricals, all of the interossei, and the adductor pollicis muscle.

IV. Arteries of the Hand (Figure 2.17)

A. Ulnar Artery

—enters the hand anterior to the flexor retinaculum and ends at the radial
side of the pisiform bone by dividing into the superficial palmar arch and
the deep palmar branch.

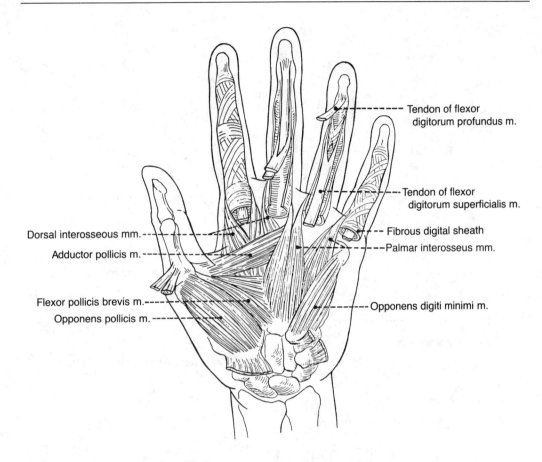

Dorsal interosseous mm.

Adductor pollicis m.

Flexor pollicis brevis m.

Opponens pollicis m.

Tendon of flexor digitorum profundus m.

Tendon of flexor digitorum superficialis m.

Fibrous digital sheath

Palmar interosseus mm.

Opponens digiti minimi m.

Figure 2.16. Deep muscles of the hand.

1. Superficial Palmar Arch

—is formed by the main termination of the ulnar artery but usually is completed by the superficial palmar branch of the radial artery.

—lies immediately under the palmar aponeurosis.

—gives off three common palmar digital arteries, each of which bifurcates into proper palmar digital arteries, which run distally to supply the adjacent sides of the fingers.

2. Deep Palmar Branch

—accompanies the deep branch of the ulnar nerve through the hypothenar muscles and anastomoses with the radial artery, thereby completing the deep palmar arch.

B. Radial Artery

—enters the palm by passing between the heads of the first dorsal interosseous muscle, and then passes between the two heads of the adductor pollicis muscle to complete the deep palmar arch.

—gives off the following arteries:

1. Princeps Pollicis Artery

—descends along the ulnar border of the first metacarpal bone under the flexor pollicis longus tendon.

—divides into two proper digital arteries for each side of the thumb.

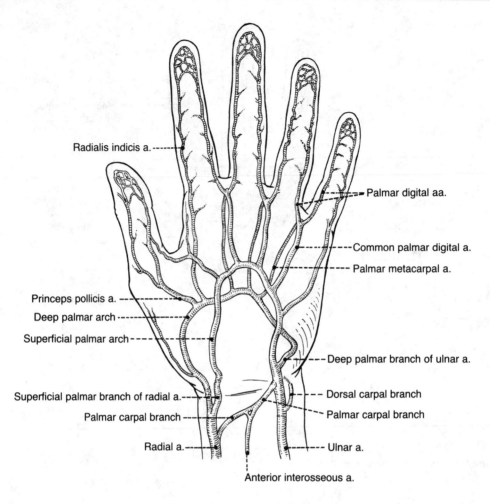

Radialis indicis a.

Palmar digital aa.

Common palmar digital a.

Palmar metacarpal a.

Princeps pollicis a.

Deep palmar arch

Superficial palmar arch

Deep palmar branch of ulnar a.

Superficial palmar branch of radial a.

Dorsal carpal branch

Palmar carpal branch

Palmar carpal branch

Radial a.

Ulnar a.

Anterior interosseous a.

Figure 2.17. Blood supply to the hand.

2. Radialis Indicis Artery

—also may arise from the deep palmar arch or the princeps pollicis artery.

3. Deep Palmar Arch

—is formed by the main termination of the radial artery and usually is completed by the deep palmar branch of the ulnar artery.

—passes between the transverse and oblique heads of the adductor pollicis muscle.

—gives off three palmar metacarpal arteries, which descend on the interossei and join the common palmar digital arteries from the superficial palmar arch.

C. Structures Inside the Carpal Tunnel

—median nerve and flexor pollicis longus, flexor digitorum profundus, and flexor digitorum superficialis muscles.

D. Structures Entering the Palm Superficial to the Flexor Retinaculum

—ulnar nerve, ulnar artery, palmaris longus tendon, and palmar cutaneous branch of the median nerve.

Clinical Considerations

I. Lesions of the Peripheral Nerves

A. Upper Trunk Injury (Erb-Duchenne Paralysis)

—is caused by a violent displacement of the head from the shoulder, such as results from a fall from a motorcycle or horse.

—results in a "waiter's tip" hand, in which the arm tends to lie in medial rotation due to paralysis of lateral rotator muscles.

B. Lower Trunk Injury (Klumpke's Paralysis)

—is caused during a difficult breech delivery (birth palsy or obstetric paralysis), or by a cervical rib (cervical rib syndrome), or by abnormal insertion or spasm of the anterior and middle scalene muscles (scalene syndrome).

—results in a "claw hand."

C. Injury to the Posterior Cord

—is caused by the pressure of the crosspiece of a crutch, resulting in paralysis of the arm called "crutch palsy."

—results in loss of the extensors of the arm, forearm, and hand.

—produces a "wrist drop."

D. Injury to the Long Thoracic Nerve

—is caused by a stab wound or during thoracic surgery.

—results in winging of the scapula due to paralysis of the serratus anterior muscle.

E. Injury to the Musculocutaneous Nerve

—results in weakness of supination (biceps) and forearm flexion (brachialis and biceps).

F. Injury to the Axillary Nerve

—is caused by a fracture of the surgical neck of the humerus or inferior dislocation of the humerus.

—results in weakness of lateral rotation and abduction of the arm (the supraspinatus can abduct the arm but not to a horizontal level).

G. Injury to the Radial Nerve

—is caused by a fracture of the midshaft of the humerus.

—results in loss of the extensors of the forearm, hand, metacarpals, and phalanges.

—results in loss of wrist extension, leading to wrist drop.

—produces a weakness of abduction and adduction of the hand.

H. Injury to the Ulnar Nerve

—is caused by a fracture of the medial epicondyle.

—results in a "claw hand," in which the ring and little fingers are hyperextended at the metacarpophalangeal joints and flexed at the interphalangeal joints.

—results in loss of abduction and adduction of the fingers and flexion of the metacarpophalangeal joints, owing to paralysis of the palmar and dorsal interossei muscles and the medial two lumbricals.

—results in loss of adduction of the thumb, owing to paralysis of the adductor pollicis muscle.

I. Injury to the Median Nerve
—may be caused by a supracondylar fracture of the humerus.

—results in loss of pronation, opposition of the thumb, flexion of the lateral two interphalangeal joints, and impairment of the medial two interphalangeal joints.

—produces a characteristic flattening of the thenar eminence, often referred to as "ape hand."

II. Fractures and Syndromes
A. Fracture of the Clavicle
—results in upward displacement of the proximal fragment, owing to the pull of the sternocleidomastoid muscle, and downward displacement of the distal fragment, owing to the pull of the deltoid muscle and gravity.

—may be caused by the obstetrician in breech (buttocks) presentation, or it may occur when the child presses against the maternal symphysis pubis during its passage through the birth canal.

—may cause an injury of the brachial plexus (lower trunk) and a fatal hemorrhage from the subclavian vein.

—is also responsible for thrombosis of the subclavian vein, leading to pulmonary embolism.

B. Colles' Fracture
—is a fracture of the lower end of the radius in which the distal fragment is displaced posteriorly.

—if the distal fragment is displaced anteriorly, it is called a reverse Colles' fracture (Smith's fracture).

C. Carpal Tunnel Syndrome
—is caused by compression of the median nerve due to the reduced size of the osseofibrous carpal tunnel, resulting from inflammation of the flexor retinaculum, anterior dislocations of the lunate bone, arthritic changes, or inflammation of the tendon and its sheath by fibers of the flexor retinaculum.

—leads to pain and paresthesia (tingling, burning, and numbness) in the hand in the area of the median nerve.

—may also cause atrophy of the thenar muscles in cases of severe compression.

D. Dupuytren's Contracture
—is a disease of the palmar fascia resulting in thickening and contracture of fibrous bands on the palmar surface of the hand and fingers.

Summary of Muscle Actions of the Upper Limb
Movement of the Scapula

Elevation—trapezius (upper part), levator scapulae

Depression—trapezius (lower part), serratus anterior, pectoralis minor

Protrusion (forward or lateral movement; abduction)—serratus anterior
Retraction (backward or medial movement; adduction)—trapezius, rhomboids
Anterior or inferior rotation of the glenoid fossa—rhomboid major
Posterior or superior rotation of the glenoid fossa—serratus anterior, trapezius

Movement at the Shoulder Joint (Ball-and-Socket Joint)
Adduction—pectoralis major, latissimus dorsi, deltoid (posterior part)
Abduction—deltoid, supraspinatus
Flexion—pectoralis major (clavicular part), deltoid (anterior part), coracobrachialis, biceps
Extension—latissimus dorsi, deltoid (posterior part)
Medial rotation—subscapularis, pectoralis major, deltoid (anterior part), latissimus dorsi, teres major
Lateral rotation—infraspinatus, teres minor, deltoid (posterior part)

Movement at the Elbow Joint (Hinge Joint)
Flexion—brachialis, biceps, brachioradialis, pronator teres
Extension—triceps (medial head), anconeus

Movement at the Radioulnar Joints (Pivot Joints)
Pronation—pronator quadratus, pronator teres
Supination—supinator, biceps brachii

Movement at the Radiocarpal and Midcarpal Joints (Ellipsoid Joints)
Adduction—flexor carpi ulnaris, extensor carpi ulnaris
Abduction—flexor carpi radialis, extensor carpi radialis longus and brevis
Flexion—flexor carpi radialis, flexor carpi ulnaris, palmaris longus, abductor pollicis longus
Extension—extensor carpi radialis longus and brevis, extensor carpi ulnaris

Movement at the Metacarpophalangeal Joint (Ellipsoid Joint)
Adduction—palmar interossei
Abduction—dorsal interossei
Flexion—lumbricals and interossei
Extension—extensor digitorum

Movement at the Interphalangeal Joint (Hinge Joint)
Flexion—flexor digitorum superficialis (proximal interphalangeal joint), flexor digitorum profundus (distal interphalangeal joint)
Extension—lumbricals and interossei (when metacarpophalangeal joint is extended by extensor digitorum)
Extension—extensor digitorum (when metacarpophalangeal joint is flexed by lumbricals and interossei)

Summary of Muscle Innervations of the Upper Limb

Muscles of the Anterior Compartment of the Arm: Musculocutaneous Nerve
Biceps brachii
Coracobrachialis
Brachialis

Muscles of the Posterior Compartment of the Arm: Radial Nerve
Triceps
Anconeus

Muscles of the Posterior Compartment of the Forearm: Radial Nerve

Superficial layer—brachioradialis; extensor carpi radialis longus; extensor carpi radialis brevis; extensor carpi ulnaris; extensor digitorum communis; extensor digiti minimi

Deep layer—supinator; abductor pollicis longus; extensor pollicis longus; extensor pollicis brevis; extensor indicis

Muscles of the Anterior Compartment of the Forearm: Median Nerve

Superficial layer—pronator teres; flexor carpi radialis; palmaris longus; flexor carpi ulnaris (ulnar nerve)*

Middle layer—flexor digitorum superficialis

Deep layer—flexor digitorum profundus (median nerve and ulnar nerve)*; flexor pollicis longus; pronator quadratus

Thenar Muscles: Median Nerve

Abductor pollicis brevis
Opponens pollicis
Flexor pollicis brevis (median and ulnar nerves)*

Abductor Pollicis Muscle: Ulnar Nerve

Hypothenar Muscles: Ulnar Nerve

Abductor digiti minimi
Opponens digiti minimi
Flexor digiti minimi

Lumbrical (Medial Two) and Interossei Muscles: Ulnar Nerve

Lumbrical Muscles (Lateral Two): Median Nerve

---

*Indicates exception or dual innervation.

Review Test

DIRECTIONS: Each of the numbered items or incomplete statements in this section is followed by answers or by completions of the statement. Select the **one** lettered answer or completion that is **best** in each case.

1. Which of the following statements concerning the brachial plexus is correct?
(A) It contains nerve fibers originating from ventral roots only and not from dorsal roots of spinal nerves.
(B) It contains nerve fibers originating from ventral primary rami only and not from dorsal primary rami of spinal nerves.
(C) Each root is formed by fibers originating from more than one ventral primary ramus of a spinal nerve.
(D) Each trunk is formed by fibers originating from more than one ventral ramus.
(E) All trunks branch to innervate muscles.

2. The vascular anastomosis around the shoulder joint involves all of the arteries below EXCEPT the
(A) dorsal scapular artery.
(B) thoracoacromial artery.
(C) subscapular artery.
(D) posterior circumflex humeral artery.
(E) superior ulnar collateral artery.

3. Each statement below concerning anatomical structures in the axilla is correct EXCEPT
(A) part of the axillary artery lies posterior to the tendon of the pectoralis minor muscle.
(B) the roots of the brachial plexus pass anterior to the middle scalene muscles.
(C) in the axilla, the axillary vein lies lateral and posterior to the axillary artery and posterior cord of the brachial plexus.
(D) the axillary sheath is a fascial extension of the prevertebral layer of cervical fascia and encloses the closely grouped great vessels and nerves of the limb.
(E) some roots of the brachial plexus give off nerves to the muscle that forms the medial wall of the axilla.

4. Each statement below concerning the mammary gland is true EXCEPT
(A) it is composed of 15–20 lobes.
(B) it is invested by deep fascia and is connected to the skin by strong connective tissue strands called suspensory ligaments.
(C) it receives some arterial blood through branches of the internal thoracic artery.
(D) each lobe has a single lactiferous duct that opens onto the nipple.
(E) it receives autonomic nerve fibers through the lateral cutaneous branches of the intercostal nerves.

5. Each statement below concerning the pectoral region or axilla is true EXCEPT
(A) normally, the areolar area of the female breast does not have adipose tissue immediately deep to it.
(B) the clavipectoral fascia, costocoracoid membrane, and suspensory ligament of the axilla are different aspects of the same fascial layer.
(C) the pectoralis major, pectoralis minor, and subclavius muscles either originate or insert on part of the humerus.
(D) the pectoralis major muscle inserts on the lateral lip of the intertubercular (bicipital) groove.
(E) the mammary gland lies in the superficial fascia.

6. Each of the following vessels is a branch of the thoracoacromial artery EXCEPT the
(A) acromial.
(B) pectoral.
(C) clavicular.
(D) deltoid branch.
(E) superior thoracic.

7. The defect known as winged scapula is caused by damage to the

(A) roots of the eighth cervical and first thoracic spinal nerves.
(B) thoracodorsal nerve.
(C) lower subscapular nerve.
(D) nerve arising from the roots of the brachial plexus.
(E) nerve arising from the upper trunk of the brachial plexus.

8. The median nerve is formed by parts of the

(A) lateral and posterior cord.
(B) posterior divisions of the upper and middle trunks.
(C) medial and lateral cords.
(D) posterior divisions of the middle and lower trunks.
(E) medial and posterior cords.

9. Which statement below concerning the muscles of the hand is correct?

(A) The adductor pollicis muscle is innervated by the median nerve.
(B) The thenar muscles are innervated by a nerve derived from the posterior cords of the brachial plexus.
(C) The lumbrical muscles arise from tendons of the flexor digitorum superficialis.
(D) The dorsal interosseous muscles assist flexion of the metacarpophalangeal joints and extension of the interphalangeal joints.
(E) The palmar interosseous muscles abduct the fingers.

10. Which of the following statements concerning the pectoralis minor muscle is true?

(A) It divides the axillary nerve into three parts.
(B) It is innervated by the medial pectoral nerve and acts to elevate the shoulder.
(C) It originates from the coracoid process.
(D) It is attached to the ribs.
(E) It forms the posterior wall of the axilla.

11. Which statement below concerning the intercostobrachial nerve is true?

(A) It is a lateral cutaneous branch of the second intercostal nerve.
(B) It typically communicates with the medial antebrachial cutaneous nerve.
(C) It pierces the coracobrachialis muscle.
(D) It is a branch of the medial cord of the brachial plexus.
(E) It contains sympathetic preganglionic fibers.

12. Each statement below concerning the medial pectoral nerve is true EXCEPT

(A) it contains general somatic efferent fibers.
(B) it contains axons that have cell bodies in the sympathetic ganglion.
(C) it supplies the pectoralis major muscle.
(D) it is connected to the lateral pectoral nerve.
(E) it contains axons that have cell bodies in the lateral horn of the spinal cord.

13. Nerves that originate in the posterior cord of the brachial plexus supply each of the following muscles EXCEPT the

(A) subscapularis muscle.
(B) teres major muscle.
(C) latissimus dorsi muscle.
(D) teres minor muscle.
(E) infraspinatus muscle.

14. A patient presents with a severely damaged radial nerve due to fracture of the lower third of the humerus. The patient will experience

(A) a loss of wrist extension, leading to wrist drop.
(B) a weakness in pronating the forearm.
(C) a sensory loss over the ventral aspect of the base of the thumb.
(D) an inability to oppose the thumb.
(E) an inability to abduct the fingers.

15. Which of the following muscles is responsible for the flexion of the proximal interphalangeal joints?

(A) Palmar interossei
(B) Flexor digitorum profundus
(C) Extensor indicis
(D) Flexor digitorum superficialis
(E) Lumbricals

16. Paralysis that impairs the flexion of the distal interphalangeal joint of the index finger will also produce each of the following conditions EXCEPT

(A) similar paralysis of the third digit.
(B) atrophy of the thenar eminence.
(C) loss of sensation over the distal part of the second digit.
(D) complete paralysis of the thumb.
(E) loss of pronation.

17. Which of the following muscles is capable of adducting the arm?

(A) Teres minor
(B) Supraspinatus
(C) Latissimus dorsi
(D) Infraspinatus
(E) Subscapularis

18. All of the muscles below attach to the middle digit EXCEPT the
(A) flexor digitorum profundus.
(B) extensor digitorum communis.
(C) palmar interossei.
(D) dorsal interossei.
(E) lumbricals.

19. In order from medial to lateral side, the major contents of the proximal portion of the cubital fossa are the
(A) biceps brachii tendon, median nerve, brachial artery, radial nerve.
(B) median nerve, biceps brachii tendon, brachial artery, radial nerve.
(C) brachial artery, median nerve, biceps brachii tendon, radial nerve.
(D) brachial artery, biceps brachii tendon, median nerve, radial nerve.
(E) median nerve, brachial artery, biceps brachii tendon, radial nerve.

20. Which combination of structures moves the shoulder joint?
(A) Both heads of the biceps and all three heads of the triceps
(B) Both heads of the biceps and two heads of the triceps
(C) Both heads of the biceps and one head of the triceps
(D) One head of the biceps and one head of the triceps
(E) One head of the biceps and two heads of the triceps

21. The extensor expansion on the middle digit is related to
(A) the extensor digitorum communis, one lumbrical, one dorsal interosseous, and one palmar interosseous muscle.
(B) the extensor digitorum communis, two lumbricals, and two dorsal interossei muscles.
(C) the extensor digitorum communis, one lumbrical, and two dorsal interossei muscles.
(D) the extensor indicis, one lumbrical, and two dorsal interossei muscles.
(E) none of the above.

22. Each statement below concerning the axillary nerve is true EXCEPT
(A) it arises from the posterior cord of the brachial plexus.
(B) it lies adjacent to the medial and posterior surface of the surgical neck of the humerus.
(C) it supplies the deltoid and the teres minor muscles.
(D) it can be damaged by inferior dislocation of the head of the humerus.
(E) it accompanies the scapular circumflex artery.

23. Which statement below concerning the radial artery is true?
(A) It passes through the carpal tunnel.
(B) It accompanies the posterior interosseous nerve in the forearm.
(C) It is the principal source of blood to the superficial palmar arch.
(D) It has the princeps pollicis artery as one of its branches.
(E) It runs distally between the flexor digitorum superficialis and the flexor digitorum profundus.

24. Which group of nerves below is intimately related to a portion of the humerus and can be affected by fractures of the humerus?
(A) Axillary, musculocutaneous, radial
(B) Axillary, median, ulnar
(C) Axillary, radial, ulnar
(D) Axillary, median, musculocutaneous
(E) Median, radial, ulnar

25. The median nerve innervates each of the following muscles EXCEPT the
(A) flexor digitorum superficialis.
(B) opponens pollicis.
(C) two medial lumbricals.
(D) pronator teres.
(E) palmaris longus.

26. Inflammation of the synovial sheath of the common flexor tendons damages the nerve passing through the carpal tunnel and can cause the clinical syndrome called carpal tunnel syndrome. Each statement below concerning this syndrome is true EXCEPT

(A) the dorsal and palmar interosseous muscles are normal.
(B) the flexor digitorum superficialis and profundus muscles are abnormal.
(C) the outer half of the thenar eminence is flattened.
(D) the medial one and one-half fingers have diminished sensitivity.
(E) the adductor pollicis muscle is not atrophied.

27. Which of the following statements concerning the position of the flexor retinaculum is correct?

(A) It lies superficial to the ulnar and median nerves.
(B) It lies deep to the ulnar and median nerves.
(C) It lies superficial to the ulnar nerve and deep to the median nerve.
(D) It lies deep to the ulnar nerve and superficial to the median nerve.
(E) It lies deep to the ulnar nerve and superficial to the ulnar artery.

28. Which of the following intrinsic muscles of the thumb attaches to the first metacarpal?

(A) Abductor pollicis brevis
(B) Flexor pollicis brevis (superficial head)
(C) Opponens pollicis
(D) Adductor pollicis
(E) Flexor pollicis brevis (deep head)

DIRECTIONS: Each group of items in this section consists of lettered options followed by a set of numbered items. For each item, select the **one** lettered option that is most closely associated with it. Each lettered option may be selected once, more than once, or not at all.

Questions 29–33

Match each statement below with the appropriate nerve.

(A) Axillary nerve
(B) Radial nerve
(C) Median nerve
(D) Musculocutaneous nerve
(E) Ulnar nerve

29. Passes through the quadrangular space of the shoulder

30. Innervates a muscle that originates from the third metacarpal and adducts the thumb

31. Innervates the flexor of the distal phalanx of the thumb; usually passes between the two heads of the pronator teres

32. Innervates muscles that abduct or adduct the fingers

33. Usually accompanied by the profunda brachii artery along part of its course

Questions 34–38

Match each statement below with the appropriate bone.

(A) Scapula
(B) Clavicle
(C) Humerus
(D) Radius
(E) Ulna

34. The brachialis muscle inserts into a rough area on the anterior surface of its coronoid process

35. Its head is at the distal end

36. Its head is at the proximal end

37. The long head of the biceps brachii muscle originates here

38. It has a trochlea with a medial margin that projects more than its lateral margin

Questions 39-43

Match each statement below with the appropriate muscle.

(A) Teres minor muscle
(B) Latissimus dorsi muscle
(C) Biceps brachii muscle
(D) Supraspinatus muscle
(E) Brachioradialis muscle

39. Supinates the forearm; originates from the scapula

40. Helps stabilize the glenohumeral joint; innervated by the axillary nerve

41. Forms part of the posterior axillary fold; innervated by a nerve that branches from the posterior cord of the brachial plexus

42. Can flex the forearm; innervated by the radial nerve

43. Abducts the arm; forms part of the rotator cuff; attaches on the greater tuberosity of the humerus

Questions 44-48

Match each statement below with the appropriate artery.

(A) Palmar metacarpal artery
(B) Anterior interosseous artery
(C) Posterior interosseous artery
(D) Radial artery
(E) Ulnar artery

44. Gives off the princeps pollicis artery

45. Lies superficial to the flexor retinaculum

46. Most superficial structure visible in the anatomical snuff-box

47. Gives off an interosseous recurrent artery

48. Descends between the flexor digitorum superficialis and the flexor digitorum profundus

Questions 49-52

Match each description below with the appropriate lettered site or structure in this radiograph of the elbow joint and its associated structures.

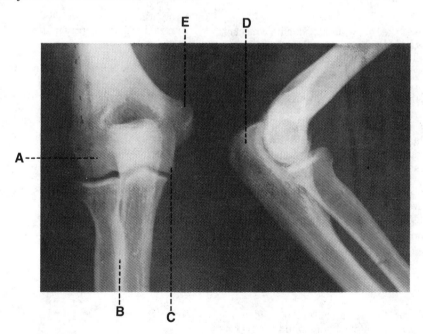

49. Site for tendinous attachment of the biceps brachii muscle

50. Site for origin of the common flexor tendon

51. The capitulum

52. The olecranon

Directions: Each question below contains four suggested answers of which **one or more** is correct. Choose answer

A if **1, 2, and 3** are correct
B if **1 and 3** are correct
C if **2 and 4** are correct
D if **4** is correct
E if **1, 2, 3, and 4** are correct

53. Which of the following muscles are innervated by the ulnar nerve?

(1) Palmar interossei
(2) Dorsal interossei
(3) Medial lumbricals
(4) Abductor pollicis brevis

54. Which of the following tendons are contained in the ulnar bursa?

(1) Tendons of the flexor digitorum superficialis
(2) Tendons of the flexor pollicis longus
(3) Tendons of the flexor digitorum profundus
(4) Tendons of the flexor carpi radialis

55. Which of the following nerves innervate the muscles that abduct the arm?

(1) Axillary
(2) Thoracodorsal
(3) Suprascapular
(4) Radial

56. Which of the following statements concerning the lymphatic drainage of the breast are true?

(1) It may constitute a path for the spread of malignant tumors to axillary nodes.
(2) It provides a rich drainage of cutaneous and glandular regions.
(3) It has paths to the venous system.
(4) It includes collecting vessels that follow the perforating blood vessels through the pectoralis major muscle.

57. Which of the following statements concerning muscles in the upper limb are true?

(1) All intrinsic muscles of the thumb except the opponens pollicis insert into the base of the proximal phalanx.
(2) All heads of the biceps brachii and the triceps brachii arise from the scapula.
(3) The little finger does not have a named adductor.
(4) The flexor digitorum profundus tendon attaches to the middle phalanx and the flexor pollicis longus attaches to the terminal phalanx of the thumb.

58. Which of the following nerves innervate the muscle that moves the metacarpophalangeal joint of the ring finger?

(1) Median
(2) Radial
(3) Musculocutaneous
(4) Ulnar

59. Which of the following muscles form the floor of the cubital fossa?

(1) Brachioradialis
(2) Brachialis
(3) Pronator teres
(4) Supinator

60. Which of the following locations contain cell bodies of nerve fibers in the medial brachial cutaneous nerve?

(1) Dorsal root ganglia
(2) Anterior horn of the gray matter of the spinal cord
(3) Sympathetic chain ganglia
(4) Lateral horn of the gray matter of the spinal cord

61. Which of the following statements concerning the musculocutaneous nerve are true?

(1) It contains postganglionic sympathetic axons.
(2) It contains afferent axons.
(3) It does not contain preganglionic parasympathetic axons.
(4) It contains somatic efferent axons.

62. Which of the following nerves innervate the abductors of the arm?

(1) Suprascapular
(2) Musculocutaneous
(3) Axillary
(4) Radial

Questions 63–66

A patient with a fracture of the clavicle at the junction of the inner and middle third of the bone exhibits overriding of the medial and lateral fragments. The arm is rotated medially.

63. The lateral portion of the fractured clavicle is displaced downward by
(1) the deltoid muscle.
(2) the pectoralis major muscle.
(3) gravity.
(4) the trapezius muscle.

64. Which of the following muscles cause the medial fragment to angle upward?
(1) Pectoralis major
(2) Deltoid
(3) Trapezius
(4) Sternocleidomastoid

65. Which of the following muscles are responsible for the medial rotation of the arm?
(1) Pectoralis major
(2) Subscapularis
(3) Teres major
(4) Latissimus dorsi

66. Which of the following conditions could occur secondary to the fractured clavicle?
(1) A fatal hemorrhage from the subclavian vein
(2) Thrombosis of the subclavian vein, causing a pulmonary embolism
(3) Thrombosis of the subclavian artery, causing an embolism in the brachial artery
(4) Damage to the lower trunk of the brachial plexus

67. Which of the following statements concerning supination are true?
(1) The palm faces backward.
(2) It is partially impaired when a nerve that lies in the spiral groove of the humerus is severed.
(3) It involves the active participation of the elbow and wrist joints.
(4) It involves the participation of the proximal and distal radioulnar joints.

Answers and Explanations

1–B. The brachial plexus is formed by the ventral primary rami of the lower four cervical nerves and first thoracic nerve. The spinal nerves are formed by the union of the dorsal and ventral roots. The root of the brachial plexus is the ventral primary ramus of the spinal nerve. The ventral primary ramus of the seventh cervical nerve remains single as the middle trunk. The upper trunk of the brachial plexus is the only trunk that sends branches (e.g., suprascapular nerve, nerve to the subclavius) to innervate muscles.

2–E. The superior ulnar collateral artery enters the anastomosis around the elbow joint.

3–C. The axillary vein ascends through the axilla along the medial side of the axillary artery and posterior cord of the brachial plexus.

4–B. The mammary gland lies in the superficial fascia.

5–C. The pectoralis minor muscle originates from the second to the fifth ribs and inserts on the coracoid process of the scapula. The subclavius muscle arises from the first rib and its costal cartilage and inserts on the clavicle.

6–E. The thoracoacromial artery has four branches: the pectoral, clavicular, acromial, and deltoid arteries.

7–D. Winged scapula is caused by paralysis of the serratus anterior muscle as a result of damage to the long thoracic nerve that arises from the roots of the brachial plexus (C5–C7).

8–C. The medial and lateral heads of the median nerve arise from the medial and lateral cords, respectively.

9–D. The palmar interosseous muscles adduct the fingers. The adductor pollicis muscle is innervated by the ulnar nerve. The thenar muscles are innervated by the median nerve, which is formed by the union of the lateral and medial heads from the lateral and medial cords, respectively. The lumbrical muscles arise from the tendons of the flexor digitorum profundus. The dorsal interosseous muscles abduct the fingers.

10–D. The pectoralis minor divides the axillary artery into three parts and forms the anterior wall of the axilla. It originates from the second to the fifth ribs and inserts on the coracoid process.

11–A. The intercostobrachial nerve arises from the lateral cutaneous branch of the second intercostal nerve. It may communicate with the medial brachial cutaneous nerve, which is a branch of the medial cord of the brachial plexus. It contains sympathetic postganglionic fibers.

12–E. The medial pectoral nerve receives communication from the lateral pectoral nerve to form a loop in front of the axillary artery. It contains somatic motor fibers that supply the pectoralis major and minor muscles. It also contains postganglionic sympathetic fibers that supply blood vessels and have cell bodies in the sympathetic chain ganglia.

13–E. The subscapularis muscle is innervated by the upper and lower subscapular nerves. The teres major muscle is innervated by the lower subscapular nerve. The latissimus dorsi muscle is innervated by the thoracodorsal nerve. The teres minor muscle is innervated by the axillary nerve. All of these nerves originate in the posterior cord of the brachial plexus. The infraspinatus is innervated by the suprascapular nerve, which originates in the upper trunk.

14–A. The pronator teres, pronator quadratus, and opponens pollicis muscles, as well as the skin over the ventral aspect of the thumb, are innervated by the median nerve. The dorsal interosseous muscles, which act to abduct the fingers, are innervated by the ulnar nerve.

15–D. The flexor digitorum superficialis muscle flexes the proximal interphalangeal joints. The flexor digitorum profundus muscle flexes the distal interphalangeal joints. The palmar interosseous muscles flex the metacarpophalangeal joints and adduct the fingers. The lumbricals can flex metacarpophalangeal joints and extend the interphalangeal joints.

16–D. Flexion of the distal interphalangeal joints of the index and middle fingers is accomplished by the flexor digitorum profundus muscle, which is innervated by the median nerve. This nerve also innervates the skin over the distal part of the second digit and the thenar muscles; however, the adductor pollicis and the deep head of the flexor pollicis brevis are innervated by the ulnar nerve. The median nerve also innervates the pronator teres and quadratus muscles.

17–C. The latissimus dorsi adducts the arm; the supraspinatus muscle abducts the arm. The infraspinatus and the teres minor rotate the arm laterally. The subscapularis rotates the arm medially.

18–C. The palmar interossei are adductors of the fingers and hence have no attachment on the middle digit (the second finger).

19–E. The contents of the cubital fossa from medial to lateral side are the median nerve, the brachial artery, the biceps brachii tendon, and the radial nerve.

20–C. Both heads of the biceps brachii and the long head of the triceps brachii are attached to the scapula.

21–C. The dorsal interossei abduct the fingers; the palmar interossei adduct the fingers. Thus, the middle finger has an abductor on each side but no adductors.

22–E. The axillary nerve arises from the posterior cord of the brachial plexus, accompanies the posterior humeral circumflex artery around the surgical neck of the humerus, and supplies the deltoid and teres minor muscles. Because it lies close to the surgical neck of the humerus, it can be damaged by inferior dislocation of the head of the humerus.

23–D. The radial artery runs distally beneath the brachioradialis with the superficial radial nerve, passes through the anatomical snuffbox, enters the palm by passing between the two heads of the first dorsal interosseous muscle, and divides into the princeps pollicis artery and the deep palmar arch.

24–C. To be injured in a fracture, the nerve must lie close to, or contact, the bone. The axillary nerve passes posteriorly around the surgical neck of the humerus; the radial nerve lies in the radial groove of the middle of the shaft of the humerus; and the ulnar nerve passes behind the medial condyle.

25–C. The median nerve innervates the two lateral lumbricals; however, the ulnar nerve innervates the two medial lumbricals.

26–D. Carpal tunnel syndrome results from injury of the median nerve. Diminished sensitivity in the medial one and one-half fingers would result from damage to the ulnar nerve.

27–D. The flexor retinaculum lies deep to the ulnar artery and the ulnar nerve and superficial to the median nerve.

28–C. The opponens pollicis inserts on the first metacarpal; all other short muscles of the thumb insert on the proximal phalanges.

29–A. The axillary nerve and the posterior humeral circumflex artery pass through the quadrangular space.

30–E. The ulnar nerve innervates the adductor pollicis muscle, which arises from the second and third metacarpals, inserts on the proximal phalanx, and draws the thumb back into the plane of the palm (i.e., adducts the thumb).

31–C. The median nerve innervates the flexor pollicis longus muscle and usually passes between the two heads of the pronator teres.

32–E. The ulnar nerve supplies the dorsal interossei (abductors) and the palmar interossei (adductors).

33–B. The radial nerve turns around in the spiral groove on the back of the humerus with the profunda brachii artery.

34–E. The brachialis muscle inserts on a rough area on the anterior surface of the coronoid process and tuberosity of the ulna.

35–E. The head of the ulna is at its distal end.

36–D. The head of the radius is at its proximal end.

37–A. The long head of the biceps brachii originates from the supraglenoid tubercle of the scapula.

38–C. The humerus has a trochlea with a medial edge that projects more inferiorly than its lateral edge. Consequently, the long axis of the humerus is oblique to the long axis of the ulna, forming an angulation called a carrying angle.

39–C. The long head of the biceps brachii muscle originates from the supraglenoid tubercle of the scapula; the short head arises from the coracoid process. This muscle supinates the forearm.

40–A. The teres minor muscle forms part of the rotator cuff, which helps stabilize the glenohumeral joint, and is innervated by the axillary nerve.

41–B. The latissimus dorsi muscle forms part of the posterior axillary fold. It is innervated by the thoracodorsal nerve, which arises from the posterior cord of the brachial plexus.

42–E. The brachioradialis muscle flexes the forearm and is innervated by the radial nerve.

43–D. The supraspinatus muscle arises from the supraspinous fossa and inserts on the superior facet of the greater tubercle of the humerus. This muscle is innervated by the suprascapular nerve and abducts the arm.

44–D. The radial artery divides into the princeps pollicis artery and the deep palmar arch.

45–E. The ulnar artery enters the hand superficial to the flexor retinaculum and lateral to the pisiform, where it divides into the superficial and deep branches.

46–D. The radial artery runs through the anatomical snuff-box and then enters the palm by passing between the two heads of the first dorsal interosseous muscle.

47–C. The posterior interosseous artery gives off an interosseous recurrent artery that anastomoses with a branch of the profunda brachii artery.

48–E. The ulnar artery descends between the flexor digitorum superficialis and the flexor digitorum profundus.

49–B. The tendon of the biceps brachii muscle inserts on the radial tuberosity.

50–E. The common flexor tendon of the arm attaches to the medial epicondyle.

51–A. The capitulum articulates with the head of the radius.

52–D. The olecranon is the curved projection on the back of the elbow.

53–A. The ulnar nerve innervates all interossei (palmar and dorsal) and the two medial lumbricals.

54–B. The ulnar bursa, or common flexor sheath, contains the tendons of both the flexor digitorum superficialis and profundus muscles. The radial bursa envelops the tendon of the flexor pollicis longus.

55–B. The abductors of the arm are the deltoid and supraspinatus muscles, which are innervated by the axillary and suprascapular nerves, respectively.

56–E. Most of the lymphatic drainage in the lateral quadrants of the breast goes to the axillary nodes, particularly the pectoral nodes. Lymphatic vessels in the medial quadrants follow the perforating vessels through the pectoralis major and enter the parasternal nodes. The remaining lymphatic drainage goes to the apical nodes, nodes of the opposite breast, and nodes of the anterior abdominal wall.

57–B. The short head of the biceps brachii arises from the coracoid process of the scapula; however, only the long head of the triceps brachii arises from the infraglenoid tubercle of the scapula. The flexor digitorum profundus tendon attaches to the distal phalanx of the fingers, and the flexor pollicis longus attaches to the distal phalanx of the thumb.

58–C. The metacarpophalangeal joint of the ring finger is flexed by the interossei and lumbricals, which are innervated by the ulnar nerve. This joint is extended by the extensor digitorum, which is innervated by the radial nerve.

59–C. The floor of the cubital fossa is formed by the brachialis and supinator muscles. The brachioradialis and pronator teres muscles form the lateral and medial boundaries of the fossa, respectively.

60–B. The medial brachial cutaneous nerve contains sensory fibers that have cell bodies in the dorsal root ganglia. It also contains postganglionic sympathetic fibers that have cell bodies in the sympathetic chain ganglia. The anterior horn contains neuronal cell bodies of skeletal motor fibers, and the lateral horn contains neuronal cell bodies of preganglionic sympathetic fibers.

61–E. The musculocutaneous nerve contains postganglionic sympathetic axons that innervate blood vessels, hair follicles, and sweat glands; afferent axons that innervate cutaneous tissues; and somatic efferent fibers that innervate skeletal muscles.

62–B. The abductors of the arm are the deltoid and supraspinatus muscles, which are innervated by the axillary and suprascapular nerves, respectively.

63–B. The lateral fragment of the clavicle is displaced downward by the pull of the deltoid muscle and gravity of the arm.

64–D. The sternocleidomastoid muscle is attached to the superior border of the medial third of the clavicle.

65–E. The pectoralis major, subscapularis, teres major, and latissimus dorsi muscles can rotate the arm medially.

66–E. The fractured clavicle may damage the subclavian vein, resulting in a pulmonary embolism; cause the thrombosis of the subclavian artery, resulting in embolism of the brachial artery; or damage the lower trunk of the brachial plexus.

67–C. In supination the palm faces forward (lateral rotation). This movement is carried out by the supinator and biceps brachii muscles, which are innervated by the radial and musculocutaneous nerves, respectively. Severing the radial nerve running in the spiral groove of the humerus partially impairs supination by affecting the action of the biceps brachii muscle.

3
Lower Limb

Cutaneous Nerves, Superficial Veins, and Lymphatics

I. Cutaneous Nerves

A. Lateral Femoral Cutaneous Nerve
—arises from the lumbar plexus (L2–L3), emerges from the lateral border of the psoas major, crosses the iliacus, and passes under the inguinal ligament near the anterior–superior iliac spine.
—supplies the skin on the anterior and lateral aspects of the thigh as far as the knee.

B. Clunial (Buttock) Nerves
—supply the skin of the gluteal region.
—consist of superior (lateral branches of the dorsal rami of the upper three lumbar nerves), middle (lateral branches of the dorsal rami of the upper three sacral nerves), and inferior (gluteal branches of the posterior femoral cutaneous nerve) nerves.

C. Posterior Femoral Cutaneous Nerve
—arises from the sacral plexus (S1–S3), passes through the greater sciatic foramen below the piriformis muscle, runs deep to the gluteus maximus muscle, and emerges from the inferior border of this muscle.
—descends in the posterior midline of the thigh deep to the fascia lata and pierces the fascia lata near the popliteal fossa.
—supplies the skin of the buttock, thigh, and calf.

D. Saphenous Nerve
—arises from the femoral nerve in the femoral triangle and descends with the femoral vessels through the femoral triangle and the adductor canal.
—descends in the leg in company with the greater saphenous vein and supplies the skin on the medial side of the leg and foot.

E. Lateral Sural Cutaneous Nerve
—arises from the common peroneal nerve in the popliteal fossa and supplies the skin on the posterolateral side of the leg.
—may have a communicating branch to join the medial sural cutaneous nerve.

F. Medial Sural Cutaneous Nerve
—arises from the tibial nerve in the popliteal fossa, may join the lateral sural nerve or its communicating branch to form the sural nerve, and supplies the medial side of the heel and foot.

71

G. Sural Nerve

—is formed by the union of the medial sural and lateral sural nerves (or the communicating branch of the lateral sural nerve).

—supplies the lateral side of the foot.

II. Superficial Veins

A. Great Saphenous Vein

—begins at the medial end of the dorsal venous arch of the foot.

—ascends in front of the medial malleolus and along the medial surface of the tibia and the femur.

—passes through the saphenous opening (fossa ovalis) in the fascia lata and pierces the femoral sheath to join the femoral vein.

—is accompanied by the saphenous nerve in the leg.

—is a suitable vessel for use in arterial bypass surgery.

B. Small (or Short) Saphenous Vein

—begins at the lateral end of the dorsal venous arch and passes upward along the lateral side of the foot with the sural nerve, behind the lateral malleolus.

—passes to the popliteal fossa, where it perforates the deep fascia and terminates in the popliteal vein.

III. Lymphatics: Inguinal Group of Lymph Nodes

—drains the superficial thigh region directly. (Lymphatic drainage of the skin and subcutaneous tissue follows that of the superficial veins.)

—drains the lower limb indirectly via deep lymphatic channels from the popliteal nodes.

—drains the anterolateral abdominal wall below the umbilicus, gluteal region, part of the vagina, anus, and external genitalia except the glans. (Lymphatic drainage from the uterus follows the round ligament.)

—consists of a superficial group (located subcutaneously near the saphenofemoral junction) and a deep group (located medial to the femoral vein).

Bones

I. Coxal (Hip) Bone (Figures 3.1 and 3.2)

—is formed by the fusion of the **ilium, pubis,** and **ischium** bones on each side of the pelvis.

—articulates with the sacrum at the sacroiliac joint to form the pelvic girdle.

A. Ilium

—forms the upper part of the acetabulum and the lateral part of the hip bone.

—presents the anterior–superior iliac spine, anterior–inferior iliac spine, posterior iliac spine, greater sciatic notch, iliac fossa, and gluteal lines.

B. Pubis

—forms the anterior part of the acetabulum and the anteromedial part of the hip bone.

—presents the body, superior and inferior rami, obturator foramen (formed by fusion of the ischium and pubis), pubic crest and tubercle, and pectineal line (or pectin pubis).

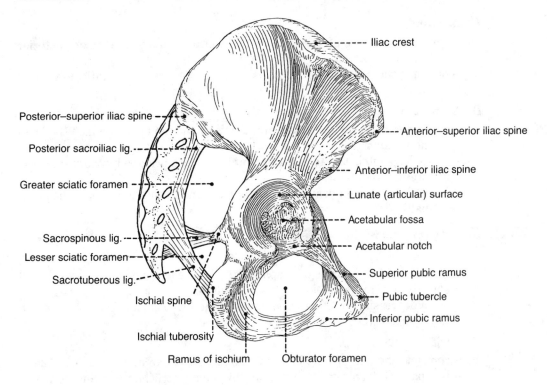

Figure 3.1. Lateral view of the hip bone.

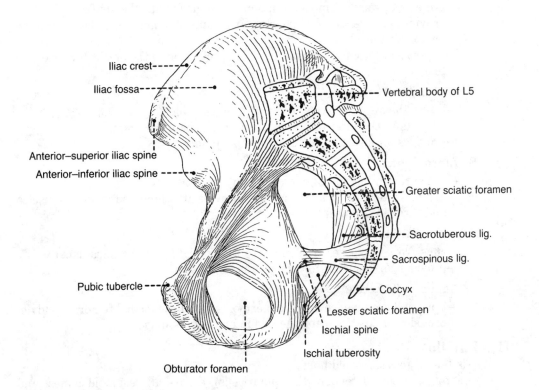

Figure 3.2. Medial view of the hip bone.

C. Ischium

—forms the posteroinferior part of the acetabulum and the lower posterior part of the hip bone.

—presents the body, ischial spine, ischial tuberosity, lesser sciatic notch, and ramus.

D. Acetabulum

—is a cup-shaped cavity on the lateral side of the hip bone in which the head of the femur fits.

—has a deep notch, the **acetabular notch,** which is bridged by the **transverse acetabular ligament.**

—is formed by the ilium superiorly, the ischium inferolaterally, and the pubis medially.

II. Femur (Figure 3.3)

A. Head

—forms about two-thirds of a sphere and is directed medially, upward, and slightly forward to fit into the acetabulum.

—has a depression in its articular surface, the fovea capitis femoris, to which the ligamentum capitis femoris is attached.

B. Neck

—connects the head to the body (shaft) and forms an angle of about 125°.

—is separated from the shaft in front by the intertrochanteric line, to which the iliofemoral ligament is attached.

C. Greater Trochanter

—projects upward from the junction of the neck with the shaft.

—provides an insertion for the gluteus medius and minimus muscles and the piriformis muscle.

—the trochanteric fossa on its medial aspect receives the obturator externus tendon.

D. Lesser Trochanter

—lies in the angle between the neck and the shaft.

—projects at the inferior end of the intertrochanteric crest.

—provides an insertion for the iliopsoas tendon.

E. Linea Aspera

—is the rough line or ridge on the body (shaft).

—exhibits lateral and medial lips that provide attachments for many muscles and three intermuscular septa.

F. Pectineal Line

—runs from the lesser trochanter to the medial lip of the linea aspera.

—provides an insertion for the pectineus muscle.

G. Adductor Tubercle

—is a small prominence at the uppermost part of the medial femoral condyle.

—provides an insertion for the adductor magnus muscle.

III. Patella

—is the largest sesamoid bone.

—is found within a tendon (the quadriceps), as are all sesamoid bones, and articulates with the femur but not with the tibia.

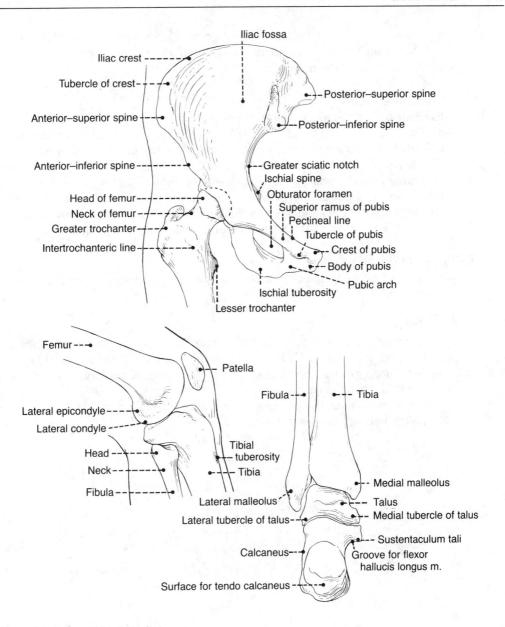

Figure 3.3. Bones of the lower limb.

—attaches to the tibial tuberosity by a continuation of the quadriceps tendon, called the patellar ligament.

—functions to increase the angle of pull of the quadriceps femoris, thereby magnifying its power.

IV. Tibia (see Figure 3.3)

—is the weight-bearing medial bone of the leg.

—has the tibial tuberosity into which the patellar ligament inserts.

—its two condyles articulate with the condyles of the femur.

—its distal end projects medially and inferiorly as the **medial malleolus,** which has the malleolar groove for the tendons of the tibialis posterior and flexor digitorum longus muscles and another groove (lateral to the malleolus groove) for the tendon of the flexor hallucis longus muscle. It also provides attachment for the deltoid ligament.

V. Fibula (see Figure 3.3)

—has little or no function in weight bearing but provides attachment for muscles.
—attaches via its head (apex) to the fibular collateral ligament of the knee joint.
—its distal end projects laterally and inferiorly as the **lateral malleolus**, which lies more inferior and posterior than does the medial malleolus. The lateral malleolus attaches via its fossa to the transverse tibiofibular and posterior talofibular ligaments and provides attachment for the anterior talofibular, posterior talofibular, and calcaneofibular ligaments. It also presents the sulcus for the tendons of the peroneus longus and brevis muscles.

VI. Tarsus (Figure 3.4)

—consists of seven tarsal bones: talus, calcaneus, navicular, cuboid, and three cuneiforms.

A. Talus

—transmits the weight of the body from the tibia to other weight-bearing bones of the foot and is the only tarsal bone without muscle attachments.
—has a deep groove, the sulcus tali, for the interosseous ligaments between the talus and the calcaneus.
—has a groove on the posterior surface of its body for the tendon of the flexor hallucis longus muscle.

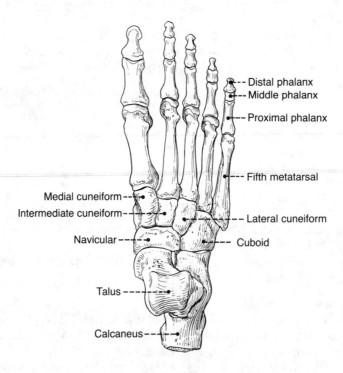

Distal phalanx
Middle phalanx
Proximal phalanx

Fifth metatarsal

Medial cuneiform
Intermediate cuneiform
Lateral cuneiform
Navicular
Cuboid

Talus

Calcaneus

Figure 3.4. Bones of the foot.

B. Calcaneus
—is the largest and strongest bone of the foot and lies below the talus.
—forms the heel and articulates with the talus superiorly and the cuboid anteriorly.
—the **sustentaculum tali,** which supports the talus, is a shelf-like medial projection from the medial surface of the calcaneus; on its inferior surface, there is a groove for the tendon of the flexor hallucis longus.

C. Navicular
—is a boat-shaped tarsal bone lying between the head of the talus and the three cuneiform bones.

D. Cuboid
—is the most laterally placed tarsal bone and has a notch and groove for the tendon of the peroneus longus muscle.

E. Cuneiform Bones
—are wedge-shaped bones that are related to the transverse arch.
—articulate with the navicular bone posteriorly and with three metatarsals anteriorly.

VII. Metatarsus
—consists of five metatarsals and has prominent medial and lateral sesamoid bones on the first metatarsal.

VIII. Phalanges
—there are 14 (2 in the first digit and 3 in each of the others).

Joints

I. Hip (Coxal) Joint (see Figure 3.3)
—is a multiaxial ball-and-socket synovial joint.
—its cavity is deepened by the acetabular rim and is completed below by the transverse ligament, which converts the acetabular notch into a foramen for passage of nutrient vessels and nerves.

A. Acetabular Labrum
—is a complete fibrocartilage rim that deepens the articular socket for the head of the femur and subsequently stabilizes the hip joint.

B. Fibrous Capsule
—is attached proximally to the margin of the acetabulum and to the transverse acetabular ligament.
—is attached distally to the neck of the femur as follows: anteriorly to the intertrochanteric line and the root of the greater trochanter and posteriorly to the intertrochanteric crest.
—encloses part of the head and most of the neck of the femur.
—is reinforced anteriorly by the **iliofemoral ligament,** posteriorly by the **ischiofemoral ligament,** and inferiorly by the **pubofemoral ligament.**

C. Ligaments

1. Iliofemoral Ligament
—is the largest and most important ligament that reinforces the fibrous capsule anteriorly and is in the form of an inverted Y.

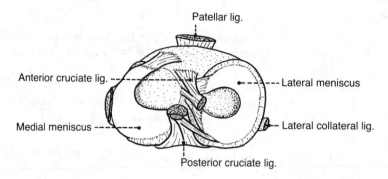

Figure 3.5. Ligaments of the knee joint.

—is attached proximally to the anterior–inferior iliac spine and the acetabular rim and distally to the intertrochanteric line and the front of the greater trochanter of the femur.

—strongly resists hyperextension at the hip joint.

2. Ischiofemoral Ligament

—reinforces the fibrous capsule posteriorly and extends from the ischial portion of the acetabular rim to the neck of the femur, medial to the base of the greater trochanter.

3. Pubofemoral Ligament

—reinforces the fibrous capsule inferiorly and extends from the pubic portion of the acetabular rim and the superior pubic ramus to the lower part of the femoral neck.

4. Ligamentum Teres (Capitis) Femoris

—arises from the floor of the acetabular fossa (more specifically, from the margins of the acetabular notch and from the transverse ligament) and attaches to the fovea of the femur.

5. Transverse Acetabular Ligament

—is a fibrous band that bridges the acetabular notch and converts it into a foramen.

D. Blood Supply

—is derived from branches of the medial and lateral femoral circumflex, superior and inferior gluteal, and obturator arteries.

E. Nerve Supply

—is derived from branches of the femoral, obturator, and superior gluteal nerves and the nerve to the quadratus femoris.

II. Knee Joint (Figures 3.5 and 3.6)

—is a hinge-type synovial joint permitting flexion, extension, and some rotation.

—is encompassed by a fibrous capsule that is rather thin, weak, and often incomplete.

—is attached via a capsule to the margins of the femoral condyles, to the patella and the patellar ligament, and to the tibia at the margins of the tibial condyles.

—is stabilized on the lateral side by the biceps and gastrocnemius (lateral head) tendons, the iliotibial tract, and the fibular collateral ligaments and on the

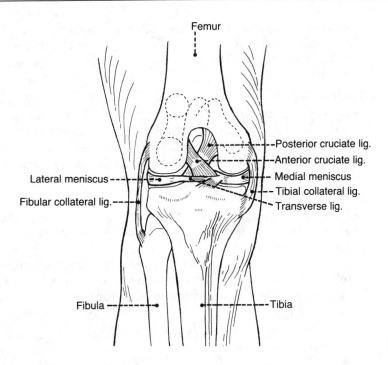

Figure 3.6. Ligaments of the knee joint (anterior view).

medial side by the sartorius, gracilis, gastrocnemius (medial head), semitendinosus, and semimembranosus muscles, and the tibial collateral ligament.

—allows rotation when the knee is flexed; full extension is accompanied by medial rotation of the femur on the tibia, pulling all ligaments taut.

A. Ligaments (see Figures 3.5 and 3.6)

1. Anterior Cruciate Ligament

—lies inside the knee joint capsule but outside the synovial cavity of the joint.

—attaches to the anterior intercondylar area of the tibia, hence the "anterior" part of its name.

—passes upward, backward, and laterally to be inserted into the posterior aspect of the medial surface of the lateral femoral condyle.

—prevents backward slipping (dislocation) of the femur on the tibia (or forward slipping of the tibia on the femur) and hyperextension of the knee joint.

—limits excessive anterior mobility of the tibia on the femur when the knee is extended.

—is lax when the knee is flexed and becomes taut when the knee is fully extended.

2. Posterior Cruciate Ligament

—lies outside the synovial cavity but within the fibrous joint capsule.

—attaches to the posterior impression of the intercondylar area.

—passes upward, forward, and medially to be inserted into the anterior part of the lateral surface of the medial femoral condyle.

—limits hyperflexion of the knee and prevents forward displacement of the femur on the tibia (or backward slipping of the tibia on the femur).

—is lax when the knee is extended and becomes taut when the knee is flexed.

3. Medial Meniscus

—lies outside the synovial cavity but within the joint capsule.

—is C-shaped (i.e., forms a semicircle) and is attached to the interarticular area of the tibia or medial collateral ligament.

—acts as a cushion or shock absorber and facilitates lubrication.

—is more frequently torn in injuries than the lateral meniscus.

4. Lateral Meniscus

—lies outside the synovial cavity but within the joint capsule.

—is approximately circular and is incompletely attached to the upper aspect of the tibia.

—is separated laterally from the fibular (or lateral) collateral ligament by the tendon of the popliteal muscle and aids in forming a more stable base for the articulation of the femoral condyle.

5. Transverse Ligament

—binds the anterior horns (ends) of the lateral and medial semilunar cartilages (menisci).

6. Tibial Collateral Ligament

—is attached to the medial meniscus as well as to the medial aspects of the articular capsule and tibial condyle.

—is a broad band separated from the capsule anteriorly by a bursa.

—prevents medial displacement of the two long bones.

—its firm attachment to the medial meniscus is of clinical significance because injury to the ligament results in concomitant damage to the medial meniscus.

7. Fibular Collateral Ligament

—extends between the lateral femoral epicondyle and the head of the fibula.

—is a rounded cord that stands well away from the capsule of the joint.

B. Bursae

1. Suprapatellar Bursa

—lies deep to the quadriceps femoris muscle and extends about 8 cm superior to the patella.

—is the major bursa communicating with the knee joint cavity (the semi-membranosus bursa also may communicate with it).

2. Prepatellar Bursa

—lies over the superficial surface of the patella.

3. Infrapatellar Bursa

—consists of a subcutaneous infrapatellar bursa, which lies over the patellar ligament, and a deep infrapatellar bursa, which lies deep to the patellar ligament.

C. Blood Supply

—is provided by the genicular branches (superior medial and lateral, inferior medial and lateral, and middle) of the popliteal artery.

—is provided also by a descending branch of the lateral femoral circumflex artery, an articular branch of the descending genicular artery, and the anterior tibial recurrent artery.

D. Nerve Supply

—is from branches of the sciatic, femoral, and obturator nerves.

E. Clinical Considerations

1. The Unhappy Triad of the Knee Joint

—is characterized by:

a. **Rupture** of the **tibial collateral ligament,** as a result of excessive abduction.

b. **Tearing** of the **anterior cruciate ligament,** as a result of forward displacement of the tibia.

c. **Injury** to the **medial meniscus,** as a result of the tibial collateral ligament attachment.

—can occur when a football player's cleated shoe is planted firmly in the turf, and the knee is struck from the lateral side.

—is indicated by a knee that is markedly swollen, particularly in the suprapatellar region.

—results in tenderness upon application of pressure along the extent of the tibial collateral ligament.

2. Knee-Jerk (or Patellar) Reflex

—occurs when the patellar ligament is tapped, resulting in a sudden contraction of the quadriceps femoris.

3. Housemaid's Knee (or Prepatellar Bursitis)

—is inflammation and swelling of the prepatellar bursa.

III. Tibiofibular Joints

A. Proximal Tibiofibular Joint

—is a plane synovial joint between the head of the fibula and the tibia.

B. Distal Tibiofibular Joint

—is a fibrous joint between the tibia and the fibula.

IV. Ankle Joint (Figure 3.7)

—is a hinge-type (ginglymus) synovial joint located between the inferior ends of the tibia and fibula and the trochlea of the talus.

A. Articular Capsule

—is a thin fibrous capsule that lies both anteriorly and posteriorly, allowing movement.

—is reinforced medially by the medial (or deltoid) ligament and laterally by the lateral ligament, which prevents anterior and posterior slipping of the tibia and fibula on the talus.

B. Medial (Deltoid) Ligament

—forms the keystone for the lateral longitudinal arch.

—has four parts: the **tibionavicular, tibiocalcaneal, anterior tibiotalar,** and **posterior tibiotalar ligaments.**

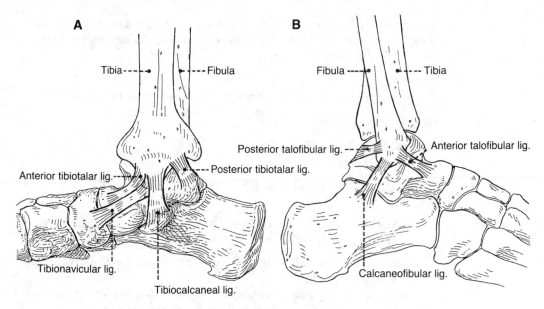

A

Tibia - - - ← → - - - Fibula

Anterior tibiotalar lig. - - → ← - - Posterior tibiotalar lig.

Tibionavicular lig.

Tibiocalcaneal lig.

B

Fibula - - - ← → - - - Tibia

Posterior talofibular lig. - - → ← - - Anterior talofibular lig.

Calcaneofibular lig.

Figure 3.7. Ligaments of the ankle joint. (*A*) Medial view; (*B*) lateral view.

—extends from the medial malleolus to the navicular, calcaneus, and talus bones.

—prevents overeversion of the foot.

C. Lateral Ligament

—consists of the **anterior talofibular, posterior talofibular,** and **calcaneofibular** (cord-like) **ligaments**.

—resists inversion of the foot and may be torn during an ankle sprain (inversion injury).

V. Intertarsal Joints

—consist of the **talocalcaneal** (or **subtalar**), **talocalcaneonavicular, calcaneocuboid,** and **transverse tarsal joints**.

A. Talocalcaneal (or Subtalar) Joint

—is part of the talocalcaneonavicular joint formed between the talus and calcaneus bones.

—allows inversion and eversion of the foot.

B. Talocalcaneonavicular Joint

—is part of the transverse tarsal joint.

—resembles a ball-and-socket joint in which the head of the talus is in contact with a socket formed by the calcaneus and navicular bones.

—is supported by the spring (plantar calcaneonavicular) ligament.

C. Calcaneocuboid Joint

—is part of the transverse tarsal joint.

—resembles a saddle joint between the calcaneus and the cuboid bones.

—is supported by the short plantar (plantar calcaneocuboid) and long plantar ligaments and by the tendon of the peroneus longus muscle.

D. Transverse Tarsal Joint
—consists of two conjoined joints, the talonavicular part of the talocalcaneo-navicular joint and the calcaneocuboid joint. These joints are separated anatomically but act together functionally.
—is important in inversion and eversion of the foot.

VI. Tarsometatarsal Joints
—are plane joints and strengthen the transverse arch.
—are united by articular capsules and are reinforced by the plantar, dorsal, and interosseous ligaments.

VII. Metatarsophalangeal Joints
—are ellipsoid (condyloid) joints.
—are joined by articular capsules and are reinforced by the plantar and collateral ligaments.

VIII. Interphalangeal Joints
—are hinge (ginglymus) joints.
—are enclosed by articular capsules and are reinforced by the plantar and collateral ligaments.

Anterior Thigh

I. Femoral Triangle
—is bounded by the inguinal ligament superiorly, the sartorius muscle laterally, and the adductor longus muscle medially.
—contains the femoral nerve and vessels.

A. Femoral Ring
—is the abdominal opening of the femoral canal.
—is bounded by the inguinal ligament anteriorly; the femoral vein laterally; the lacunar ligament medially; and the pectineal ligament posteriorly.

B. Femoral Canal
—lies medial to the femoral vein in the femoral sheath.
—contains fat, areolar connective tissue, and lymph nodes.
—transmits lymphatics from the lower limb and perineum to the peritoneal cavity.
—is a potential weak area and a site of femoral herniation.

C. Femoral Sheath
—is a funnel-shaped extension of the transversalis and iliopsoas fasciae into the thigh, deep to the inguinal ligament.
—contains (from lateral to medial) the femoral artery, femoral vein, and femoral canal. (The femoral nerve lies outside the femoral sheath, lateral to the femoral artery.)
—its distal end reaches the level of the proximal end of the saphenous opening.

D. Femoral Hernia
—lies lateral to the pubic tubercle and deep to the inguinal ligament.
—its sac is formed by the parietal peritoneum.
—passes through the femoral ring and canal.
—is more common in women than in men.

E. Adductor Canal

—begins at the apex of the femoral triangle and ends at the tendinous adductor hiatus.

—lies between the adductor magnus and longus muscles and the vastus medialis muscle and is covered by the sartorius muscle and fascia.

—contains the femoral vessels, the saphenous nerve, and the nerve to the vastus medialis.

II. Anterior Muscles of the Thigh

Muscle	Origin	Insertion	Nerve	Action
Iliacus	Iliac fossa; ala of sacrum	Lesser trochanter	Femoral n.	Flexes and rotates thigh medially (with psoas major)
Sartorius	Anterior–superior iliac spine	Upper medial side of tibia	Femoral n.	Flexes and rotates thigh laterally; flexes and rotates leg medially
Rectus femoris	Anterior–inferior iliac spine; posterior–superior rim of acetabulum	Base of patella; tibial tuberosity	Femoral n.	Flexes thigh; extends leg
Vastus medialis	Intertrochanteric line; linea aspera; medial intermuscular septum	Medial side of patella; tibial tuberosity	Femoral n.	Extends leg
Vastus lateralis	Intertrochanteric line; greater trochanter; linea aspera; gluteal tuberosity; lateral intermuscular septum	Lateral side of patella; tibial tuberosity	Femoral n.	Extends leg
Vastus intermedius	Upper shaft of femur; lower lateral intermuscular septum	Upper border of patella; tibial tuberosity	Femoral n.	Extends leg

III. Medial Muscles of the Thigh

Muscle	Origin	Insertion	Nerve	Action
Adductor longus	Body of pubis below its crest	Middle third of linea aspera	Obturator n.	Adducts, flexes, and rotates thigh laterally
Adductor brevis	Body and inferior pubic ramus	Pectineal line; upper part of linea aspera	Obturator nn.	Adducts, flexes, and rotates thigh laterally
Adductor magnus	Ischiopubic ramus; ischial tuberosity	Linea aspera; medial supracondylar line; adductor tubercle	Obturator and sciatic nn.	Adducts, flexes, and extends thigh

(Continued on next page)

Muscle	Origin	Insertion	Nerve	Action
Pectineus	Pectineal line of pubis	Pectineal line of femur	Obturator and femoral nn.	Adducts and flexes thigh
Gracilis	Body and inferior pubic of ramus	Medial surface of upper quarter of tibia	Obturator n.	Adducts and flexes thigh; flexes and rotates leg medially
Obturator externus	Margin of obturator foramen and obturator membrane	Intertrochanteric fossa of femur	Obturator n.	Rotates thigh laterally

IV. Blood Vessels of the Thigh (Figure 3.8)

A. Obturator Artery

—arises from the internal iliac artery in the pelvis and passes through the obturator foramen, where it divides into **anterior** and **posterior branches**.

1. Anterior Branch

—descends in front of the adductor brevis muscle and gives off muscular branches.

2. Posterior Branch

—descends behind the adductor brevis muscle and supplies the adductor muscles.

—gives off the acetabular branch, which passes through the acetabular notch and provides a small artery to the head of the femur.

B. Femoral Artery

—begins as the continuation of the external iliac artery distal to the inguinal ligament, descends through the femoral triangle, and enters the adductor canal.

1. Superficial Epigastric Artery

—runs subcutaneously upward toward the umbilicus.

2. Superficial Circumflex Iliac Artery

—runs laterally toward the anterior–superior iliac spine, almost parallel with the inguinal ligament.

3. External Pudendal Artery

—emerges through the saphenous ring, runs medially over the spermatic cord (or the round ligament of the uterus), and sends inguinal branches to the skin above the pubic and anterior scrotal (or labial) branches to the skin of the scrotum (or labium majus).

4. Profunda Femoris (Deep Femoral) Artery

—arises from the femoral artery within the femoral triangle.

—descends in front of the pectineus, adductor brevis, and adductor magnus muscles, but behind the adductor longus muscle.

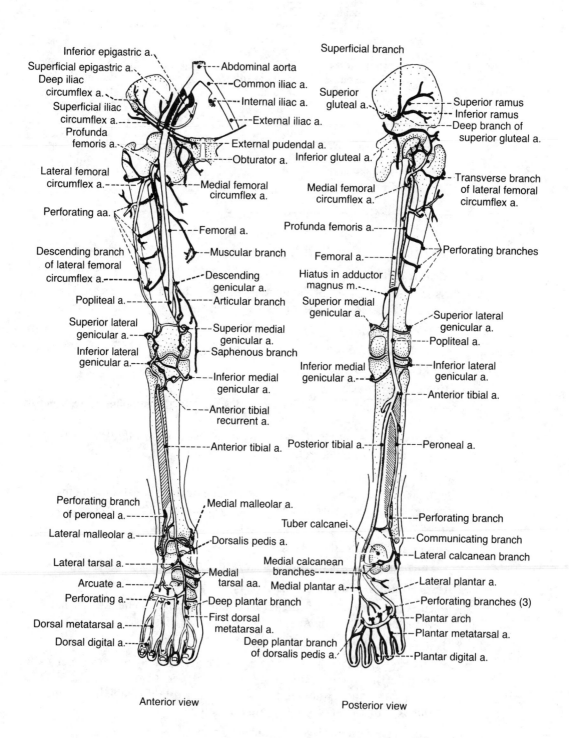

Figure 3.8. Blood supply to the lower limb (anterior and posterior views).

—gives off the medial and lateral femoral circumflex and muscular branches.

—provides, in the adductor canal, four perforating arteries that perforate and supply the adductor magnus and hamstring muscles.

—its first perforating branch anastomoses with the inferior gluteal and the transverse branches of the medial and lateral femoral circumflex arteries. This is known as the **cruciate anastomosis** of the buttock.

5. **Medial Femoral Circumflex Artery**

—arises from the femoral, or profunda femoris, artery in the femoral triangle.

—runs between the pectineus and iliopsoas muscles, continues between the obturator externus and adductor brevis muscles, and enters the gluteal region between the adductor magnus and quadratus femoris muscles.

—gives off muscular branches and an acetabular branch to the hip joint and then divides into an ascending branch, which anastomoses with the gluteal arteries, and a transverse branch, which joins the cruciate anastomosis.

—is clinically important because it supplies most of the blood to the head and neck of the femur.

6. **Lateral Femoral Circumflex Artery**

—arises from the femoral or profunda femoris artery.

—passes laterally deep to the sartorius and rectus femoris muscles.

—divides into ascending, transverse, and descending branches.

   a. **Ascending Branch**

   —ascends deep to the tensor fasciae latae muscle and then between the gluteus medius and minimus muscles to anastomose with the superior gluteal artery.

   b. **Transverse Branch**

   —penetrates the vastus lateralis muscle, winds around the femur, and joins the cruciate anastomosis.

   c. **Descending Branch**

   —descends with the nerve to the vastus lateralis muscle beneath the rectus femoris.

   —anastomoses with the superior lateral genicular branch of the popliteal artery.

7. **Descending Genicular Artery**

—arises from the femoral artery just before it passes through the adductor canal.

—divides into the articular branch, which enters the anastomosis around the knee, and the saphenous branch, which supplies the superficial tissue and skin on the medial side of the knee.

V. Nerves (Figure 3.9)

   A. **Obturator Nerve**

   —arises from the lumbar plexus (L3–L4) and enters the thigh through the obturator foramen.

   —divides into **anterior** and **posterior branches**.

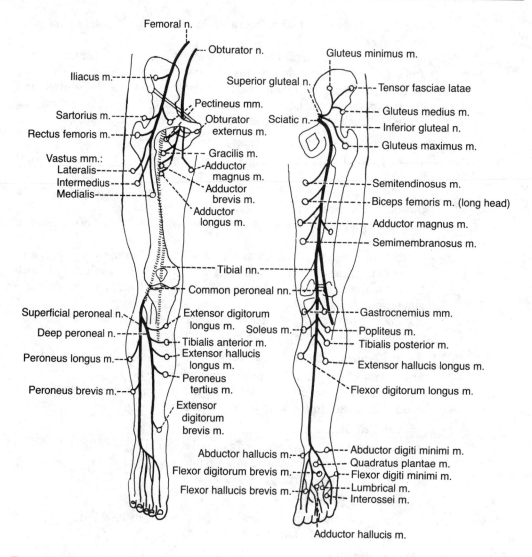

Figure 3.9. Innervation of the lower limb.

1. **Anterior Branch**

    —descends between the adductor longus and adductor brevis muscles and supplies the adductor longus, adductor brevis, gracilis, and pectineus muscles.

2. **Posterior Branch**

    —descends between the adductor brevis and adductor magnus muscles and supplies the obturator externus and adductor magnus muscles.

B. **Femoral Nerve**

    —arises from the lumbar plexus (L2–L4) within the substance of psoas major muscle, emerges between the iliacus and psoas major muscles, and enters the thigh by passing deep to the inguinal ligament and lateral to the femoral sheath.

    —gives off **muscular branches**; **articular branches** to the hip and knee joints; and **cutaneous branches**, including the anterior femoral cutaneous nerve and the saphenous nerve, which descends through the femoral triangle and accompanies the femoral vessels in the adductor canal.

Gluteal Region and Posterior Thigh

I. Bony Landmarks and Ligaments

A. Sacrotuberous Ligament
—extends from the ischial tuberosity to the posterior iliac spines, lower sacrum, and coccyx and converts, with the sacrospinous ligament, the lesser sciatic notch into the lesser sciatic foramen.

B. Sacrospinous Ligament
—extends from the ischial spine to the lower sacrum and the coccyx and converts the greater sciatic notch into the greater sciatic foramen.

C. Greater Sciatic Foramen
—transmits the piriformis muscle; superior and inferior gluteal vessels and nerves; internal pudendal vessels and pudendal nerve; sciatic nerve; posterior femoral cutaneous nerve; and the nerves to the obturator internus and quadratus femoris muscles.

D. Lesser Sciatic Foramen
—transmits the tendon of the obturator internus; the nerve to the obturator internus; and the internal pudendal vessels and pudendal nerve.

E. Structures That Pass Through Both the Greater and Lesser Sciatic Foramina
—are the pudendal nerve, the internal pudendal vessels, and the nerve to the obturator internus.

F. Iliotibial Tract
—is a particularly strong lateral portion of the fascia lata.
—is derived from the insertion of the gluteus maximus and tensor fasciae latae muscles.
—helps to form the fibrous capsule of the knee joint.

G. Fascia Lata
—is a strong membranous fascia covering muscles of the thigh and forms the lateral and medial intermuscular septa by its inward extension to the femur.
—is attached to the pubic symphysis, crest, and ramus, the ischial tuberosity, the inguinal and sacrotuberous ligaments, and the sacrum and coccyx.

II. Muscles of the Gluteal Region

Muscle	Origin	Insertion	Nerve	Action
Gluteus maximus	Ilium; sacrum; coccyx; sacrotuberous ligament	Gluteal tuberosity; iliotibial tract	Inferior gluteal n.	Extends and rotates thigh laterally
Gluteus medius	Ilium between iliac crest, and anterior and posterior gluteal lines	Greater trochanter	Superior gluteal n.	Abducts and rotates thigh medially

(Continued on next page)

Muscle	Origin	Insertion	Nerve	Action
Gluteus minimus	Ilium between anterior and inferior gluteal lines	Greater trochanter	Superior gluteal n.	Abducts and rotates thigh medially
Tensor fasciae latae	Iliac crest; anterior–superior iliac spine	Iliotibial tract	Superior gluteal n.	Flexes, abducts, and rotates thigh medially
Piriformis	Pelvic surface of sacrum; sacrotuberous ligament	Upper end of greater trochanter	Sacral n. (S1–S2)	Rotates thigh laterally
Obturator internus	Ischiopubic rami; obturator membrane	Greater trochanter	N. to obturator internus	Abducts and rotates thigh laterally
Superior gemellus	Ischial spine	Obturator internus tendon	N. to obturator internus	Rotates thigh laterally
Inferior gemellus	Ischial tuberosity	Obturator internus tendon	N. to quadratus femoris	Rotates thigh laterally
Quadratus femoris	Ischial tuberosity	Intertrochanteric crest	N. to quadratus femoris	Rotates thigh laterally

III. Posterior Muscles of the Thigh*

Muscle	Origin	Insertion	Nerve	Action
Semitendinosus	Ischial tuberosity	Medial surface of upper part of tibia	Tibial portion of sciatic n.	Extends thigh; flexes and rotates leg medially
Semimembranosus	Ischial tuberosity	Medial condyle of tibia	Tibial portion of sciatic n.	Extends thigh; flexes and rotates leg medially
Biceps femoris	Long head from ischial tuberosity; short head from linea aspera and upper supracondylar line	Head of fibula	Tibial (long head) and common peroneal (short head) divisions of sciatic n.	Extends thigh; flexes and rotates leg medially

*These three muscles collectively are called hamstrings.

IV. Blood Vessels (see Figure 3.8)

A. Superior Gluteal Artery

—arises from the internal iliac artery, passes between the lumbosacral trunk and the first sacral nerve, and enters the buttock through the greater sciatic foramen above the piriformis muscle.

—runs deep to the gluteus maximus muscle and divides into a superficial branch, which forms numerous branches to supply the gluteus maximus, and a deep branch, which runs between the gluteus medius and minimus muscles and supplies these muscles and the tensor fasciae latae.

B. Inferior Gluteal Artery
—arises from the internal iliac artery, usually passes between the first and second sacral nerves, and enters the buttock through the greater sciatic foramen below the piriformis.
—enters the deep surface of the gluteus maximus and descends on the medial side of the sciatic nerve, in company with the posterior femoral cutaneous nerve.

V. Nerves (see Figure 3.9)

A. Superior Gluteal Nerve
—arises from the sacral plexus (L4–S1) and enters the buttock through the greater sciatic foramen above the piriformis.
—passes between the gluteus medius and minimus muscles and divides into numerous branches to innervate the gluteus medius and minimus, the tensor fasciae latae, and the hip joint.

B. Inferior Gluteal Nerve
—arises from the sacral plexus (L5–S2) and enters the buttock through the greater sciatic foramen below the piriformis.
—divides into numerous branches to supply the overlying gluteus maximus.

C. Posterior Femoral Cutaneous Nerve
—arises from the sacral plexus (S1–S3) and enters the buttock through the greater sciatic foramen below the piriformis.
—runs deep to the gluteus maximus and emerges from the inferior border of this muscle.
—descends on the posterior thigh and supplies the skin of the buttock, thigh, and calf.

D. Sciatic Nerve
—arises from the sacral plexus (L4–S3) and is the largest nerve in the body, consisting of the tibial and common peroneal components.
—enters the buttock through the greater sciatic foramen below the piriformis.
—descends over the obturator internus gemelli and quadratus femoris muscles between the ischial tuberosity and the greater trochanter.
—innervates the hamstring muscles by its tibial division, except for the short head of the biceps femoris, which is innervated by its common peroneal division.
—provides articular branches to the hip and knee joints.

VI. Popliteal Fossa
—is bounded superomedially by the semitendinosus and semimembranosus muscles and superolaterally by the biceps muscle.
—is bounded inferolaterally by the lateral head of the gastrocnemius muscle and inferomedially by the medial head of the gastrocnemius muscle.
—contains the popliteal vessels, the common peroneal and tibial nerves, and the small saphenous vein.

—its floor is composed of the femur, the oblique popliteal ligament, and the popliteus muscle.

A. Popliteal Artery (see Figure 3.8)

—is a continuation of the femoral artery at the adductor hiatus and runs through the popliteal fossa.

—terminates at the lower border of the popliteus muscle by dividing into the anterior and posterior tibial arteries.

—gives off five genicular arteries:

1. Superior Lateral Genicular Artery

—passes deep to the biceps femoris tendon.

2. Superior Medial Genicular Artery

—passes deep to the semimembranosus and semitendinosus muscles and enters the substance of the vastus medialis.

3. Inferior Lateral Genicular Artery

—passes laterally above the head of the fibula and then deep to the fibular collateral ligament.

4. Inferior Medial Genicular Artery

—passes medially along the upper border of the popliteus muscle, deep to the popliteus fascia.

5. Middle Genicular Artery

—pierces the oblique popliteal ligament and enters the knee joint, where it supplies the cruciate ligaments and the synovial membrane of the joint cavity.

B. Popliteal Vein

—ascends through the popliteal fossa behind the popliteal artery.

—receives the small saphenous vein and those veins corresponding to the branches of the popliteal artery.

C. Ligaments

1. Arcuate Popliteal Ligament

—arises from the head of the fibula and passes upward and medially over the tendon of the popliteus muscle on the back of the knee joint.

2. Oblique Popliteal Ligament

—is an oblique expansion of the tendon of the semimembranosus muscle and passes upward obliquely across the posterior surface of the knee joint from the medial condyle of the tibia.

Leg and Foot

I. Anterior and Lateral Muscles of the Leg

Muscle	Origin	Insertion	Nerve	Action
Anterior:				
Tibialis anterior	Lateral tibial condyle; interosseous membrane	First cuneiform; first metatarsal	Deep peroneal n.	Dorsiflexes and inverts foot

(Continued on next page)

Muscle	Origin	Insertion	Nerve	Action
Extensor hallucis longus	Middle half of anterior surface of fibula; interosseous membrane	Base of distal phalanx of big toe	Deep peroneal n.	Extends big toe; dorsiflexes and inverts foot
Extensor digitorum longus	Lateral tibial condyle; upper two-thirds of fibula; interosseous membrane	Bases of middle and distal phalanges	Deep peroneal n.	Extends toes; dorsiflexes foot
Peroneus tertius	Distal one-third of fibula; interosseous membrane	Base of fifth metatarsal	Deep peroneal n.	Dorsiflexes and everts foot
Lateral:				
Peroneus longus	Lateral tibial condyle; head and upper lateral side of fibula	Base of first metatarsal; medial cuneiform	Superficial peroneal n.	Everts and plantar flexes foot
Peroneus brevis	Lower lateral side of fibula; intermuscular septa	Base of fifth metatarsal	Superficial peroneal n.	Everts and plantar flexes foot

II. Posterior Muscles of the Leg

Muscle	Origin	Insertion	Nerve	Action
Superficial group:				
Gastrocnemius	Lateral (lateral head) and medial (medial head) femoral condyles	Posterior aspect of calcaneus via tendo calcaneus	Tibial n.	Flexes knee; plantar flexes foot
Soleus	Upper fibula head; soleal line on tibia	Posterior aspect of calcaneus via tendo calcaneus	Tibial n.	Plantar flexes foot
Plantaris	Lower lateral supracondylar line	Posterior surface of calcaneus	Tibial n.	Flexes and rotates leg medially
Deep group:				
Popliteus	Lateral condyle of femur; popliteal ligament	Upper posterior side of tibia	Tibial n.	Flexes and rotates leg medially
Flexor hallucis longus	Lower two-thirds of fibula; interosseous membrane; intermuscular septa	Base of distal phalanx of big toe	Tibial n.	Flexes distal phalanx of big toe

(Continued on next page)

Muscle	Origin	Insertion	Nerve	Action
Flexor digitorum longus	Middle posterior aspect of tibia	Distal phalanges of lateral four toes	Tibial n.	Flexes lateral four toes; plantar flexes foot
Tibialis posterior	Interosseous membrane; upper parts of tibia and fibula	Tuberosity of navicular; sustentacula tali; three cuneiforms; cuboid; bases of metatarsals 2–4	Tibial n.	Plantar flexes and inverts foot

III. Blood Vessels (see Figure 3.8)

A. Posterior Tibial Artery

—arises from the popliteal artery at the lower border of the popliteus, between the tibia and the fibula.

—is accompanied by two venae comitantes and the tibial nerve.

—gives off the **peroneal (fibular) artery,** which descends between the tibialis posterior and the flexor hallucis longus muscles and supplies the lateral muscles in the posterior compartment. The peroneal artery passes behind the lateral malleolus, gives off the posterior lateral malleolar branch, and ends in branches to the ankle and heel.

—also gives off the posterior medial malleolar, perforating, and muscular branches and terminates by dividing into the **medial** and **lateral plantar arteries.**

B. Medial Plantar Artery

—is the smaller terminal branch of the posterior tibial artery.

—runs between the abductor hallucis and the flexor digitorum brevis muscles.

—gives off a superficial branch, which supplies the big toe, and a deep branch, which forms three superficial digital branches.

C. Lateral Plantar Artery

—is the larger terminal branch of the posterior tibial artery.

—runs forward laterally in company with the lateral plantar nerve between the quadratus plantae and the flexor digitorum brevis muscles and then between the flexor digitorum brevis and the adductor digiti minimi muscles.

—forms the plantar arch by joining the deep plantar branch of the dorsalis pedis artery. The plantar arch gives off four plantar metatarsal arteries.

D. Anterior Tibial Artery

—arises from the popliteal artery and enters the anterior compartment by passing through the gap at the upper end of the interosseous membrane.

—descends on the interosseous membrane between the tibialis anterior and extensor digitorum longus muscles.

—gives off the anterior tibial recurrent artery, which ascends to the knee joint, and the anterior medial and lateral malleolar arteries at the ankle.

—runs distally across the ankle midway between the lateral and medial malleoli and continues onto the dorsum of the foot as the dorsalis pedis artery.

E. Dorsalis Pedis Artery

—begins anterior to the ankle joint midway between the two malleoli as the continuation of the anterior tibial artery.

—gives off the **medial** and **lateral tarsal arteries** and divides into a **deep plantar artery** to the deep plantar arch and an **arcuate artery,** which gives off the second, third, and fourth **dorsal metatarsal arteries.**

IV. Nerves (see Figure 3.9)

A. Common Peroneal Nerve

—is separated from the tibial portion at the apex of the popliteal fossa and descends through the fossa.

—superficially crosses the lateral head of the gastrocnemius muscle, and then turns laterally around the neck of the fibula deep to the peroneus longus, where it divides into the **deep peroneal** (or **anterior tibial**) and **superficial peroneal nerves.**

—gives off the **lateral sural cutaneous nerve,** which supplies the skin on the lateral part of the back of the leg, and the **recurrent articular branch** to the knee joint.

B. Superficial Peroneal Nerve

—arises from the common peroneal nerve between the peroneus longus and the neck of the fibula, descends in the lateral compartment, and supplies the peroneus longus and brevis muscles.

—emerges between the peroneus longus and brevis muscles by piercing the deep fascia at the lower third of the leg to become subcutaneous and supplies the skin of the lower leg and foot.

C. Deep Peroneal (Anterior Tibial) Nerve

—arises from the common peroneal nerve between the peroneus longus and the neck of the fibula.

—gives off a recurrent branch to the knee joint.

—passes around the neck of the fibula and through the extensor digitorum longus muscle.

—descends on the interosseous membrane between the extensor digitorum longus and the tibialis anterior and then between the extensor digitorum longus and the extensor hallucis longus muscles.

—supplies the anterior muscles of the leg and divides into medial and lateral branches.

D. Tibial Nerve

—descends through the popliteal fossa and then lies on the popliteus muscle.

—gives off three articular branches, which accompany the medial superior genicular, middle genicular, and medial inferior genicular arteries to the knee joint.

—supplies the muscular branches to the posterior muscles of the leg.

—gives rise to the **medial sural cutaneous nerve,** the medial calcaneal branch to the skin of the heel and sole, and the articular branches to the ankle joint.

—terminates beneath the flexor retinaculum by dividing into the **medial** and **lateral plantar nerves.**

1. **Medial Plantar Nerve**

   —arises beneath the flexor retinaculum, deep to the posterior portion of the abductor hallucis muscles, as the larger terminal branch from the tibial nerve.

   —passes distally between the abductor hallucis and flexor digitorum brevis and supplies these muscles.

   —gives off **common digital branches** that divide into **proper digital branches,** which supply the flexor hallucis brevis and the first lumbrical and the skin of the medial three and one-half toes.

2. **Lateral Plantar Nerve**

   —is the smaller terminal branch of the tibial nerve.

   —runs distally and laterally between the quadratus plantae and the flexor digitorum brevis, supplying the quadratus plantae and the abductor digiti minimi muscles.

   —divides into a superficial branch, which supplies the flexor digiti minimi brevis, and a deep branch, which supplies the plantar and dorsal interossei, the lateral three lumbricals, and the adductor hallucis.

V. Muscles of the Foot

Muscle	Origin	Insertion	Nerve	Action
Dorsum of foot:				
Extensor digitorum brevis	Dorsal surface of calcaneus	Tendons of extensor digitorum longus	Deep peroneal n.	Extends toes
Extensor hallucis brevis	Dorsal surface of calcaneus	Base of proximal phalanx of big toe	Deep peroneal n.	Extends big toe
Sole of foot:				
Abductor hallucis	Medial tubercle of calcaneus	Base of proximal phalanx of big toe	Medial plantar n.	Abducts big toe
Flexor digitorum brevis	Medial tubercle of calcaneus	Middle phalanges of lateral four toes	Medial plantar n.	Flexes middle phalanges of lateral four toes
Abductor digiti minimi	Medial and lateral tubercles of calcaneus	Proximal phalanx of little toe	Lateral plantar n.	Abducts little toe
Quadratus plantae	Medial and lateral side of calcaneus	Tendons of flexor digitorum longus	Lateral plantar n.	Aids in flexing toes
Lumbricals (4)	Tendons of flexor digitorum longus	Proximal phalanges; extensor expansion	First by medial plantar n.; lateral three by lateral plantar n.	Flex metatarsophalangeal joints and extend interphalangeal joints

(Continued on next page)

Muscle	Origin	Insertion	Nerve	Action
Flexor hallucis brevis	Cuboid; third cuneiform	Proximal phalanx of big toe	Medial plantar n.	Flexes big toe
Adductor hallucis:				
Oblique head	Bases of metatarsals 2–4	Proximal phalanx of big toe	Lateral plantar n.	Adducts big toe
Transverse head	Capsule of lateral four metatarsophalangeal joints			
Flexor digiti minimi brevis	Base of metatarsal 5	Proximal phalanx of little toe	Lateral plantar n.	Flexes little toe
Plantar interossei (3)	Medial sides of metatarsals 3–5	Medial sides of base of proximal phalanges 3–5	Lateral plantar n.	Adduct toes; flex proximal, and extend distal phalanges
Dorsal interossei (4)	Adjacent shafts of metatarsals	Proximal phalanges of second toes (medial and lateral sides), and third and fourth toes (lateral sides)	Lateral plantar n.	Abduct toes; flex proximal, and extend distal phalanges

VI. Retinacula, Ligaments, and Arches (Figure 3.10)

A. Flexor Retinaculum
—is a deep fascial band that passes between the medial malleolus and the medial surface of the calcaneus.
—holds three tendons in place beneath it: the tibialis posterior, flexor digitorum longus, and flexor hallucis longus.
—transmits the tibial nerve and posterior tibial artery beneath it.

B. Plantar Aponeurosis
—is a thick fascia investing the plantar muscles.
—radiates from the calcaneal tuberosity (or tuber calcanei) toward the toes and gives attachment to the short flexor muscles of the toes.

C. Transverse (Metatarsal) Arch
—is formed by the navicular, the three cuneiforms, the cuboid, and the five metatarsal bones of the foot.
—is maintained anteriorly by the transverse head of the adductor hallucis.

D. Lateral Longitudinal Arch
—is formed by the calcaneus, cuboid, and lateral two metatarsal bones.
—its "keystone" is the cuboid bone.
—is supported by the peroneus longus tendon and the long and short plantar ligaments.

E. Medial Longitudinal Arch
—is formed by the talus, calcaneus, navicular, cuneiforms, and medial three metatarsals and is supported by the spring ligament.

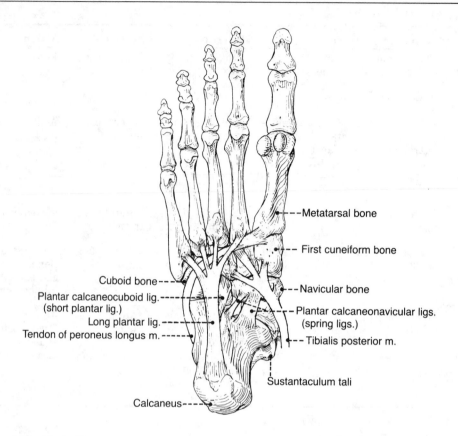

Figure 3.10. Plantar ligaments.

F. Long Plantar Ligament

—extends from the calcaneus to the tuberosity of the cuboid and the base of the metatarsals and forms a canal for the tendon of the peroneus longus.

—supports the lateral side of the longitudinal arch of the foot.

G. Short Plantar (Plantar Calcaneocuboid) Ligament

—extends from the front of the inferior surface of the calcaneus to the plantar surface of the cuboid bone.

—lies deep to the long plantar ligament and supports the lateral side of the longitudinal arch.

H. Spring (Plantar Calcaneonavicular) Ligament

—passes from the sustentaculum tali of the calcaneus bone to the navicular bone.

—supports the head of the talus and the medial side of the longitudinal arch.

VII. Clinical Considerations

A. Gluteal Gait (Gluteus Medius Limp)

—is a waddling gait, characterized by the pelvis falling toward the unaffected side at each step due to paralysis of the gluteus medius muscle. This muscle normally functions to stabilize the pelvis when the opposite foot is off the ground.

B. **Trendelenburg's Symptom**
—is seen in a fracture of the femoral neck, dislocated hip joint, and weakness and paralysis of the gluteus medius muscle. When the patient stands on the affected limb, the pelvis on the sound side will sag, whereas, if normal, the pelvis will rise.

C. **Damage to the Common Peroneal Nerve**
—results in "foot drop" and loss of sensation from the dorsum of the foot.
—causes paralysis of all of the dorsiflexor and evertor muscles of the foot.

D. **Damage to the Deep Peroneal Nerve**
—results in "foot drop" and hence a characteristic high-stepping gait.

E. **Damage to the Tibial Nerve Above the Ankle**
—results in a characteristic "clawing" of the toes and secondary loss on the sole of the foot, affecting posture and locomotion of the lower limb.

F. **Anterior Tibial Compartment Syndrome**
—is characterized by ischemic necrosis of the muscles of the anterior tibial compartment of the leg. Presumably this happens as a result of compression of arteries by swollen muscles, following excessive exertion.
—is accompanied by extreme tenderness and pain on the anterolateral aspect of the leg.

G. **Ankle Jerk**
—is a reflex twitch of the triceps surae (i.e., the medial and lateral heads of the gastrocnemius and the soleus muscles).
—is induced by tapping the tendo calcaneus.
—its reflex center is in the fifth lumbar and first sacral segments of the spinal cord.

H. **Talipes Planus (Flat Foot)**
—is characterized by a waddling gait with the feet turned out.
—results in disappearance of the medial portion of the longitudinal arch, which appears completely flattened.
—causes greater wear on the inner border of the soles and heels of the shoes than on the outer border.
—causes pain as a result of stretching of the spring ligament and the long and short plantar ligaments.

I. **Talipes Equinovarus (Clubfoot)**
—is characterized by plantar flexion, inversion, and adduction of the foot.
—is a congenitally deformed foot that is twisted from its natural position.

Summary of Muscle Actions of the Lower Limb

Movements at the Hip Joint (Ball-and-Socket Joint)
Flexion—iliopsoas, tensor fasciae latae, rectus femoris, adductors, sartorius, pectineus
Extension—hamstrings, gluteus maximus, adductor magnus
Adduction—adductor magnus, adductor longus, adductor brevis, pectineus, gracilis
Abduction—gluteus medius, gluteus minimus
Medial rotation—tensor fasciae latae, gluteus medius, gluteus minimus

Lateral rotation—obturator internus, obturator externus, gemelli, piriformis, quadratus femoris, gluteus maximus

Movements at the Knee Joint (Hinge Joint)
Flexion—hamstrings, gracilis, sartorius, gastrocnemius, popliteus
Extension—quadriceps femoris
Medial rotation—semitendinosus, popliteus
Lateral rotation—biceps femoris

Movements at the Ankle Joint (Hinge Joint)
Dorsiflexion—anterior tibialis, extensor digitorum longus, extensor hallucis longus, peroneus tertius
Plantar flexion—triceps surae, plantaris, posterior tibialis, peroneus longus, flexor digitorum longus, flexor hallucis longus (when the knee is fully flexed)

Movements at the Intertarsal Joint (Talocalcanean, Transverse Tarsal Joint)
Inversion—tibialis posterior, tibialis anterior, triceps surae
Eversion—peroneus longus, peroneus brevis, peroneus tertius

Movements at the Metatarsophalangeal Joint (Ellipsoid Joint)
Flexion—lumbricals, interossei, flexor hallucis brevis, flexor digiti minimi brevis
Extension—extensor digitorum longus and brevis, extensor hallucis longus

Movements at the Interphalangeal Joint (Hinge Joint)
Flexion—flexor digitorum longus and brevis, flexor hallucis longus
Extension—extensor digitorum longus and brevis, extensor hallucis longus

---

Summary of Muscle Innervations of the Lower Limb

Muscles of the Thigh

Muscles of the Anterior Compartment: Femoral Nerve
Sartorius
Quadriceps femoris–rectus femoris; vastus medialis; vastus intermedius; and vastus lateralis

Muscles of the Medial Compartment: Obturator Nerve
Adductor longus; adductor brevis; adductor magnus (obturator and tibial nerves)*; gracilis; obturator externus; pectineus (femoral and obturator nerves)*

Muscles of the Posterior Compartment: Tibial Part of Sciatic Nerve
Semitendinosus; semimembranosus; biceps femoris, long head; biceps femoris, short head (common peroneal part of sciatic nerve)*; adductor magnus (tibial part of sciatic and obturator nerve)*

Muscles of the Lateral Compartment
Gluteus maximus (inferior gluteal nerve)
Gluteus medius (superior gluteal nerve)
Gluteus minimus (superior gluteal nerve)
Tensor fasciae latae (superior gluteal nerve)
Piriformis (nerve to piriformis)
Obturator internus (nerve to obturator internus)
Superior gemellus (nerve to obturator internus)

Inferior gemellus (nerve to quadratus femoris)
Quadratus femoris (nerve to quadratus femoris)

Muscles of the Leg

Muscles of the Anterior Compartment: Deep Peroneal Nerve
Tibialis anterior; extensor digitorum longus; extensor hallucis longus; peroneus tertius

Muscles of the Lateral Compartment: Superficial Peroneal Nerve
Peroneus longus; peroneus brevis

Muscles of the Posterior Compartment: Tibial Nerve
Superficial layer—gastrocnemius; soleus; plantaris
Deep layer—popliteus; tibialis posterior; flexor digitorum longus; flexor hallucis longus

---

* Indicates exception

Review Test

DIRECTIONS: Each of the numbered items or incomplete statements in this section is followed by answers or by completions of the statement. Select the **one** lettered answer or completion that is **best** in each case.

1. The medial plantar nerve is a branch of the
(A) common peroneal nerve.
(B) tibial nerve.
(C) superficial peroneal nerve.
(D) deep peroneal nerve.
(E) anterior tibial nerve.

2. The dorsalis pedis artery is a continuation of the
(A) posterior tibial artery.
(B) popliteal artery.
(C) femoral artery.
(D) peroneal artery.
(E) anterior tibial artery.

3. Which of the following ligaments of the hip joint transmits medial epiphysial vessels to the head of the femur?
(A) Transverse acetabular ligament
(B) Pubofemoral ligament
(C) Ischiofemoral ligament
(D) Iliofemoral ligament
(E) Ligamentum teres femoris

4. Which of the following ligaments prevents forward displacement of the femur on the tibia when the knee is flexed?
(A) Anterior cruciate ligament
(B) Fibular collateral ligament
(C) Patellar ligament
(D) Posterior cruciate ligament
(E) Tibial collateral ligament

5. The femoral nerve innervates the
(A) psoas major muscle.
(B) skin on the lateral side of the foot.
(C) skin over the greater trochanter.
(D) sartorius muscle.
(E) tensor fasciae latae.

6. Each of the following muscles contributes directly to the stability of the knee joint EXCEPT the
(A) soleus muscle.
(B) semimembranosus muscle.
(C) sartorius muscle.
(D) biceps femoris muscle.
(E) gastrocnemius muscle.

7. Which of the following arteries sends a branch to the ligamentum teres femoris?
(A) Superficial circumflex iliac
(B) Inferior gluteal
(C) Superficial epigastric
(D) Superficial external pudendal
(E) Obturator

8. Each of the following statements concerning muscles in the leg is true EXCEPT
(A) the tendon of the flexor digitorum longus passes deep to the tendon of the tibialis posterior as they pass around the medial malleolus.
(B) the tendon of the flexor hallucis longus passes deep to the tendon of the flexor digitorum longus across the sole of the foot.
(C) the flexor digitorum longus arises from the tibia.
(D) the flexor hallucis longus arises from the fibula.
(E) the tibialis posterior inverts the foot.

9. Each structure below passes deep to the inferior or superior extensor retinaculum of the ankle EXCEPT the
(A) anterior tibial nerve.
(B) extensor digitorum longus muscle.
(C) dorsalis pedis artery.
(D) peroneus tertius muscle.
(E) superficial peroneal nerve.

10. Which of the following statements concerning muscles in the foot is true?
(A) The peroneus longus inserts on the proximal end of the first metatarsus and the medial cuneiform bone.
(B) The flexor digitorum brevis inserts into the base of the proximal phalanx of each toe except the first toe.
(C) The quadratus plantae inserts into the medial border of the flexor digitorum brevis.
(D) The flexor hallucis longus inserts into the base of the proximal phalanx of the first toe.
(E) The tibialis anterior inserts on the calcaneus.

11. Which of the following statements concerning the longitudinal arches of the foot is true?

(A) They include the proximal phalanges distally.
(B) They are primarily supported by muscles.
(C) The lateral longitudinal arch includes the navicular bone.
(D) The talus transmits weight from the tibia to the longitudinal arches.
(E) They are supported in part by a ligament that extends from the calcaneus to the sustentaculum tali.

12. Which of the following statements concerning the knee joint is true?

(A) The medial meniscus is circular.
(B) The knee joint is supported medially by a collateral ligament that is free of the capsule.
(C) The anterior cruciate ligament prevents anterior displacement of the femur on the tibia.
(D) The medial collateral ligament is firmly attached to the medial meniscus.
(E) The synovial cavity usually is continuous with the prepatellar bursa.

13. Each statement below concerning the femoral triangle is true EXCEPT

(A) it is covered superficially by fascia lata.
(B) it contains the femoral artery, femoral vein, and femoral nerve.
(C) it is bounded medially by the adductor longus muscle.
(D) the femoral nerve is covered by the femoral sheath.
(E) it is bounded superiorly by the inguinal ligament.

14. Which of the following arteries usually gives off the branch that follows the ligament of the head of the femur?

(A) Medial circumflex femoral artery
(B) Inferior gluteal artery
(C) Lateral circumflex femoral artery
(D) Obturator artery
(E) Superior gluteal artery

15. A patient presents with a slight weakness in the lateral arch of the foot on one side and cannot dorsiflex the foot. These signs probably indicate damage to the

(A) superficial peroneal nerve.
(B) lateral plantar nerve.
(C) deep peroneal nerve.
(D) sural nerve.
(E) posterior tibial nerve.

16. When the common peroneal nerve is severed in the popliteal fossa but the tibial nerve is spared, the foot will be

(A) plantar flexed and inverted.
(B) dorsiflexed and everted.
(C) dorsiflexed and inverted.
(D) plantar flexed and everted.
(E) dorsiflexed only.

17. To avoid damaging the sciatic nerve during an intramuscular injection in the gluteal region, the needle should be inserted

(A) in the area over the sacrospinous ligament.
(B) midway between the ischial tuberosity and the lesser trochanter.
(C) at the midpoint of the gemelli muscles.
(D) in the upper right quadrant of the gluteal region.
(E) in the lower right quadrant of the gluteal region.

18. Which of the following statements concerning the peroneal artery is true?

(A) It ends in branches to the ankle and heel.
(B) It continues into the foot as the dorsalis pedis artery.
(C) It gives rise to the medial plantar artery in the foot.
(D) It is a branch of the anterior tibial artery.
(E) It continues into the foot as the lateral plantar artery.

19. Which muscle below has a tendon that occupies the groove in the lower surface of the cuboid bone?

(A) Peroneus tertius muscle
(B) Peroneus brevis muscle
(C) Peroneus longus muscle
(D) Tibialis anterior muscle
(E) Tibialis posterior muscle

20. Which muscle below has a tendon that occupies the groove on the undersurface of the sustentaculum tali?

(A) Flexor digitorum brevis muscle
(B) Flexor digitorum longus muscle
(C) Flexor hallucis brevis muscle
(D) Flexor hallucis longus muscle
(E) Tibialis posterior muscle

DIRECTIONS: Each group of items in this section consists of lettered options followed by a set of numbered items. For each item, select the **one** lettered option that is most closely associated with it. Each lettered option may be selected once, more than once, or not at all.

Questions 21–25

Match each description below with the most appropriate ligament.

(A) Calcaneofibular ligament
(B) Long plantar ligament
(C) Plantar calcaneonavicular ligament
(D) Short plantar ligament
(E) Deltoid ligament

21. The thickened medial part of the ankle joint capsule

22. Forms a canal for the tendon of the peroneus longus muscle

23. The cord-like ligament that reinforces the lateral side of the ankle joint

24. Extends from the front of the inferior surface of the calcaneus to the plantar surface of the cuboid bone; supports the longitudinal arch

25. Supports the head of the talus in maintaining the medial longitudinal arch

Questions 26–30

Match each description below with the most appropriate muscle.

(A) Biceps femoris muscle
(B) Gluteus medius muscle
(C) Iliopsoas muscle
(D) Rectus femoris muscle
(E) Tensor fasciae latae muscle

26. Rotates the leg laterally

27. The chief flexor of the thigh

28. Flexes the thigh and extends the leg

29. Can flex and medially rotate the thigh during running and climbing

30. Its paralysis causes the pelvis to tilt away from the paralyzed side when the body is supported by the leg on the same side

Questions 31–35

Match each bone below with the appropriate lettered structure in this radiograph of the foot.

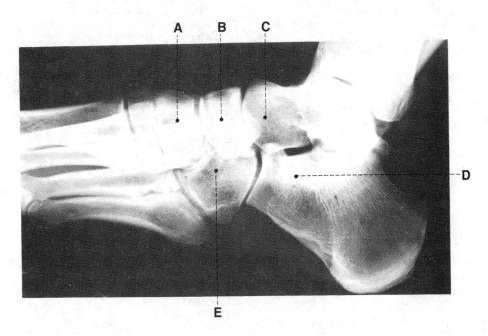

31. Navicular bone

32. Cuboid bone

33. Cuneiform bones

34. Calcaneus bone

35. Talus bone

DIRECTIONS: Each question below contains four suggested answers of which **one or more** is correct. Choose answer

> A if **1, 2, and 3** are correct
> B if **1 and 3** are correct
> C if **2 and 4** are correct
> D if **4** is correct
> E if **1, 2, 3, and 4** are correct

36. Which of the following muscles bound the adductor canal in the lower extremity?
(1) Sartorius
(2) Vastus medialis
(3) Adductor longus and adductor magnus
(4) Gracilis

37. Which of the following statements concerning the great saphenous vein are true?
(1) It ascends posterior to the medial malleolus.
(2) It empties into the femoral vein.
(3) It courses anterior to the medial condyles of the tibia and femur.
(4) It passes superficial to the fascia lata of the thigh.

38. The adductor canal contains the
(1) femoral vessels.
(2) saphenous nerve.
(3) nerve to the vastus medialis.
(4) saphenous vein.

39. The inability to extend the leg at the knee joint indicates paralysis of the
(1) semitendinosus and semimembranosus muscles.
(2) sartorius muscle.
(3) gracilis muscle.
(4) quadriceps femoris muscle.

40. Which of the following statements concerning the superior gluteal nerve are true?
(1) It innervates the gluteus medius and gluteus minimus muscles.
(2) It leaves the pelvis through the greater sciatic foramen, above the piriformis.
(3) It runs primarily in the plane between the gluteus medius and gluteus minimus muscles.
(4) It innervates the gluteus maximus muscle.

Questions 41–45

Answer questions 41–45 using the diagram below.

42. Which of the following structures pass through *B* in the diagram?
(1) Inferior epigastric artery
(2) Iliopsoas muscle
(3) External pudendal artery
(4) Femoral vein

43. Which of the following structures pass through *C* in the diagram?
(1) Iliolumbar artery
(2) Lumbosacral trunk
(3) Ilioinguinal nerve
(4) Obturator nerve

44. Which of the following structures pass through *D* in the diagram?
(1) Pudendal nerve
(2) Posterior cutaneous nerve of the thigh
(3) Obturator internus tendon
(4) Inferior gluteal artery

45. Which of the following structures pass through both *A* and *D* in the diagram?
(1) Nerve to the quadratus femoris muscle
(2) Pudendal nerve
(3) Sciatic nerve
(4) Nerve to the obturator internus muscle

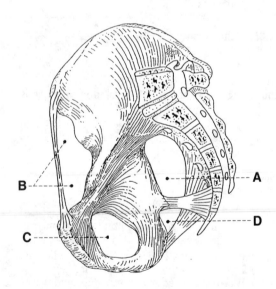

41. Which of the following structures pass through *A* in the diagram?
(1) Piriformis muscle
(2) Superior gluteal nerve
(3) Sciatic nerve
(4) Pudendal nerve

Answers and Explanations

1-B. The medial plantar nerve is a branch of the tibial nerve.

2-E. The dorsalis pedis artery is a continuation of the anterior tibial artery.

3-E. The ligamentum teres femoris transmits medial epiphysial vessels to the head of the femur. These vessels originate from the posterior branch of the obturator artery.

4-D. The posterior cruciate ligament is important for preventing forward displacement of the femur on the tibia when the knee is flexed.

5-D. The femoral nerve lies outside the femoral sheath, lateral to the femoral artery, and innervates the quadratus femoris and sartorius muscles. The second and third lumbar nerves innervate the psoas major muscle. The sural nerve innervates the skin on the lateral side of the foot. The iliohypogastric nerve and superior clunial nerves supply the skin over the greater trochanter. The superior gluteal nerve innervates the tensor fasciae latae.

6-A. The soleus muscle arises from the fibula and tibia below the knee joint and, therefore, does not contribute to the stability of the knee joint.

7-E. The posterior branch of the obturator artery sends a branch to the ligamentum teres femoris.

8-A. The tendon of the flexor digitorum longus muscle passes superficial to the tendon of the tibialis posterior muscle as they pass around the medial malleolus. The tendon of the flexor hallucis longus muscle passes deep to the tendon of the flexor digitorum longus muscle across the sole of the foot. The flexor digitorum longus arises mainly from the middle half of the posterior shaft of the tibia. The flexor hallucis longus arises from the middle half of the posterior shaft of the fibula. The tibialis posterior inverts the foot.

9-E. The superficial peroneal nerve emerges between the peroneus longus and peroneus brevis muscles and descends superficial to the extensor retinaculum of the ankle, supplying the skin of the lower leg and foot.

10-A. The peroneus longus muscle inserts on the proximal end of the first metatarsus and the medial cuneiform bone. The flexor digitorum brevis muscle inserts into the base of the middle phalanges of the lateral four toes. The quadratus plantae muscle inserts into the tendons of the flexor digitorum longus muscle. The flexor hallucis longus muscle inserts into the base of the distal phalanx of the first (great) toe. The tibialis anterior muscle inserts on the first cuneiform and the first metatarsal.

11-D. The talus transmits weight from the tibia to the longitudinal arch. The longitudinal arches include the metatarsals distally. The medial longitudinal arch includes the navicular bone and is supported by the spring ligament, which extends from the sustentaculum tali of the calcaneus to the navicular. The lateral longitudinal arch is supported by the long and short plantar ligaments.

12-D. The medial collateral ligament of the knee joint is firmly attached to the medial meniscus, the articular capsule, and the tibia. The medial meniscus is semicircular (i.e., it is C-shaped). The knee joint is strengthened internally by the paired cruciate ligaments. The anterior cruciate ligament prevents posterior displacement of the femur on the tibia. It is unusual for the synovial cavity of the knee joint to be continuous with the prepatellar bursa.

13-D. The femoral nerve lies outside the femoral sheath, lateral to the femoral artery. From lateral to medial, the femoral sheath contains the femoral artery, femoral vein, and femoral canal.

14-D. All of the arteries listed are involved in the vascular anastomosis around the hip joint. The obturator artery usually gives off a branch that follows the ligament of the head of the femur.

15-C. The deep peroneal nerve supplies the anterior muscles of the leg, including the tibialis anterior, extensor hallucis longus, extensor digitorum longus, and the peroneus tertius, which dorsiflex the foot.

16–A. The common peroneal nerve supplies the anterior and lateral muscles of the leg; the tibial nerve supplies the posterior muscles of the leg. Severance of the common peroneal nerve paralyzes the muscles for dorsiflexion and eversion of the foot; thus, the foot is plantar flexed and inverted.

17–D. To avoid damaging the sciatic nerve during an intramuscular injection, the needle should be inserted in the upper right quadrant of the gluteal region.

18–A. The peroneal artery is a branch of the posterior tibial artery, supplies the lateral muscles of the leg, passes behind the lateral malleolus, and ends in branches to the ankle and heel.

19–C. The groove in the lower surface of the cuboid bone is occupied by the tendon of the peroneus longus muscle.

20–D. The groove on the undersurface of the sustentaculum tali is occupied by the tendon of the flexor hallucis longus muscle.

21–E. The deltoid ligament is the thickened medial part of the ankle joint capsule. It consists of the anterior tibiotalar, tibionavicular, tibiocalcaneal, and posterior tibiotalar ligaments.

22–B. The long plantar ligament forms a canal for the tendon of the peroneus longus muscle.

23–A. The calcaneofibular ligament is the cord-like ligament that reinforces the lateral side of the ankle joint.

24–D. The short plantar ligament extends from the front of the inferior surface of the calcaneus to the plantar surface of the cuboid bone and supports the longitudinal arch.

25–C. The plantar calcaneonavicular (spring) ligament supports the head of the talus in maintaining the medial longitudinal arch.

26–A. The biceps femoris muscle rotates the leg laterally when the knee is flexed.

27–C. The iliopsoas is the primary flexor of the thigh.

28–D. The rectus femoris muscle crosses the hip and knee joints; thus, it can flex the thigh and extend the leg.

29–E. The tensor fasciae latae can flex and medially rotate the thigh during running and climbing.

30–B. The gluteus medius muscle supports the pelvis. When the opposing leg is raised, for example during walking, the gluteus medius muscle swings the pelvis forward and prevents it from tilting to the opposite side.

31–B. The navicular bone is a boat-shaped tarsal bone between the head of the talus and the three cuneiform bones.

32–E. The cuboid bone is the most lateral tarsal bone. It has a notch and groove for the tendon of the peroneus longus muscle.

33–A. The cuneiform bones are wedge-shaped bones that articulate posteriorly with the navicular bone and anteriorly with three metatarsals.

34–D. The calcaneus bone forms the heel. It articulates superiorly with the talus bone and anteriorly with the cuboid bone.

35–C. The talus transmits the weight of the body from the tibia to other weight-bearing bones of the foot and is the only tarsal bone without muscle attachments. It articulates anteriorly with the navicular bone and posteriorly with the calcaneus bone. It has a groove on the posterior surface for the tendon of the flexor hallucis longus muscle.

36–A. The adductor canal is bounded by the sartorius, vastus medialis, and adductor longus and adductor magnus muscles.

37–C. The greater saphenous vein courses anterior to the medial malleolus and medial to the medial condyles of the tibia and femur, ascends superficial to the fascia lata, and terminates in the femoral vein.

38–A. The adductor canal contains the femoral vessels, the saphenous nerve, and the nerve to the vastus medialis.

39–D. The quadriceps femoris muscle consists of the rectus femoris and the vastus medialis, intermedialis, and lateralis. They extend the leg at the knee joint. The semitendinosus and semimembranosus muscles extend the thigh and flex the leg. The sartorius and gracilis muscles can flex the leg.

40–A. The superior gluteal nerve leaves the pelvis through the greater sciatic foramen above the piriformis and runs between the gluteus medius and gluteus minimus muscles. It innervates the gluteus medius and gluteus minimus muscles and the tensor fasciae latae.

41–E. Space *A* in the diagram is the greater sciatic foramen. The piriformis muscle and the superior gluteal, sciatic, and pudendal nerves pass through the greater sciatic foramen.

42–C. The space deep to the inguinal ligament is separated by the iliopectineal arcus (ligament) into the lateral muscular and the medial vascular lacunae, which transmit the iliopsoas muscle and the femoral vessels, respectively.

43–D. Space *C* in the diagram is the obturator foramen. The obturator nerve passes through the obturator foramen, where it divides into anterior and posterior branches.

44–B. Space *D* in the diagram is the lesser sciatic foramen. The pudendal nerve and the obturator internus tendon pass through the lesser sciatic foramen.

45–C. The pudendal nerve and the nerve to the obturator internus muscle pass through both the greater (*A*) and lesser (*D*) sciatic foramina.

4
Thorax

Thoracic Wall

I. Skeleton of the Thorax (Figure 4.1)

A. Sternum

1. Manubrium

—has a superior margin, the jugular notch, that can be readily palpated at the root of the neck.

—has a clavicular notch on each side for articulation with the clavicle.

—also articulates with the cartilage of the first rib, the upper half of the second rib, and the body of the sternum at the manubriosternal joint or at the sternal angle.

2. Sternal Angle (of Louis)

—is the junction between the manubrium and the body of the sternum.

—is located at the level where:

a. The second rib articulates with the sternum.

b. The aortic arch begins and ends.

c. The trachea bifurcates into the right and left bronchi.

d. The inferior border of the superior mediastinum is demarcated.

e. A transverse plane can pass through the vertebral column between T4 and T5.

3. Body

—articulates with the second to seventh costal cartilages.

—also articulates with the xiphoid process at the xiphosternal joint, which is level with the ninth thoracic vertebra.

4. Xiphoid Process

—is a flat, cartilaginous process at birth and ossifies slowly from the central core and unites with the body of the sternum after middle age.

—can be palpated in the epigastrium.

—is attached via its pointed caudal end to the linea alba.

B. Ribs

—consist of 12 pairs that form the main part of the thoracic cage, extending from the vertebrae to or toward the sternum.

—are divided into head, neck, tubercle, and body (shaft).

—the head articulates with the corresponding vertebral bodies and intervertebral disks and suprajacent vertebral bodies.

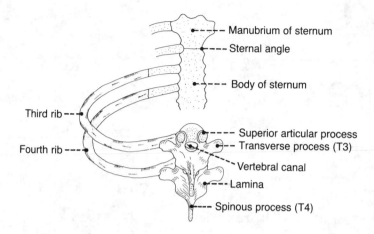

Figure 4.1. Articulations of the ribs with the vertebrae and the sternum.

—the tubercle articulates with the transverse processes of the corresponding vertebrae, with the exception of ribs 11 and 12.

1. **True Ribs**

—are the first seven ribs (ribs 1 to 7), which are attached to the sternum by their costal cartilages.

2. **False Ribs**

—are the lower five ribs (ribs 8 to 12); ribs 8 to 10 are connected to the costal cartilages immediately above them to form the anterior costal margin.

3. **Floating Ribs**

—are the last two ribs (ribs 11 and 12), which are connected only to the vertebrae.

4. **First Rib**

—is the broadest and shortest of the true ribs.
—has a single articular facet on its head, which articulates with the first thoracic vertebra.
—has a scalene tubercle for the insertion of the anterior scalene muscle.

5. **Second Rib**

—has two articular facets on its head, which articulate with the bodies of the first and second thoracic vertebrae.
—is about twice as long as the first rib.

6. **Tenth Rib**

—has a single articular facet on its head, which articulates with the tenth thoracic vertebra.

7. **Eleventh and Twelfth Ribs**

—have a single articular facet on the head.
—have no neck or tubercle.

II. Articulation of the Thorax (see Figures 2.6 and 4.1)

A. Sternoclavicular Joint

—provides the only bony attachment between the appendicular and axial skeletons.

—is a saddle-type synovial joint but has the movements of a ball-and-socket joint.

—has a fibrocartilaginous articular surface and contains two separate synovial cavities.

B. Sternocostal (Sternochondral) Joints

—are synchondroses in which the sternum articulates with the first seven costal cartilages.

C. Costochondral Joints

—are synchondroses in which the ribs articulate with their respective costal cartilages.

III. Muscles of the Thoracic Wall

Muscle	Origin	Insertion	Nerve	Action
External intercostals	Lower border of ribs	Upper border of rib below	Intercostal n.	Elevate ribs in inspiration
Internal intercostals	Lower border of ribs	Upper border of rib below	Intercostal n.	Depress ribs; interchondral part elevates ribs
Innermost intercostals	Lower border of ribs	Upper border of rib below	Intercostal n.	Elevate ribs
Transverse thoracic	Posterior surface of lower sternum and xiphoid	Inner surface of costal cartilages 2–6	Intercostal n.	Depresses ribs
Subcostalis	Inner surface of lower ribs near their angles	Upper borders of ribs 2 or 3 below	Intercostal n.	Elevates ribs
Levator costarum	Transverse processes of T7–T11	Subjacent ribs between tubercle and angle	Dorsal primary rami of C8–T11	Elevates ribs

IV. Blood Vessels

A. Internal Thoracic Artery (see Figure 2.7)

—usually arises from the first part of the subclavian artery and descends directly behind the first six costal cartilages, just lateral to the sternum.

—gives off two anterior intercostal arteries in each of the upper six intercostal spaces and terminates at the sixth intercostal space by dividing into the superior epigastric and musculophrenic arteries.

1. **Pericardiacophrenic Artery**

   —accompanies the phrenic nerve between the pleura and the pericardium to the diaphragm.

   —supplies the pleura, pericardium, and diaphragm (upper surface).

2. **Anterior Intercostal Arteries**

   —supply the upper six intercostal spaces and anastomose with the posterior intercostal arteries.

   —provide muscular branches to the intercostal, serratus anterior, and pectoral muscles.

3. **Anterior Perforating Branches**

   —perforate the internal intercostal muscles in the upper six intercostal spaces, course with the anterior cutaneous branches of the intercostal nerves, and supply the pectoralis major muscle and the skin and subcutaneous tissue over it.

   —provide **medial mammary branches** by their second, third, and fourth perforating branches.

4. **Musculophrenic Artery**

   —follows the costal arch on the inner surface of the costal cartilages.

   —gives off two anterior arteries in the seventh, eighth, and ninth spaces; perforates the diaphragm; and ends in the tenth intercostal space, where it anastomoses with the deep circumflex iliac artery.

   —supplies the pericardium, diaphragm, and muscles of the abdominal wall.

5. **Superior Epigastric Artery**

   —descends on the deep surface of the rectus abdominis muscle within the rectus sheath; supplies this muscle and anastomoses with the inferior epigastric artery.

   —supplies the diaphragm, peritoneum, and anterior abdominal wall.

B. **Thoracoepigastric Vein**

   —is a venous connection between the lateral thoracic vein and the superficial epigastric vein.

V. Lymphatic Drainage

A. **Sternal or Parasternal (Internal Thoracic) Nodes**

   —are placed along the internal thoracic artery.

   —receive lymph from the medial portion of the breast, intercostal spaces, diaphragm, and supraumbilical region of the abdominal wall.

   —drain into the junction of the internal jugular and subclavian veins.

B. **Intercostal Nodes**

   —lie near the heads of the ribs.

   —receive lymph from the intercostal spaces and the pleura.

   —drain into the cisterna chyli or the thoracic duct.

C. **Phrenic Nodes**

   —lie on the thoracic surface of the diaphragm.

   —receive lymph from the pericardium, diaphragm, and liver.

   —drain into the sternal and posterior mediastinal nodes.

VI. Nerves

A. Intercostal Nerves

—are the anterior primary rami of the first 11 thoracic spinal nerves.

—are called typical intercostal nerves (for the third to sixth nerves).

—run between the internal and innermost layers of muscles, with the intercostal veins and arteries above (**van:** veins-arteries-nerves).

—are lodged in the costal grooves on the inferior surface of the ribs.

—give off lateral and anterior cutaneous branches and muscular branches.

B. Subcostal Nerves

—are the anterior rami of the twelfth thoracic nerve (see p 176).

Mediastinum, Pleura, and Organs of Respiration

I. Mediastinum (Figure 4.2)

—is an **interpleural space** (area between the pleural cavities) in the thorax and is bounded laterally by the pleural cavities, anteriorly by the sternum, and posteriorly by the vertebral column (does not contain the lungs).

—consists of the superior mediastinum above the pericardium and the three lower divisions: anterior, middle, and posterior.

A. Superior Mediastinum

—is bounded superiorly by the oblique plane of the first rib and inferiorly by the imaginary line running from the sternal angle to the intervertebral disk between the fourth and fifth thoracic vertebrae.

—contains the superior vena cava, brachiocephalic veins, arch of the aorta, thoracic duct, trachea, esophagus, thymus, vagus nerve, left recurrent laryngeal nerve, and phrenic nerve.

B. Anterior Mediastinum

—lies anterior to the pericardium and contains the remnants of the thymus gland, lymph nodes, fat, and connective tissue.

C. Middle Mediastinum

—lies between the right and left pleural cavities and contains the heart, pericardium, phrenic nerves, roots of the great vessels, arch of the azygos vein, and main bronchi.

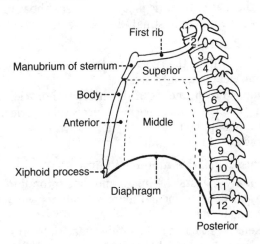

Figure 4.2. Mediastinum.

D. Posterior Mediastinum

—lies posterior to the pericardium between the mediastinal pleurae and contains the esophagus, thoracic aorta, azygos and hemiazygos veins, thoracic duct, vagus nerves, sympathetic trunk, and splanchnic nerves.

II. Trachea and Bronchi

A. Trachea

—begins at the inferior border of the cricoid cartilage (C6).
—has 16 to 20 incomplete hyaline cartilaginous rings that prevent the trachea from collapsing and that open posteriorly.
—is about 9 to 15 cm in length and bifurcates into the right and left main bronchi at the level of the sternal angle (junction of T4 and T5).
—has a **carina,** which is a downward and backward projection of the last tracheal cartilage, forming a prominent semilunar ridge separating the openings of the right and left main bronchi.

B. Right Main Stem (Primary) Bronchus

—is shorter, wider, and more vertical than the left main stem bronchus and therefore receives more foreign bodies through the trachea.
—runs under the arch of the azygos vein and divides into the superior, middle, and inferior lobar (secondary) bronchi.

C. Left Main Stem (Primary) Bronchus

—crosses anterior to the esophagus and divides into two lobar (secondary) bronchi, the upper and lower.

D. Right Superior Lobar (Secondary) Bronchus

—is known as the **eparterial** bronchus because it passes above the level of the pulmonary artery. (All others are the hyparterial bronchi.)
—is more superior than any other bronchus.

III. Pleurae and Pleural Cavities (Figures 4.3 and 4.4)

A. Pleura

—is a thin serous membrane that lines the inner surface of the thoracic wall and the mediastinum.
—is innervated by the intercostal nerves for the costal and peripheral portions of the diaphragmatic pleura, whereas the remainder of the diaphragmatic pleura and mediastinal pleura are supplied by the phrenic nerve.

1. Parietal Pleura

—lines the wall of the thorax and has costal, diaphragmatic, mediastinal, and cervical (cupola) parts.
—is supplied by branches of the internal thoracic, superior phrenic, posterior intercostal, and superior intercostal arteries. However, the visceral pleura is supplied by the bronchial arteries.
—is very sensitive to pain.

2. Visceral Pleura (Pulmonary Pleura)

—intimately invests the lungs and dips into all of the fissures.
—is supplied by bronchial arteries, but its venous blood is drained by pulmonary veins.
—is insensitive to pain but contains vasomotor fibers and sensory endings of vagal origin, which may be involved in respiratory reflexes.

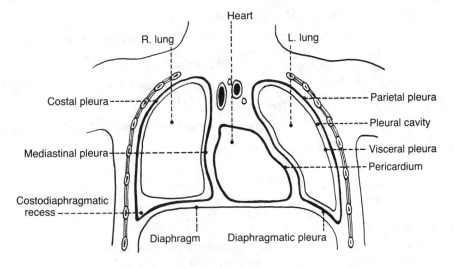

Figure 4.3. Frontal section of the thorax.

3. Cervical Pleura (Cupola of Pleura)
—is the dome of the pleura, projecting into the neck above the neck of the first rib.
—is strengthened by the suprapleural membrane (Sibson's fascia), which is a thickening of the endothoracic fascia.

4. Pulmonary Ligament
—is a vertical fold of mediastinal pleura, extending from the hilus of the lung to the base (diaphragmatic surface) on the medial surface of the lungs. It supports the lung in the plural sac.

B. Pleural Cavity
—is a potential space between the parietal and visceral pleurae.
—contains a film of fluid that lubricates the surface of the pleurae and facilitates the movement of the lungs.

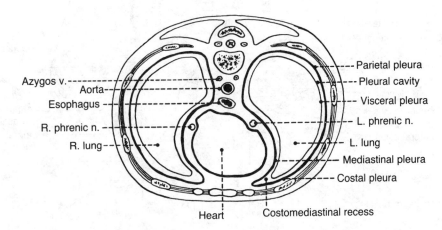

Figure 4.4. Horizontal section through the thorax.

1. **Costodiaphragmatic Recesses**

   —are the pleural recesses formed by the reflection of the costal and diaphragmatic pleurae.

   —can accumulate fluid when in the erect position.

   —allow the lungs to be pulled in and expanded during inspiration.

2. **Costomediastinal Recesses**

   —are part of the pleural cavity where the costal and mediastinal pleurae meet.

3. **Endothoracic Fascia**

   —is a small amount of loose connective tissue that separates the parietal pleurae from the thoracic wall.

IV. Lungs (Figure 4.5)

—are the essential organs of respiration and are attached to the heart and trachea by their roots and the pulmonary ligaments.

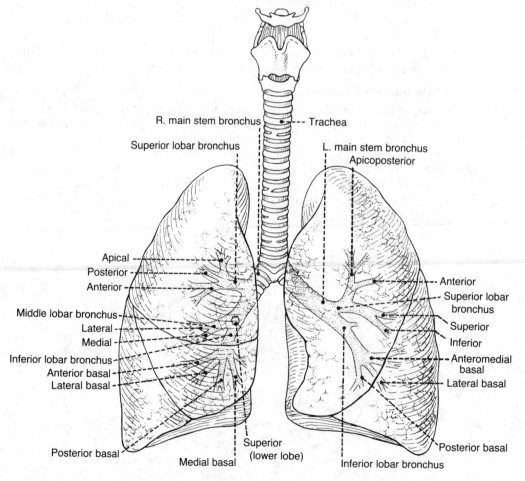

Figure 4.5. Anterior view of the trachea, bronchi, and lungs.

—contain nonrespiratory tissues, which are nourished by the bronchial arteries and drained by the pulmonary veins.

—receive parasympathetic fibers that supply the smooth muscle and glands of the bronchial tree and probably are excitatory to these structures.

—receive sympathetic fibers that supply blood vessels, smooth muscle, and glands of the bronchial tree and probably are inhibitory.

—have some sensory endings of vagal origin, which are stimulated by the stretching of the lung during inspiration and are concerned in the reflex control of respiration.

—their bases rest upon the convex surface of the diaphragm, descend during inspiration, and ascend during expiration.

A. Right Lung

—its apex projects into the root of the neck and is smaller than that of the left lung.

—is heavier and shorter than the left lung, due to the higher right dome of the diaphragm, and is wider because the heart bulges more to the left.

—is divided into upper, middle, and lower lobes by the **oblique** and **horizontal fissures,** but usually receives a single bronchial artery.

—has 3 lobar (secondary) bronchi and 10 segmental (tertiary) bronchi.

B. Left Lung

—has two lobes, upper and lower, and usually receives two bronchial arteries.

—has an oblique fissure that follows the line of the sixth rib.

—has a **lingula,** a tongue-shaped portion of its upper lobe that corresponds to the middle lobe of the right lung and is formed by the more pronounced **cardiac notch** in the left lung.

—has 2 lobar (secondary) bronchi and 8 (or possibly 9 or 10) segmental bronchi.

C. Bronchopulmonary Segment

—is the anatomical, functional, and surgical unit (subdivision) of the lungs.

—consists of a tertiary (lobular or segmental) bronchus, a branch of the pulmonary artery, and a portion of the lung tissue, surrounded by a delicate connective-tissue septum.

—refers to the portion of the lung supplied by each segmental bronchus and segmental artery and vein.

—is clinically important because of the ability to surgically remove a segment of the lung without seriously disrupting the surrounding lung tissue.

V. Lymphatic Vessels of the Lung

—drain the bronchial tree, pulmonary vessels, and connective-tissue septa.

—run along the bronchiole and bronchi toward the hilus, where they end in the pulmonary and bronchopulmonary nodes, which in turn drain into the tracheobronchial nodes.

—are not present in the walls of the pulmonary alveoli.

VI. Blood Vessels (Figure 4.6)

A. Pulmonary Trunk

—extends upward from the conus arteriosus of the right ventricle of the heart.

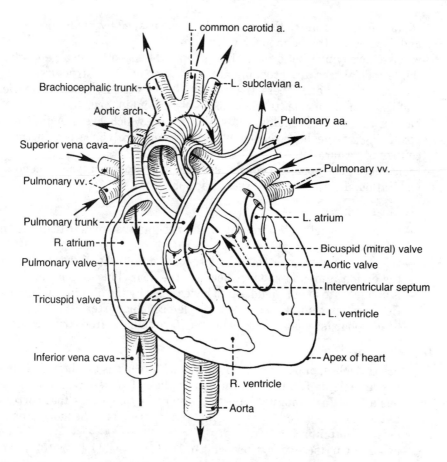

Figure 4.6. Pulmonary circulation and circulation through the heart chambers.

—passes superiorly and posteriorly from the front of the ascending aorta to its left side for about 5 cm and bifurcates into the right and left pulmonary arteries.

—is partially invested with fibrous pericardium.

—has much lower blood pressure than that in the aorta.

B. Left Pulmonary Artery

—carries unoxygenated blood to the left lung and is shorter and narrower than the right artery.

—is connected to the arch of the aorta by the **ligamentum arteriosum,** the fibrous remains of the ductus arteriosus.

C. Right Pulmonary Artery

—runs horizontally toward the hilus of the right lung under the arch of the aorta behind the ascending aorta and superior vena cava and anterior to the right bronchus.

D. Pulmonary Veins

—are intersegmental in drainage (do not accompany the bronchi or the segmental artery within the parenchyma of the lungs).

—are five in number as they leave the lung, consisting of one from each lobe of the lungs, although the right upper and middle veins usually join so that only four veins enter the left atrium.

—carry oxygenated blood and enter the left atrium.

—collect arterial blood from the respiratory part of the lung and venous blood from the visceral pleura and from the bronchi.

E. Bronchial Arteries

—arise from the thoracic aorta; usually there is one artery for the right lung and two for the left lung.

—supply oxygenated blood to the nonrespiratory conducting tissues of the lungs. Anastomoses occur between the capillary beds of the bronchial and the pulmonary arterial systems.

F. Bronchial Veins

—receive blood from the larger subdivisions of the bronchi and empty into the azygos vein on the right and into the accessory hemiazygos vein on the left. (Other venous blood is drained by the pulmonary veins.)

—may receive twigs from the tracheobronchial lymph nodes.

VII. Respiration

—is the vital exchange of oxygen and carbon dioxide that occurs in the lungs.

A. During Inspiration

—the diaphragm contracts, pulling the dome inferiorly into the abdomen, thereby increasing the vertical diameter of the thorax and decreasing intrathoracic and intrapulmonary pressures.

—the pleural cavities and the lungs enlarge, the intra-alveolar pressure is reduced, and the air passively rushes into the lungs as a result of atmospheric pressure.

—the ribs are elevated and the abdominal pressure is increased with decreased abdominal volume.

B. During Expiration

—the diaphragm, the intercostal muscle, and other muscles relax; the thoracic volume is decreased; and the intrathoracic pressure is increased.

—the stretched elastic tissue of the lungs recoils, and much of the air is expelled.

—the abdominal pressure is decreased, and the ribs are depressed.

C. Muscles of Inspiration

—include the diaphragm, external intercostal, sternocleidomastoid, levator scapulae, serratus anterior, scalenus, pectoralis major and minor, erector spinae, and serratus posterior–superior muscles.

D. Muscles of Expiration

—include the muscles of the abdominal wall, internal intercostals, and serratus posterior–inferior muscles.

VIII. Nerve Supply

A. Pulmonary Plexus

—receives afferent and efferent (parasympathetic) fibers from the vagus nerve, joined by branches from the sympathetic trunk and cardiac plexus.

—is divided into the **anterior pulmonary plexus,** which lies in front of the root of the lung, and the **posterior pulmonary plexus,** which lies behind the root of the lung.

—its branches accompany the blood vessels and bronchi into the lung.

B. Phrenic Nerve

—arises from the third through fifth cervical nerves (C3–C5) and lies in front of the anterior scalene muscle.

—enters the thorax by passing deep to the subclavian vein and superficial to the subclavian arteries.

—runs anterior to the root of the lung, whereas the vagus nerve runs posterior to the root of the lung.

—is accompanied by the pericardiacophrenic vessels of the internal thoracic vessels and descends between the mediastinal pleura and the pericardium.

—supplies the pericardium, the mediastinal and diaphragmatic pleurae, and the diaphragm.

—its sectioning or crushing in the neck may not by itself produce complete paralysis of the corresponding half of the diaphragm because the accessory phrenic nerve, derived from the fifth cervical nerve as a branch of the nerve to the subclavius, usually joins the phrenic nerve in the root of the neck or in the upper part of the thorax.

IX. Clinical Considerations

A. Pneumothorax

—is the presence of air or gas in the pleural cavity.

B. Emphysema

—is the accumulation of air in the terminal bronchioles and alveolar sacs.

—reduces the surface area available for gas exchange and thereby reduces oxygen absorption.

C. Pneumonia (Pneumonitis)

—is an inflammation of the lungs.

D. Pleural Tap (Thoracentesis or Pleuracentesis)

—is a surgical puncture of the thoracic wall into the pleural cavity for aspiration of fluid.

—is performed posterior to the midaxillary line at one or two intercostal spaces below the fluid level but not below the ninth intercostal space.

Pericardium and Heart

I. Pericardium

—is a fibroserous sac that encloses the heart and roots of the great vessels and occupies the middle mediastinum.

—is composed of the fibrous pericardium and serous pericardium.

—receives blood from the pericardiacophrenic, bronchial, and esophageal arteries.

—is supplied with vasomotor and sensory fibers from the phrenic and vagus nerves and the sympathetic trunks.

A. Fibrous Pericardium

—is a strong, dense, fibrous layer that blends with the adventitia of the roots of the great vessels and the central tendon of the diaphragm.

B. Serous Pericardium

—consists of the parietal layer, which lines the inner surface of the fibrous pericardium, and the visceral layer, which forms the outer layer (epicardium) of the heart wall and the roots of the great vessels.

C. Pericardial Cavity

—is a potential space between the visceral layer of the serous pericardium (epicardium) and the parietal layer of the serous pericardium lining the inner surfaces of the fibrous pericardium.

D. Pericardial Sinuses

1. Transverse Sinus

—is a subdivision of the pericardial sac, lying posterior to the aorta and pulmonary trunk, anterior to the superior vena cava and the left atrium, and superior to the right and left pulmonary veins.

2. Oblique Sinus

—is a subdivision of the pericardial sac behind the heart.

—is surrounded by the reflection of the serous pericardium around the right and left pulmonary veins and the inferior vena cava.

II. Heart

A. General Characteristics

—its apex lies in the left fifth intercostal space slightly medial to the nipple line and is useful clinically for determining the left border of the heart and for the auscultation of the mitral valve.

—its wall consists of three layers: inner **endocardium**, middle **myocardium**, and outer **epicardium**.

B. Internal Anatomy (Figure 4.7)

1. Right Atrium

—its walls are relatively smooth except for the presence of the pectinate muscles.

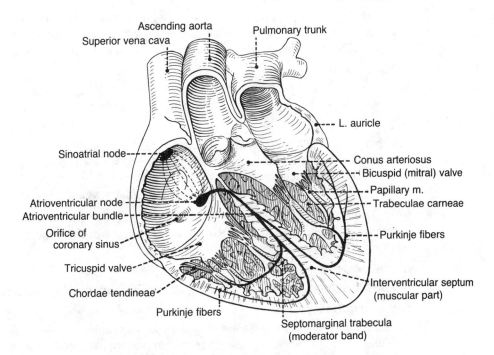

Figure 4.7. Internal anatomy and conducting system of the heart.

—is larger than the left atrium, but its wall is thinner.

—its pressure is normally slightly lower than the left atrial pressure.

a. Right Auricle

—is the conical muscular pouch of the superior extremity of the right atrium.

—covers the first part of the right coronary artery.

b. Sinus Venarum (Cavarum)

—is a posteriorly situated, smooth-walled area that is separated from the more muscular atrium proper by the **crista terminalis.**

—represents the embryonic sinus venosus and receives the superior vena cava, inferior vena cava, coronary sinus, and anterior cardiac veins.

c. Pectinate Muscles

—are prominent ridges of atrial myocardium located in the interior of both auricles and the right atrium. The inner surface of the left atrium is smooth.

d. Crista Terminalis

—is a vertical muscular ridge running anteriorly along the right atrial wall from the opening of the superior vena cava to the inferior vena cava.

—represents the junction between the primitive sinus venosus and the right atrium proper and is indicated externally by the sulcus terminalis.

—provides the origin of the pectinate muscles, which run across the right atrial wall.

e. Venae Cordis Minimae

—are the smallest cardiac veins that begin in the substance of the heart (endocardium and innermost layer of the myocardium) and end chiefly in the atria at the **foramina venarum minimarum.**

f. Fossa Ovalis

—represents the position of the foramen ovale, through which blood runs from the right to the left atrium before birth.

2. Left Atrium

—is smaller and has thicker walls than the right atrium.

—is the most posterior of the four chambers, and its walls are smooth, except for a few pectinate muscles in the auricle.

—receives oxygenated blood through four pulmonary veins.

3. Right Ventricle

—makes up the major portion of the anterior surface of the heart.

—contains the following structures:

a. Trabeculae Carneae

—are anastomosing muscular ridges of myocardium in the ventricles.

b. Papillary Muscles

—are cone-shaped muscles enveloped by endocardium.

—extend from the anterior and posterior ventricular walls and the septum, and their apices are attached to the chordae tendineae.

—contract to tighten the chordae tendineae, preventing the cusps of the valve from being everted into the atrium by the pressure developed by the pumping action of the heart.

c. Chordae Tendineae
—extend from one papillary muscle to more than one cusp.
—prevent eversion of the valve during the ventricular contractions.

d. Conus Arteriosus (Infundibulum)
—is the upper end of the right ventricle.
—has smooth walls and leads into the pulmonary orifice.

e. Septomarginal Trabecula (Moderator Band)
—is an isolated trabecula of the bridge type, extending from the interventricular septum to the base of the anterior papillary muscle of the right ventricle.
—carries the right limb of the atrioventricular bundle from the septum to the sternocostal wall of the ventricle.

f. Interventricular Septum
—gives an attachment to the septal cup of the tricuspid valve.
—is mostly muscular but has a small membranous upper part, which is a common site of ventricular septal defects.

4. Left Ventricle
—is divided into the left ventricle proper and the aortic vestibule, which is the upper and anterior part of the left ventricle and leads into the aorta.
—contains two **papillary muscles** (anterior and posterior) with their chordae tendineae and a meshwork of muscular ridges called the **trabeculae carneae.**
—performs more work than the right ventricle; its wall is usually twice as thick.
—is longer, narrower, and more conical-shaped than the right ventricle.

C. Valves (Figures 4.8 and 4.9)

1. Pulmonary Valve
—lies behind the medial end of the third left costal cartilage and adjoining part of the sternum.
—its sound is most audible over the left second intercostal space.

2. Aortic Valve
—lies behind the left half of the sternum opposite the third intercostal space.
—its sound is most audible over the right second intercostal space.

3. Tricuspid (Right Atrioventricular) Valve
—lies behind the right half of the sternum opposite the fourth intercostal space and is covered by endocardium.
—its sound is most audible over the right lower part of the body of the sternum.
—has three cusps: anterior, posterior, and septal.
—is attached by the chordae tendineae to three papillary muscles that keep the valves against the pressure developed by the pumping action of the heart.

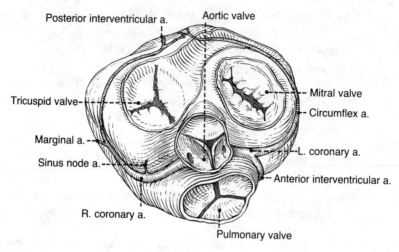

Figure 4.8. Superior view of the heart.

4. Mitral (Left Atrioventricular) Valve
—lies behind the left half of the sternum opposite the fourth costal cartilage.
—its sound is most audible over the left fifth intercostal space (apical region at the midclavicular line).
—has two cusps: anterior and posterior.

D. Conducting System (see Figure 4.7)

1. Sinoatrial Node
—is a small mass of specialized cardiac muscle fibers.
—is the "pacemaker" of the heart and initiates the heartbeat, which can be altered by nervous stimulation (sympathetic stimulation speeds it up and vagal stimulation slows it down).

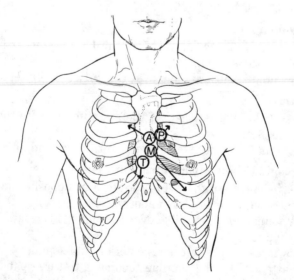

Figure 4.9. Positions of the valves of the heart and heart sounds. *P* = pulmonary valve; *A* = aortic valve; *M* = mitral valve; *T* = tricuspid valve. Arrows indicate positions of the heart sounds.

—lies in the myocardium at the upper end of the crista terminalis, near the opening of the superior vena cava in the right atrium.

—is supplied by the sinus node artery, which is usually a branch of the right coronary artery.

2. **Atrioventricular Node**

—lies beneath the endocardium in the septal wall, above the opening of the coronary sinus in the right atrium.

—is supplied by the posterior interventricular branch of the right coronary artery.

—is supplied by autonomic nerve fibers, although the cardiac muscle fibers lack motor endings.

3. **Atrioventricular Bundle (Bundle of His)**

—begins at the atrioventricular node and runs along the membranous septum.

—splits into right and left branches, which descend into the interventricular septum and spread out into the ventricular walls.

—breaks up into terminal conducting fibers (Purkinje fibers).

E. **Coronary Arteries** (Figure 4.10)

1. **Right Coronary Artery**

—arises from the anterior (right) aortic sinus and runs between the root of the pulmonary trunk and the right auricle.

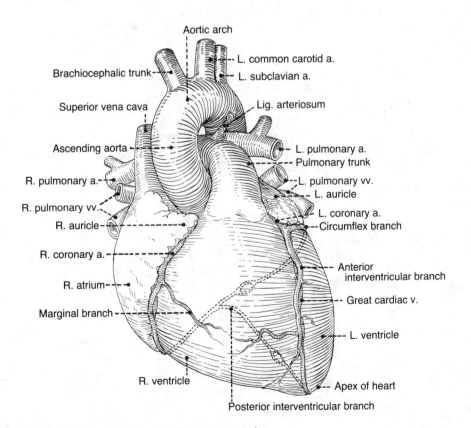

Figure 4.10. Anterior view of the heart with coronary arteries.

—gives off branches to the right atrium and ventricle and also a **marginal artery** to the right ventricular wall.

—gives off a branch termed the **sinus node artery**, which passes between the right atrium and the opening of the superior vena cava and supplies the sinoatrial node.

—gives off the **posterior interventricular artery** (right posterior descending artery).

—is supplied by sensory and autonomic nerve fibers of the coronary plexuses.

2. Left Coronary Artery

—arises from the left aortic sinus, just above the aortic semilunar valve.

—is shorter than the right coronary artery and usually is distributed to more of the myocardium.

—gives rise to the **circumflex branch** and the **anterior interventricular branch** (left anterior descending branch).

3. Blood Flow in the Coronary Arteries

—is maximal during diastole and minimal during systole, owing to the compression of the arterial branches in the myocardium during systole.

F. Cardiac Veins and Coronary Sinus (Figures 4.11 and 4.12)

1. Coronary Sinus

—is the largest vein draining the heart and lies in the coronary sulcus.

—opens into the right atrium between the opening of the inferior vena cava and the atrioventricular opening.

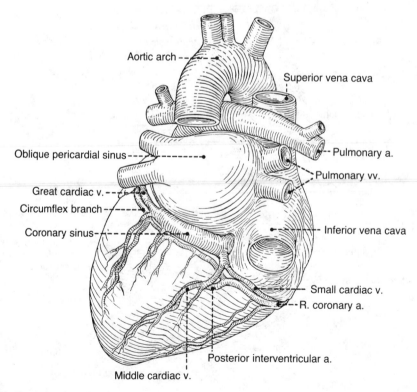

Aortic arch

Superior vena cava

Oblique pericardial sinus

Pulmonary a.

Pulmonary vv.

Great cardiac v.

Circumflex branch

Coronary sinus

Inferior vena cava

Small cardiac v.

R. coronary a.

Posterior interventricular a.

Middle cardiac v.

Figure 4.11. Posterior view of the heart.

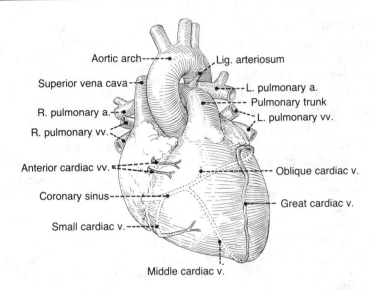

Figure 4.12. Anterior view of the heart.

—has a one-cusp valve at the right margin of its aperture.
—receives the great, middle, and small cardiac veins; the oblique vein of the left atrium; and the posterior vein of the left ventricle.

2. Great Cardiac Vein

—begins at the apex of the heart, ascends in the anterior interventricular groove, and drains upward alongside the anterior interventricular branch of the left coronary artery.
—turns to the left to lie in the coronary sinus, where it continues as the coronary sinus.

3. Middle Cardiac Vein

—begins at the apex of the heart and ascends in the posterior interventricular groove, accompanying the posterior interventricular branch of the right coronary artery.
—drains into the right end of the coronary sinus.

4. Small Cardiac Vein

—runs along the right margin of the heart in company with the marginal artery and then runs posteriorly in the coronary sulcus to end in the right end of the coronary sinus.

5. Oblique Vein of the Left Atrium

—descends to empty into the coronary sinus, near its left end.

6. Anterior Cardiac Vein

—drains the anterior right ventricle, crosses the coronary groove, and ends directly in the right atrium.

7. Smallest Cardiac Veins (Venae Cordis Minimae)

—begin in the wall of the heart and empty directly into its chambers.

G. Lymphatic Vessels

—receive lymph from the myocardium and epicardium.

—follow the right coronary artery to empty into the anterior mediastinal nodes and follow the left coronary artery to empty into a node of the tracheobronchial group.

H. Cardiac Plexus

—receives the superior, middle, and inferior cervical and thoracic cardiac nerves from the sympathetic trunks and vagus nerves.

—is divisible into the **superficial cardiac plexus,** which lies beneath the arch of the aorta, in front of the pulmonary artery, and the **deep cardiac plexus,** which lies posterior to the arch of the aorta, in front of the bifurcation of the trachea.

—gives branches to the atria, pulmonary plexuses, and coronary plexuses that distribute to the areas supplied by the arteries.

—richly supplies the conducting system of the heart. (The cardiac muscle fibers are devoid of motor endings and are activated only by the conducting system.)

—supplies the heart with sympathetic fibers, which increase the heart rate, and parasympathetic fibers, which decrease the heart rate.

I. Clinical Considerations

1. Angina Pectoris

—is characterized by attacks of chest pain originating in the heart and felt beneath the sternum, often radiating to the left shoulder and down the arm.

—is caused by an insufficient supply of oxygen to the heart muscle, owing to coronary arterial disease.

—its pain impulses travel in visceral afferent fibers through the middle and inferior cervical and thoracic cardiac branches of the sympathetic nervous system before entering the thoracic segment of the spinal cord.

2. Coronary Atherosclerosis

—is characterized by the presence of sclerotic plaques containing cholesterol and lipoid material, which impair myocardial blood flow, leading to ischemia and myocardial infarction.

3. Pericardial Tamponade

—is an acute compression of the heart due to pericardial effusion (passage of fluid from the pericardial capillaries into the pericardial sac).

—results in decreased diastolic capacity, reduced cardiac output with an increased heart rate, increased venous pressure with jugular vein distension, hepatic enlargement, and peripheral edema.

4. Pericarditis

—is an inflammation of the parietal serous pericardium, which may result in cardiac tamponade and precordial and epigastric pain.

—causes the surfaces of the pericardium to become rough, and the resulting friction, which sounds like the rustle of silk, can be heard on auscultation.

5. Pericardiocentesis

—is a surgical puncture of the pericardial cavity for the aspiration of fluid, which is necessary to relieve the pressure of accumulated fluid on the heart. A needle is inserted into the pericardial sac through the fifth or sixth intercostal spaces adjacent to the sternum.

III. Fetal Circulation (Figure 4.13)

A. Foramen Ovale

—is an opening in the septum secundum.

—usually closes functionally at birth, but anatomical closure occurs later.

—shunts blood from the right atrium into the left atrium before birth, without passing through the lungs (pulmonary circulation).

B. Ductus Arteriosus

—is derived from the sixth aortic arch and connects the bifurcation of the pulmonary trunk.

—closes functionally soon after birth, but anatomical closure requires several weeks.

—becomes the ligamentum arteriosum, which connects the left pulmonary artery (at its origin from the pulmonary trunk) to the concavity of the arch of the aorta.

—shunts blood from the pulmonary trunk to the aorta before birth, bypassing the pulmonary circulation.

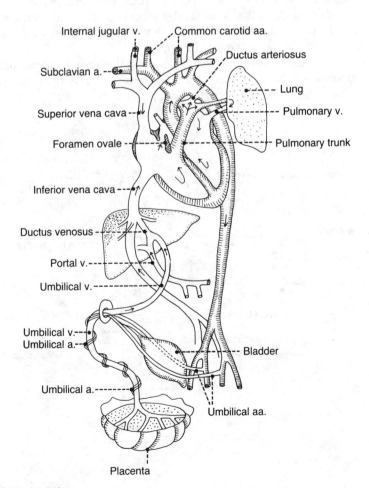

Figure 4.13. Fetal circulation.

C. Ductus Venosus

—shunts blood from the umbilical vein to the inferior vena cava before birth, bypassing the liver.

—becomes the ligamentum venosum (fibrous cord), which joins the left branch of the portal vein to the inferior vena cava.

Structures in the Posterior Mediastinum

I. Esophagus

—is a muscular tube that is continuous with the pharynx in the neck and enters the thorax behind the trachea.

—has three constrictions: one at the level of the sixth cervical vertebra, where it begins; one at the crossing of the left bronchus; and one at the tenth thoracic vertebra, where it pierces the diaphragm.

—receives blood from three branches of the aorta (the inferior thyroid, bronchial, and esophageal arteries) and from the left gastric and inferior phrenic arteries.

II. Blood Vessels and Lymphatic Vessels (Figure 4.14)

A. Thoracic Aorta

—begins at the level of the fourth thoracic vertebra.

—descends on the left side of the vertebral column and then approaches the median plane to end in front of the vertebral column by passing through the aortic hiatus of the diaphragm.

—gives off nine pairs of posterior intercostal arteries and one pair of subcostal arteries. The first two intercostal arteries arise from the highest intercostal arteries of the costocervical trunks.

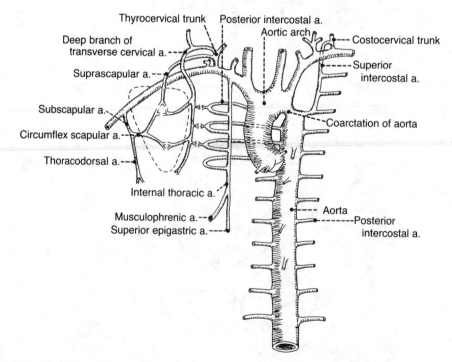

Figure 4.14. Coarctation of the aorta.

—also gives off the pericardial, bronchial (one right and two left), esophageal, mediastinal, and superior phrenic branches.

B. Coarctation of the Aorta

—takes place distal to the point of the entrance of the ductus arteriosus into the aorta, in which case an adequate collateral circulation develops before birth. If this condition occurs proximal to the origin of the left subclavian artery of the ductus arteriosus, adequate collateral circulation does not develop.

—results in tortuous and enlarged blood vessels, especially the internal thoracic, intercostal, epigastric, and scapular arteries.

—results in elevated blood pressure in the radial artery and decreased pressure in the femoral artery.

—causes the femoral pulse to occur after the radial pulse. (Normally, the femoral pulse occurs slightly before the radial pulse and is under about the same pressure.)

—leads to the development of the important collateral circulation over the thorax.

C. Collateral Circulation

—occurs between the:

1. Anterior intercostal branches of the internal thoracic artery and the posterior intercostal arteries.

2. Superior epigastric branch of the internal thoracic artery and the inferior epigastric artery.

3. Posterior–superior intercostal branch of the costocervical trunk and the third posterior intercostal artery.

4. Posterior intercostal arteries and the descending scapular (or dorsal scapular) artery, which anastomoses with the suprascapular and circumflex scapular arteries around the scapula.

D. Azygos Venous System (Figure 4.15)

1. Azygos Vein

—is formed by the union of the right ascending lumbar and right subcostal veins.

—enters the thorax through the aortic opening of the diaphragm.

—receives the right intercostal veins and the right posterior–superior intercostal vein.

—arches over the root of the right lung and empties into the superior vena cava, of which it is the first tributary.

2. Hemiazygos Vein

—is formed by the junction of the left subcostal and ascending lumbar veins.

—ascends on the left side of the vertebral bodies behind the thoracic aorta, receiving the lower four posterior intercostal veins.

3. Accessory Hemiazygos Vein

—begins at the fourth intercostal space, receives the fourth to seventh or eighth intercostal veins, descends in front of the posterior intercostal arteries, and terminates in the azygos vein.

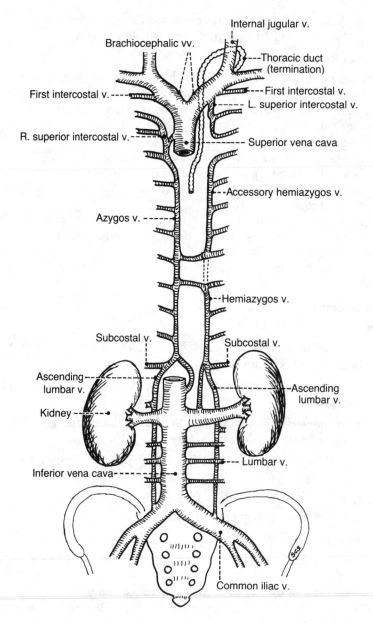

Figure 4.15. Azygos venous system.

4. Posterior Intercostal Vein

—its first intercostal vein on each side drains into the corresponding brachiocephalic vein.

—its second, third, and often fourth intercostal veins join to form the **superior intercostal vein,** which drains into the azygos vein on the right and into the brachiocephalic vein on the left.

—the rest of the vein drains into the azygos vein on the right and into the hemiazygos or accessory hemiazygos veins on the left.

E. Thoracic Duct (Figure 4.16; see Figure 4.15)

—begins in the abdomen at the **cisterna chyli,** which is the dilated junction of the intestinal, lumbar, and descending intercostal trunks.

—drains the lower limbs, pelvis, abdomen, left thorax, left upper limb, and left side of the head and neck.

—passes through the aortic opening of the diaphragm and ascends through the posterior mediastinum between the aorta and the azygos vein.

—arches over the cupula of the left pleura to lie posterior to the left subclavian vein and usually empties into the junction of the left internal jugular and subclavian veins.

III. Autonomic Nerves in the Thorax

A. Sympathetic Trunk (see Figure 5.18)

—descends in front of the neck of the ribs and the posterior intercostal vessels.

—shows the cervicothoracic (or stellate) ganglion, which is formed by fusion of the inferior cervical ganglion with the first thoracic ganglion.

—enters the abdomen through the crus of the diaphragm or behind the medial lumbocostal arch.

—gives rise to cardiac, pulmonary, mediastinal, and splanchnic branches.

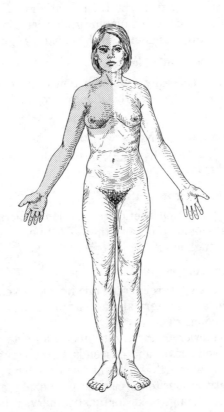

Figure 4.16. All areas except the shaded area (upper right quadrant) are drained by the thoracic duct.

—is connected to the thoracic spinal nerves by gray and white rami communicantes.

—contains the cell bodies of the postganglionic sympathetic (GVE) fibers.

1. Rami Communicantes

a. Gray Rami Communicantes

—contain postganglionic sympathetic (GVE; unmyelinated) fibers that supply the blood vessels, sweat glands, and arrector pili muscles of hair follicles.

—are connected to every spinal nerve and contain fibers with cell bodies located in the sympathetic trunk.

b. White Rami Communicantes

—contain preganglionic sympathetic (GVE; myelinated) fibers with cell bodies located in the lateral horn (intermediolateral cell column) of the spinal cord, and GVA fibers with cell bodies located in the dorsal root ganglia.

—are limited to the spinal cord segments between T1 and L2.

2. Splanchnic Nerves

—contain largely preganglionic sympathetic (GVE) fibers with cell bodies located in the lateral horn of the spinal cord, and GVA fibers with cell bodies located in the dorsal root ganglia.

a. Greater Splanchnic Nerve

—arises usually from the fifth through ninth thoracic ganglia, perforates the crus of the diaphragm or occasionally passes through the aortic hiatus, and ends in the celiac ganglion.

b. Lesser Splanchnic Nerve

—is derived usually from the tenth and eleventh thoracic ganglia, pierces the crus of the diaphragm, and ends in the aorticorenal ganglion.

c. Least Splanchnic Nerve

—is derived usually from the twelfth thoracic ganglion, pierces the crus of the diaphragm, and ends in the renal plexus.

B. Vagus Nerve (see Figure 5.18)

—contains the parasympathetic preganglionic fibers with cell bodies located in the medulla oblongata, and the GVA fibers with cell bodies located in the nodose ganglion.

1. Right Vagus Nerve

—gives off the **right recurrent laryngeal nerve,** which hooks around the right subclavian artery and ascends into the neck between the trachea and the esophagus.

—crosses anterior to the right subclavian artery, runs posterior to the superior vena cava, and descends at the right surface of the trachea and then posterior to the right main bronchus.

—gives contributions to the cardiac, pulmonary, and esophageal plexuses.

—forms the vagal trunks (or gastric nerves) at the lower part of the esophagus and enters the abdomen through the esophageal hiatus.

2. Left Vagus Nerve

—enters the thorax between the left common carotid and subclavian arteries and behind the left brachiocephalic vein and descends on the arch of the aorta.

—gives off the **left recurrent laryngeal nerve,** which hooks around the arch of the aorta to the left of the ligamentum arteriosum. It ascends through the superior mediastinum in the groove between the trachea and the esophagus.

—gives off the thoracic cardiac branches, breaks up into the pulmonary plexuses, and then continues into the esophageal plexus.

3. Clinical Considerations

—the vagus nerves can be cut at the lower portion of the esophagus (vagotomy) in an attempt to reduce gastric secretion in the treatment of peptic ulcer disease.

—the recurrent laryngeal nerve may be damaged on the left side by an aneurysm of the aortic arch, leading to coughing, hoarseness, and paralysis of the ipsilateral vocal cord because it innervates almost all of the muscles of the larynx.

Review Test

DIRECTIONS: Each of the numbered items or incomplete statements in this section is followed by answers or by completions of the statement. Select the **one** lettered answer or completion that is **best** in each case.

1. The right ventricle contains all of the following structures EXCEPT the
(A) papillary muscles.
(B) chordae tendineae.
(C) septomarginal trabeculae.
(D) pectinate muscles.
(E) trabeculae carneae.

2. Which of the following statements concerning the ligamentum arteriosum is true?
(A) It is located on the visceral surface of the liver.
(B) It connects the left umbilical vein to the left portal vein.
(C) It connects the left pulmonary artery to the aortic arch.
(D) It becomes the median umbilical ligament in adults.
(E) It becomes the round ligament of the liver in adults.

3. On the surface of the chest, the apex of the heart is located
(A) at the level of the xiphoid process of the sternum.
(B) at the level of the sternal angle.
(C) in the left fifth intercostal space.
(D) in the right fifth intercostal space.
(E) in the left fourth intercostal space.

4. Normal, quiet expiration is achieved by contraction of the
(A) elastic tissue in the thoracic wall and lungs.
(B) serratus posterior–superior muscles.
(C) pectoralis minor muscles.
(D) serratus anterior muscles.
(E) diaphragm.

5. Which of the following groups of fibers best describes fibers in the greater splanchnic nerves?
(A) Somatic afferent and preganglionic visceral efferent fibers
(B) Visceral afferent and postganglionic visceral efferent fibers
(C) Visceral afferent and preganglionic visceral efferent fibers
(D) Somatic efferent and postganglionic visceral efferent fibers
(E) Visceral afferent and somatic efferent fibers

6. Which of the following statements concerning the intercostal arteries is true?
(A) Some anterior intercostal arteries are branches of the costocervical trunk.
(B) All posterior intercostal arteries branch directly from the aorta.
(C) The upper two posterior intercostal arteries are branches of the internal thoracic artery.
(D) The anterior intercostal arteries branch directly or indirectly from the internal thoracic artery.
(E) The highest posterior intercostal arteries are branches of the thyrocervical trunk of the subclavian artery.

7. Which of the following groups of fibers best describes fibers in white rami communicantes?
(A) Preganglionic sympathetic efferent and somatic efferent fibers
(B) Postganglionic sympathetic efferent fibers
(C) Preganglionic sympathetic efferent and postganglionic parasympathetic efferent fibers
(D) Preganglionic sympathetic efferent and visceral afferent fibers
(E) Preganglionic parasympathetic efferent, postganglionic sympathetic efferent, and visceral afferent fibers

8. The sound of the mitral valve can be heard best
(A) in the left fifth intercostal space at the midclavicular line.
(B) over the medial end of the second left intercostal space.
(C) over the medial end of the second right intercostal space.
(D) over the right half of the lower end of the body of the sternum.
(E) in the left fourth intercostal space at the midclavicular line.

9. Which of the following statements concerning the cardiac veins is true?

(A) The great cardiac vein accompanies the posterior descending interventricular artery.
(B) The middle cardiac vein accompanies the anterior descending interventricular artery.
(C) The anterior cardiac vein ends in the right atrium.
(D) The small cardiac vein accompanies the circumflex branch of the left coronary artery.
(E) The oblique veins of the left atrium end in the left atrium.

10. The largest portion of the sternocostal surface of the heart is composed of the

(A) left atrium.
(B) right atrium.
(C) left ventricle.
(D) right ventricle.
(E) base of the heart.

11. Which of the following structures guards the opening between the right atrium and right ventricle?

(A) Pulmonary semilunar valves
(B) Mitral valve
(C) Valve of the coronary sinus
(D) Tricuspid valve
(E) Aortic semilunar valves

12. Which of the following terms is used to describe the conductive tissue of the heart known as the cardiac pacemaker?

(A) Atrioventricular bundle
(B) Atrioventricular node
(C) Sinoatrial node
(D) Purkinje fiber
(E) Moderator band

13. The heart is situated in

(A) the superior mediastinum.
(B) the anterior mediastinum.
(C) the middle mediastinum.
(D) the posterior mediastinum.
(E) none of the above.

14. Which of the following bronchopulmonary segments connects to the right middle lobe bronchus?

(A) Medial and lateral
(B) Anterior and posterior
(C) Anterior basal and medial basal
(D) Anterior basal and posterior basal
(E) None of the above

15. The eparterial bronchus is the

(A) left superior bronchus.
(B) left inferior bronchus.
(C) right superior bronchus.
(D) right middle bronchus.
(E) right inferior bronchus.

16. Which portion of the heart musculature is most likely to be ischemic if a blood clot blocks the circumflex branch of the left coronary artery?

(A) Anterior portion of the left ventricle
(B) Anterior interventricular region
(C) Posterior interventricular region
(D) Posterior portion of the left ventricle
(E) Anterior portion of the right ventricle

17. Each of the following statements concerning the thoracic duct is true EXCEPT

(A) it originates from the cisterna chyli in the abdomen.
(B) it passes upward through the aortic opening in the diaphragm.
(C) it drains into the junction of the left internal jugular vein and the left subclavian vein.
(D) it receives drainage from the right cervical lymph nodes.
(E) it ascends between the aorta and the azygos vein.

18. Each statement below concerning the right lung is true EXCEPT

(A) it has ten bronchopulmonary segments.
(B) it usually receives one bronchial artery.
(C) it has three lobar (secondary) bronchi.
(D) its upper lobe has a tongue-shaped portion called a lingula.
(E) it has a slightly larger capacity than the left lung.

19. Which of the following muscles functions to elevate the ribs?

(A) Sternocleidomastoid muscle
(B) Anterior scalene muscle
(C) External intercostal muscles
(D) Internal intercostal muscles
(E) All of the above

20. Each structure below is found in the superior mediastinum EXCEPT

(A) the brachiocephalic veins.
(B) the trachea.
(C) part of the left common carotid artery.
(D) the arch of the aorta.
(E) the hemiazygos vein.

21. Each of the following veins receives drainage from the posterior intercostal veins EXCEPT

(A) the hemiazygos vein.
(B) the azygos vein.
(C) the subclavian veins.
(D) the brachiocephalic veins.
(E) the superior intercostal vein.

22. Which of the following statements concerning the phrenic nerve is true?

(A) It enters the thorax by passing in front of the subclavian vein.
(B) It passes posterior to the root of the lung as it courses through the thorax.
(C) It contains only somatic motor nerve fibers.
(D) It supplies the pericardium and diaphragm.
(E) It supplies the visceral pleura.

23. Each of the following statements concerning the left vagus nerve is true EXCEPT

(A) it passes in front of the left subclavian artery as it enters the thorax.
(B) it contributes to the anterior esophageal plexus.
(C) it forms the anterior vagal trunk at the lower part of the esophagus.
(D) it can be cut on the lower part of the esophagus to reduce gastric secretion.
(E) it contains parasympathetic postganglionic fibers.

DIRECTIONS: Each group of items in this section consists of lettered options followed by a set of numbered items. For each item, select the **one** lettered option that is most closely associated with it. Each lettered option may be selected once, more than once, or not at all.

Questions 24–28

Match each of the following statements with the most appropriate nerve.
(A) Vagus nerve
(B) Phrenic nerve
(C) Sympathetic trunk
(D) Left recurrent laryngeal nerve
(E) Greater splanchnic nerve

24. Contributes to the formation of the esophageal plexus superior to the diaphragm

25. Loops around the arch of the aorta near the ligamentum arteriosum

26. Arises in the posterior mediastinum and enters the celiac ganglion

27. Lies between the fibrous pericardium and parietal mediastinal pleura along part of its course

28. Is connected to the intercostal nerves by white and gray rami

Questions 29–33

Match each of the following descriptions with the appropriate lettered structure in this superior cross-sectional view of the heart.

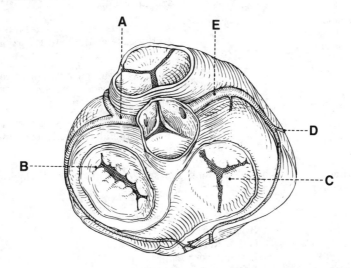

29. Runs toward the apex of the heart; supplies atrial blood to the anterior right ventricle

30. Supplies arterial blood to the sinoatrial node

31. Arises from the left aortic sinus

32. Provides the anterior interventricular artery

33. Is best heard through a stethoscope in the left fifth intercostal space at the midclavicular line

Questions 34–38

Match each statement below with the most appropriate chamber of the heart.
(A) Right atrium
(B) Left atrium
(C) Right ventricle
(D) Left ventricle
(E) None of the above

34. Blood leaves this chamber through a valve that can be auscultated over the second left intercostal space, just lateral to the sternum

35. Receives blood from the anterior cardiac veins

36. Chamber associated with the apex of the heart

37. Enlarges briefly in response to coarctation of the aorta

38. The sinoatrial node is in the wall of this chamber

Questions 39–43

Match each statement to the right with the appropriate structure.

(A) Crista terminalis
(B) Septomarginal trabecula
(C) Chordae tendineae
(D) Pectinate muscle
(E) Anulus fibrosus

39. Maintains constant tension on the cusps of the arterioventricular valve

40. Attachment site of the cusps of the atrioventricular valves

41. Extends from the interventricular septum to the base of the anterior papillary muscle of the right ventricle

42. A vertical muscular ridge representing the junction between the sinus venarum and the atrium proper

43. Contains Purkinje fibers from the right limb of the atrioventricular bundle

DIRECTIONS: Each question below contains four suggested answers of which **one or more** is correct. Choose answer

A if **1, 2, and 3** are correct
B if **1 and 3** are correct
C if **2 and 4** are correct
D if **4** is correct
E if **1, 2, 3, and 4** are correct

44. Which of the following anatomical features occur at the level of the sternal angle?

(1) The second rib articulates with the sternum.
(2) The trachea bifurcates into the right and left bronchi.
(3) The aortic arch both begins and ends at the level of the sternal angle.
(4) The inferior border of the superior mediastinum is demarcated.

45. Which of the following statements concerning the ductus arteriosus are true?

(1) It is anatomically closed approximately three months after birth.
(2) It carries poorly oxygenated blood during prenatal life.
(3) It is derived from the sixth aortic arch.
(4) It connects the left pulmonary vein to the aorta.

46. Which of the following conditions can cause right ventricular hypertrophy?

(1) A constricted pulmonary artery
(2) An abnormally small left atrioventricular opening
(3) Improper closing of the pulmonary valves
(4) An abnormally large right atrioventricular opening

47. The third rib articulates posteriorly with the

(1) transverse process of the third thoracic vertebra.
(2) body of the second thoracic vertebra.
(3) body of the third thoracic vertebra.
(4) transverse process of the second thoracic vertebra.

48. Which of the following statements concerning the right primary bronchus are true?

(1) It has a larger diameter than the left primary bronchus.
(2) It receives more foreign bodies through the trachea than the left.
(3) It gives rise to the eparterial bronchus.
(4) It is longer than the left.

49. Which of the following statements concerning the azygos vein are true?

(1) It receives the left superior intercostal vein.
(2) It receives the right superior intercostal vein.
(3) It empties directly into the right brachiocephalic vein.
(4) It empties directly into the superior vena cava.

50. Which of the following statements concerning the left recurrent laryngeal nerve are true?

(1) It hooks below the aortic arch, lateral to the ligamentum arteriosum.
(2) It ascends into the neck, passing in front of the subclavian artery.
(3) It can be damaged by an aortic aneurysm, causing hoarseness.
(4) It forms the majority of the esophageal plexus.

51. Which of the following conditions could result from myocardial infarction limited to the interventricular septum?

(1) Tricuspid valve insufficiency
(2) Pulmonary valve insufficiency
(3) Defective conduction of cardiac impulses
(4) Mitral valve insufficiency

52. Which of the following conditions represent the normal changes that occur in the circulation at, or soon after, birth?

(1) Increased blood flow through the lungs
(2) Closure of the ductus arteriosus
(3) Increased left atrial pressure
(4) Closure of the foramen ovale

Answers and Explanations

1–D. The pectinate muscles are prominent ridges of atrial myocardium located in the interior of both auricles and the right atrium. The inner surface of the left atrium is smooth.

2–C. The ligamentum arteriosum connects the aortic arch to the left pulmonary artery; the ductus arteriosus connects the aortic arch to the bifurcation of the pulmonary trunk. The ligamentum venosum is the fibrous remnant of the ductus venosus. It is located between the left lobe and the caudate lobe on the ventral surface of the liver. The left umbilical vein becomes the round ligament of the liver in the adult.

3–C. On the surface of the chest, the apex of the heart can be located in the left fifth intercostal space.

4–A. Normal, quiet expiration is achieved by contraction of extensible tissue in the lungs and the thoracic wall. The serratus posterior–superior, diaphragm, pectoralis major, and serratus anterior muscles are muscles of inspiration.

5–C. The greater splanchnic nerves contain visceral afferent and preganglionic visceral efferent sympathetic fibers.

6–D. The first two posterior intercostal arteries are branches of the highest (superior) intercostal artery of the costocervical trunk; the remaining nine branches are from the aorta. The internal thoracic arteries give rise to anterior intercostal arteries in the first six intercostal spaces and end at the sixth intercostal space by dividing into the superior epigastric artery and the musculophrenic artery. The musculophrenic artery gives off anterior intercostal arteries in the seventh, eighth, ninth, and tenth intercostal spaces.

7–D. The white rami communicantes contain preganglionic sympathetic efferent fibers and general visceral afferent fibers.

8–A. The mitral valve (left atrioventricular valve) produces the apex beat of the heart, which is most audible over the left fifth intercostal space at the midclavicular line.

9–C. The great cardiac vein accompanies the anterior interventricular artery. The middle cardiac vein accompanies the posterior interventricular artery. The anterior cardiac veins drain into the right atrium. The small cardiac vein accompanies the right marginal artery. The oblique veins of the left atrium end in the coronary sinus, near its beginning.

10–D. The right ventricle forms a large part of the sternocostal surface of the heart.

11–D. The tricuspid valve guards the opening between the right atrium and right ventricle.

12–C. The sinoatrial node initiates the impulse of contraction and is known as the pacemaker of the heart.

13–C. The heart is situated in the middle mediastinum.

14–A. The right middle lobe bronchus leads to the medial and lateral bronchopulmonary segments.

15–C. The eparterial bronchus is the right superior lobar (secondary) bronchus; all of the other bronchi are hyparterial bronchi.

16–D. The circumflex branch of the left coronary artery supplies the posterior portion of the left ventricle. The right coronary artery gives rise to the marginal branch, which supplies the anterior portion of the right ventricle, and the posterior interventricular artery, which supplies the septum and the adjacent ventricles.

17–D. The thoracic duct drains most of the lymph in the body; however, it does not drain lymph from the right side of the head and neck, right upper limb, and right thorax. The lower, dilated end of the thoracic duct is called the cisterna chyli. The thoracic duct passes through the aortic opening of the diaphragm, and then ascends between the aorta and the azygos vein. It empties into the junction of the left internal jugular and subclavian veins.

18–D. The lingula is the tongue-shaped portion of the upper lobe of the left lung. The oblique and horizontal fissures divide the right lung into upper, middle, and lower lobes; thus, the right lung has three lobar (secondary) bronchi and ten lobar bronchopulmonary segments. The right lung receives one bronchial artery and has a slightly larger capacity than the left.

19–E. Sternocleidomastoid, anterior scalene and external intercostal muscles elevate the ribs during inspiration. The interchondral parts of the internal intercostal muscles also elevate the ribs.

20–E. The brachiocephalic vein, trachea, part of the left common carotid artery, and the arch of the aorta are located in the superior mediastinum. The hemiazygos vein is located in the inferior mediastinum.

21–C. The first posterior intercostal vein drains into the corresponding brachiocephalic vein. The second, third, and fourth posterior intercostal veins join to form the superior intercostal vein, which drains into the azygos vein on the right and the brachiocephalic vein on the left. The rest of the posterior intercostal veins drain into the azygos vein on the right and into the hemiazygos vein or accessory hemiazygos vein on the left.

22–D. The phrenic nerve descends behind the subclavian vein, passes anterior to the root of the lung, and supplies the pericardium and diaphragm. It contains somatic efferent, somatic afferent, and visceral efferent (postganglionic sympathetic) fibers.

23–E. The left vagus nerve enters the thorax in front of the left subclavian artery and behind the left brachiocephalic vein. It passes behind the left bronchus, forms the pulmonary plexus, and continues to form an esophageal plexus. The vagus nerves lose their identity in the esophageal plexus. At the lower end of the esophagus, branches of the plexus reunite to form an anterior vagal trunk (anterior gastric nerve), which can be cut (vagotomy) to reduce gastric secretion. The vagus nerve carries parasympathetic preganglionic fibers to the thoracic and abdominal viscerae.

24–A. The vagus nerve forms the esophageal plexus after they leave the pulmonary plexus.

25–D. The left recurrent laryngeal nerve loops inferior to the arch of the aorta, left of the ligamentum arteriosum.

26–E. The greater splanchnic nerve arises from the fifth through ninth thoracic sympathetic ganglia in the posterior mediastinum, pierces the crus of the diaphragm, and ends in the celiac ganglion.

27–B. The phrenic nerve descends between the fibrous pericardium and the mediastinal pleura.

28–C. The gray rami connect the sympathetic trunk to every spinal nerve. The white rami are limited to the spinal cord segments between T1 and L2. The intercostal nerves are ventral primary rami of the thoracic nerves.

29–D. The marginal branch of the right coronary artery supplies the anterior wall of the right ventricle.

30–E. The right coronary artery gives off the sinus node artery, which supplies the sinoatrial node. The posterior interventricular artery, a branch of the right coronary artery, gives off a branch to supply the atrioventricular node.

31–A. The left coronary artery arises from the left aortic sinus.

32–A. The left coronary artery branches to form the anterior interventricular artery.

33–B. The mitral valve is best heard through a stethoscope in the left fifth intercostal space at the midclavicular line.

34–C. Blood passes through the pulmonary valve as it leaves the right ventricle. The pulmonary valve can be auscultated over the second left intercostal space, just lateral to the sternum.

35–A. The right atrium receives its blood supply from the anterior cardiac veins.

36–D. The apex of the heart is formed by the tip of the left ventricle and is located in the left fifth intercostal space.

37–D. The left ventricle briefly becomes enlarged in response to coarctation of the aorta.

38–A. The sinoatrial node is in the myocardium of the posterior wall of the right atrium, near the opening of the superior vena cava.

39–C. The chordae tendineae are tendinous strands that extend from the papillary muscles to the cusps of the artrioventricular valve. The papillary muscles and chordae tendineae prevent the cusps from being everted into the atrium during the ventricular contraction.

40–E. The anulus fibrosus is a fibrous ring surrounding the atrioventricular orifice, which is the attachment site of the cusps of the atrioventricular valves.

41–B. The septomarginal trabecula is a moderator band that extends from the interventricular septum to the base of the anterior papillary muscle of the right ventricle.

42–A. The crista terminalis is a muscular ridge that represents the junction between the sinus venarum and the remainder of the right atrium. It is indicated externally by the sulcus terminalis.

43–B. The septomarginal trabecula carries the right branch of the atrioventricular bundle from the septum to the opposite wall of the ventricle.

44–E. The sternal angle is the junction of the manubrium and the body of the sternum. It is located at the level where the second rib articulates with the sternum, the trachea bifurcates into the right and left bronchi, the aortic arch both begins and ends, and the inferior border of the superior mediastinum is demarcated.

45–A. The ductus arteriosus is derived from the sixth aortic arch. Before birth, the ductus arteriosus connects the bifurcation of the pulmonary trunk with the aorta and carries poorly oxygenated blood. It is functionally closed shortly after birth; however, anatomical closure requires several weeks or months. After birth, the ductus arteriosus becomes the ligamentum arteriosum, which connects the arch of the aorta to the left pulmonary artery.

46–E. Right ventricular hypertrophy can occur as a result of pulmonary stenosis, pulmonary and tricuspid valve defects, or mitral valve stenosis.

47–A. The third rib articulates posteriorly with the transverse process and the body of the third thoracic vertebra and the body of the second thoracic vertebra.

48–A. The right primary bronchus is shorter than the left primary bronchus and has a larger diameter. It receives more foreign bodies through the trachea and gives rise to the eparterial bronchus.

49–C. The right superior intercostal vein drains into the azygos vein, which empties into the superior vena cava. The left superior intercostal vein drains into the left brachiocephalic vein.

50–B. The left recurrent laryngeal nerve hooks around the aortic arch lateral to the ligamentum arteriosum; therefore, it can be damaged by an aneurysm of the aortic arch, causing paralysis of the laryngeal muscles. This nerve ascends behind the subclavian artery and lies in a groove between the trachea and the esophagus. The majority of the esophageal plexus is formed by the vagus nerves.

51–B. The three cusps of the tricuspid valve are the anterior, posterior, and septal cusps. The septal cusp lies on the margin of the orifice adjacent to the interventricular septum. The atrioventricular bundle descends into the interventricular septum. The pulmonary and mitral valves are not closely associated with the interventricular septum.

52–E. The changes in circulation that occur at, or soon after, birth include obliteration of the umbilical arteries, umbilical vein, and ductus venosus; functional closure of the ductus arteriosus and the foramen ovale (anatomical closure requires weeks or months); increased blood flow through the lungs; and increased left atrial pressure.

5

Abdomen

Anterior Abdominal Wall

I. Abdomen (Figure 5.1)

—is divided topographically by two transverse and two longitudinal planes into nine regions: right and left hypochondriac; epigastric; right and left lumbar; umbilical; right and left inguinal (iliac); and hypogastric (pubic).

—is divided also by vertical and horizontal planes through the umbilicus into four quadrants: right and left upper quadrants and right and left lower quadrants.

II. Muscles of the Anterior Abdominal Wall

Muscle	Origin	Insertion	Nerve	Action
External oblique	External surface of lower eight ribs (5–12)	Anterior half of iliac crest; anterior–superior iliac spine; pubic tubercle; linea alba	Intercostal n. (T7–T11); subcostal n. (T12)	Compresses abdomen; flexes trunk; active in forced expiration
Internal oblique	Lateral two-thirds of inguinal ligament; iliac crest; thoracolumbar fascia	Lower four costal cartilages; linea alba; pubic crest; pectineal line	Intercostal n. (T7–T11); subcostal n. (T12); iliohypogastric and ilioinguinal nn. (L1)	Compresses abdomen; flexes trunk; active in forced expiration
Transversus abdominis	Lateral one-third of inguinal ligament; iliac crest; thoracolumbar fascia; lower six costal cartilages	Linea alba; pubic crest; pectineal line	Intercostal n. (T7–T12); subcostal n. (T12); iliohypogastric and ilioinguinal nn. (L1)	Compresses abdomen; depresses ribs
Rectus abdominis	Pubic crest and pubic symphysis	Xiphoid process and costal cartilages 5–7	Intercostal n. (T7–T11); subcostal n. (T12)	Depresses ribs; flexes trunk
Pyramidalis	Pubic body	Linea alba	Subcostal n. (T12)	Tenses linea alba

(Continued on next page)

Muscle	Origin	Insertion	Nerve	Action
Cremaster	Middle of inguinal ligament; lower margin of internal oblique muscle	Tubercle and pubic crest	Genitofemoral n.	Retracts testis

III. Fasciae and Ligaments

—are organized into superficial (tela subcutanea) and deep layers, the superficial having a thin fatty layer (Camper's fascia) and the deep having a membranous layer (Scarpa's fascia).

A. Superficial Layer of the Superficial Fascia (Camper's Fascia)

—continues over the inguinal ligament as the superficial fascia of the thigh.
—continues over the pubis and perineum and the superficial layer of the superficial perineal fascia.

B. Deep Layer of the Superficial Fascia (Scarpa's Fascia)

—is attached to the fascia lata just below the inguinal ligament.
—continues over the pubis and perineum as the membranous layer (Colles' fascia) of the superficial perineal fascia.
—continues over the penis as the superficial fascia of the penis and over the scrotum as the Dartos tunic, which contains smooth muscle.
—may contain extravasated urine (by a rupture of the spongy urethra) between this fascia and the deep fascia of the abdomen.

C. Deep Fascia

—covers the muscles and continues over the spermatic cord at the superficial inguinal ring as the external spermatic fascia.

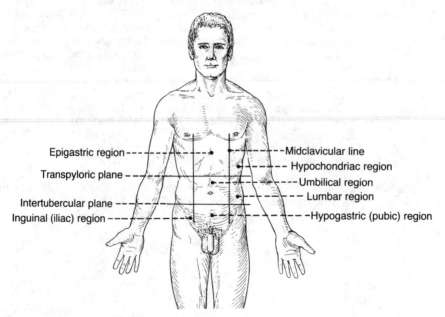

Epigastric region
Transpyloric plane
Intertubercular plane
Inguinal (iliac) region

Midclavicular line
Hypochondriac region
Umbilical region
Lumbar region
Hypogastric (pubic) region

Figure 5.1. Planes of subdivision of the abdomen.

—continues over the penis as the deep fascia of the penis (Buck's fascia) and over the pubis and perineum as the deep perineal fascia.

D. Linea Alba

—is a tendinous median raphe between the two rectus abdominis muscles, extending from the xiphoid process to the pubic symphysis.

—is formed by the fusion of the aponeuroses of the external oblique, internal oblique, and transversus abdominis muscles.

E. Linea Semilunaris

—is a curved line along the lateral border of the rectus abdominis muscle.

F. Linea Semicircularis (Arcuate Line)

—is a crescent-shaped line marking the termination of the posterior sheath of the rectus abdominis muscle just below the level of the iliac crest.

G. Lacunar Ligament

—represents the medial triangular expansion of the inguinal ligament to the pecten pubis.

—forms the medial border of the femoral ring and the floor of the inguinal canal.

H. Pectineal (Cooper's) Ligament

—is a strong fibrous band that extends laterally from the lacunar ligament along the pectineal line of the pubis.

I. Inguinal Ligament

—is the folded lower border of the aponeurosis of the external oblique muscle, extending between the anterior–superior iliac spine and the pubic tubercle.

J. Iliopectineal Arcus or Ligament

—is a fascial partition that separates the muscular (lateral) and vascular (medial) lacunae deep to the inguinal ligament. The **muscular lacuna** transmits the iliopsoas muscle, whereas the **vascular lacuna** transmits the femoral sheath and its contents, including the femoral vessels, a femoral branch of the genitofemoral nerve, and the femoral canal.

K. Reflected Inguinal Ligament

—is formed by certain fibers of the inguinal ligament reflected from the pubic tubercle upward toward the linea alba.

—also has some reflection from the lacunar ligament.

L. Conjoint Tendon (Falx Inguinalis)

—is formed by the aponeuroses of the internal oblique and transverse abdominis muscles.

—strengthens the posterior wall of the medial half of the inguinal canal.

M. Rectus Sheath (Figure 5.2)

—is formed by fusion of the aponeuroses of the external oblique, internal oblique, and transversus abdominis muscles.

—encloses the rectus abdominis muscle and sometimes the pyramidalis muscle.

—also contains the superior and inferior epigastric vessels and the ventral primary rami of thoracic nerves 7 to 12.

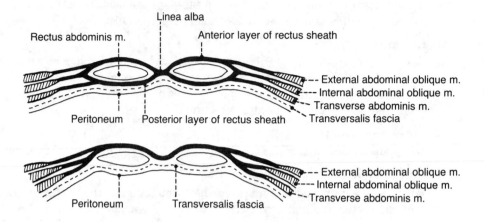

Figure 5.2. Arrangement of the rectus sheath above the umbilicus (upper) and below the arcuate line (lower).

1. **Anterior Layer**
   a. **Above the arcuate line:** aponeuroses of the external and internal oblique abdominis muscles.
   b. **Below the arcuate line:** aponeuroses of the external oblique, internal oblique, and transversus abdominis muscles.

2. **Posterior Layer**
   a. **Above the arcuate line:** aponeuroses of the internal oblique and transversus abdominis muscles.
   b. **Below the arcuate line:** rectus muscle is in contact with the transversalis fascia.

IV. Inguinal Region

A. Inguinal (Hesselbach's) Triangle

—is bounded medially by the lateral edge of the rectus abdominis muscle (the linea semilunaris), laterally by the inferior epigastric vessels, and inferiorly by the inguinal ligament.

—is an area of potential weakness and hence often the site of a direct inguinal hernia.

B. Inguinal Rings

1. **Superficial Inguinal Ring**

   —is a triangular opening in the aponeurosis of the external oblique abdominis muscle and lies just lateral to the pubic tubercle.

   —transmits the spermatic cord in the male and the round ligament of the uterus in the female.

2. **Deep Inguinal Ring**

   —lies in the transversalis fascia, just lateral to the inferior epigastric vessels.

   —is formed by embryonic extension of the processus vaginalis through the abdominal wall and subsequent passage of the testes through the transversalis fascia during descent of the testes into the scrotum.

C. **Inguinal Canal**
—begins at the deep inguinal ring and ends at the superficial ring.
—is much smaller in the female than in the male.
—transmits the spermatic cord (or round ligament of the uterus) and the ilioinguinal nerve.
—its walls include:
   1. **Anterior wall:** the aponeurosis of the external oblique and internal oblique muscles.
   2. **Posterior wall:** the aponeurosis of the transversus abdominis muscle and transversalis fascia.
   3. **Superior wall (roof):** arching fibers of the internal oblique and transversus abdominis muscles.
   4. **Inferior wall (floor):** the inguinal and lacunar ligaments.

D. **Inguinal Hernia**
—occurs superior to the inguinal ligament, whereas the femoral hernia occurs inferior to the ligament.
—occurs medial to the pubic tubercle, whereas the femoral hernia occurs lateral to the tubercle.
   1. **Indirect Inguinal Hernia**
      —passes through the deep inguinal ring, inguinal canal, and superficial inguinal ring and descends into the scrotum.
      —lies lateral to the inferior epigastric vessels.
      —is congenital (present at birth) and is associated with the persistence of the processus vaginalis.
      —is found more often on the right side in men and is more common than the direct inguinal hernia.
      —is covered by the peritoneum and the coverings of the spermatic cord.
   2. **Direct Inguinal Hernia**
      —occurs through the posterior wall of the inguinal canal (in the region of the inguinal triangle) but does not descend into the scrotum.
      —lies medial to the inferior epigastric vessels and protrudes forward to (but rarely through) the superficial inguinal ring.
      —is acquired (develops after birth) and is associated with weakness in the posterior wall of the inguinal canal lateral to the conjoint tendon.
      —its sac is formed by the peritoneum.

V. Spermatic Cord, Scrotum, and Testis

A. **Spermatic Cord**
—contains the ductus deferens, deferential vessels, testicular artery, pampiniform plexus of veins, lymphatics, and autonomic nerves of the testes.
—has several fasciae (coverings):
   1. **External spermatic fascia** is derived from the external oblique aponeurosis.
   2. **Cremasteric fascia** (cremaster muscle and fascia) originates from the internal oblique abdominis muscle.
   3. **Internal spermatic fascia** is derived from the transversalis fascia.

B. **Processus Vaginalis**
—is a peritoneal diverticulum (outpouching) in the fetus, extends into the scrotum before the testis descends, and forms the visceral and parietal layers of the tunica vaginalis.

—its neck normally fuses at birth or shortly thereafter.
—it normally loses its connection with the peritoneal cavity.
—its occlusion may cause fluid accumulation (hydrocele processus vaginalis).
—its persistence may result in an indirect inguinal hernia.

C. Tunica Vaginalis

—is a double serous membrane, a peritoneal sac (a remnant of the processus vaginalis) that covers the front and sides of the testis and epididymis.
—is derived from the abdominal peritoneum and forms the innermost layer of the scrotum.

D. Gubernaculum Testis

—is homologous to the ovarian ligament and the round ligament of the uterus.
—is the fetal ligamentous cord that connects the bottom of the fetal testis to the developing scrotum.
—appears to be important in testicular descent (pulls the testis down as it migrates).

VI. Inner Surface of the Anterior Abdominal Wall (Figure 5.3)

A. Supravesical Fossa

—is a depression on the anterior abdominal wall between the medial and lateral umbilical folds of the peritoneum.

B. Medial Inguinal Fossa

—is a depression on the anterior abdominal wall between the medial and lateral umbilical folds of the peritoneum.
—lies lateral to the superior vesical fossa.
—is the fossa where most direct inguinal hernias occur.

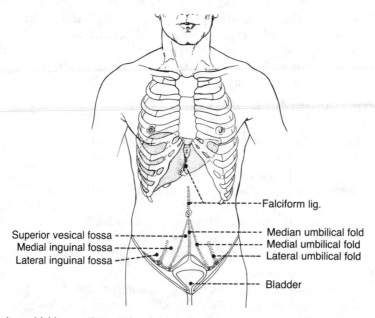

Superior vesical fossa
Medial inguinal fossa
Lateral inguinal fossa

Falciform lig.

Median umbilical fold
Medial umbilical fold
Lateral umbilical fold

Bladder

Figure 5.3. Peritoneal folds over the anterior abdominal wall.

C. Lateral Inguinal Fossa

—is a depression on the anterior abdominal wall, lateral to the lateral umbilical fold of the peritoneum.

D. Umbilical Folds or Ligaments

1. Median Umbilical Fold

—extends from the apex of the bladder to the umbilicus and contains the urachus (the fibrous remains of the fetal allantois).

2. Medial Umbilical Fold

—extends from the side of the bladder to the umbilicus and contains the obliterated umbilical artery (a branch of the internal iliac artery).

3. Lateral Umbilical Fold

—extends from the medial side of the deep inguinal ring to the arcuate line and contains inferior epigastric vessels.

E. Transversalis Fascia

—is a layer of connective tissue that lines the abdominal wall and lies between the parietal peritoneum and the abdominal muscles.

—continues with the diaphragmatic, psoas, iliac, pelvic, and quadratus lumborum fasciae.

—forms the deep inguinal ring and gives rise to the femoral sheath and the internal spermatic fascia.

—is directly in contact with the rectus abdominis muscle, below the arcuate line.

VII. Nerves

A. Subcostal Nerve

—is the ventral ramus of the twelfth thoracic nerve and supplies the muscles of the anterior abdominal wall (see p 176).

B. Iliohypogastric Nerve

—arises from the first lumbar nerve and supplies the internal oblique and transversus abdominis muscles.

—divides into a lateral cutaneous branch to supply the skin of the lateral side of the buttocks and an anterior cutaneous branch to supply the skin above the pubis (see p 176).

C. Ilioinguinal Nerve

—arises from the first lumbar nerve, pierces the internal oblique muscle near the deep inguinal ring, and runs medially through the inguinal canal and then through the superficial inguinal ring.

—supplies the internal oblique and transversus abdominis muscles (see p 177).

—gives off the femoral branch, which supplies the upper and medial parts of the thigh, and the anterior scrotal nerve, which supplies the skin of the root of the penis (or the skin of the mons pubis) and the anterior part of the scrotum (or the labium majus).

VIII. Lymphatic Drainage

A. Lymphatics in the Region Above the Umbilicus

—drain into the axillary lymph nodes.

B. Lymphatics in the Region Below the Umbilicus
  —drain into the superficial inguinal nodes.

C. Superficial Inguinal Lymph Nodes
  —receive lymph from the lower abdominal wall, buttocks, penis, scrotum, labium majus, and the lower parts of the vagina and anal canal.

IX. Blood Vessels

A. Superior Epigastric Artery
  —arises from the internal thoracic artery, enters the rectus sheath, and descends on the posterior surface of the rectus abdominis muscle.
  —anastomoses with the inferior epigastric artery within the rectus abdominis muscle.

B. Inferior Epigastric Artery
  —arises from the external iliac artery above the inguinal ligament, enters the rectus sheath, and ascends between the rectus muscle and the posterior layer of the rectus sheath.
  —anastomoses with the superior epigastric artery, providing collateral circulation between the subclavian and external iliac arteries.
  —gives off the cremasteric artery, which accompanies the spermatic cord.

C. Deep Circumflex Iliac Artery
  —arises from the external iliac artery and runs laterally along the inguinal ligament and the iliac crest between the transversus and the internal oblique muscles.
  —its ascending branch anastomoses with the musculophrenic artery.

D. Superficial Epigastric Artery
  —arises from the femoral artery and runs superiorly toward the umbilicus over the inguinal ligament.
  —anastomoses with branches of the inferior epigastric artery.

E. Superficial Circumflex Iliac Artery
  —arises from the femoral artery and runs laterally upward, parallel to the inguinal ligament.
  —anastomoses with the deep circumflex iliac and lateral femoral circumflex arteries.

F. Superficial (External) Pudendal Artery
  —arises from the femoral artery, pierces the cribriform fascia, and runs medially to supply the skin above the pubis.

G. Thoracoepigastric Vein
  —is a longitudinal venous connection between the lateral thoracic vein and the superficial epigastric vein.
  —provides a collateral route for venous return if a caval or portal obstruction occurs.

Peritoneum

I. Peritoneum
  —is a serous membrane lined by mesothelial cells.
  —consists of the **parietal peritoneum,** which lines the abdominal and pelvic walls and the inferior surface of the diaphragm, and the **visceral peritoneum,** which covers the viscerae.

—its **parietal layer** is supplied by the phrenic, lower intercostal, subcostal, iliohypogastric, and ilioinguinal nerves; however, its **visceral layer** is innervated by visceral nerves, which travel along autonomic pathways, and is relatively insensitive to pain.

II. Peritoneal Reflections (Figures 5.4 and 5.5)

A. Omentum

1. Lesser Omentum

—is a double layer of peritoneum extending from the porta hepatis of the liver to the lesser curvature of the stomach and the beginning of the duodenum.

—consists of the hepatogastric and hepatoduodenal ligaments and forms the anterior wall of the lesser sac of the peritoneal cavity.

—its right free margin contains the hepatic artery, bile duct, and portal vein.

—contains the left and right gastric vessels, which run between its two layers.

2. Greater Omentum

—hangs down like an apron from the greater curvature of the stomach, covering the transverse colon and other abdominal viscerae.

—is often referred to by surgeons as the "abdominal policeman" because it plugs the neck of a hernial sac, preventing the entrance of coils of the small intestine.

—adheres to areas of inflammation and wraps itself around the inflamed organs, thus preventing serious diffuse peritonitis.

B. Mesenteries

1. Mesentery of the Small Intestine (Mesentery Proper)

—its root extends from the duodenojejunal flexure to the right iliac fossa and is about 15 cm (6 inches) long.

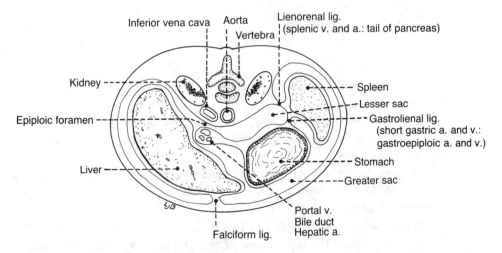

Figure 5.4. Horizontal section of the upper abdomen.

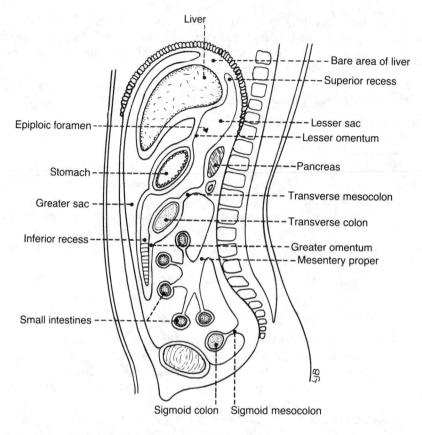

Liver

Bare area of liver

Superior recess

Epiploic foramen

Lesser sac

Lesser omentum

Pancreas

Stomach

Transverse mesocolon

Greater sac

Transverse colon

Inferior recess

Greater omentum

Mesentery proper

Small intestines

Sigmoid colon   Sigmoid mesocolon

Figure 5.5. Sagittal section of the abdomen.

—is a double fold of peritoneum that suspends the jejunum and ileum from the posterior abdominal wall and transmits the nerves and blood vessels to and from the small intestine.

—its free border encloses the small intestine, which is about 6 m (20 feet) long.

—contains the superior mesenteric and intestinal (jejunal and ileal) vessels, nerves, and lymphatics.

2. Transverse Mesocolon

—connects the posterior surface of the transverse colon to the posterior body wall.

—fuses with the greater omentum to form the **gastrocolic ligament**.

—contains the middle colic vessels, nerves, and lymphatic vessels.

3. Sigmoid Mesocolon

—is an inverted V-shaped peritoneal fold that connects the sigmoid colon to the pelvic wall and contains the sigmoid vessels.

4. Mesoappendix

—connects the appendix to the mesentery of the ileum and contains the appendicular vessels.

C. Peritoneal Ligaments

1. Lienogastric (Gastrosplenic) Ligament

—extends from the left portion of the greater curvature of the stomach to the hilus of the spleen.

—contains the short gastric vessels and the left gastroepiploic vessels.

2. Lienorenal Ligament
—runs from the hilus of the spleen to the left kidney.
—contains the splenic vessels and the tail of the pancreas.

3. Gastrophrenic Ligament
—runs from the upper part of the greater curvature of the stomach to the diaphragm.

4. Gastrocolic Ligament
—runs from the greater curvature of the stomach to the transverse colon.

5. Phrenicocolic Ligament
—runs from the colic flexure to the diaphragm.

6. Falciform Ligament
—is a sickle-shaped peritoneal fold, connecting the liver to the diaphragm and the anterior abdominal wall.
—appears to demarcate the right lobe from the left lobe of the liver on the diaphragmatic surface.
—contains the ligamentum teres hepatis, which is the fibrous remnant of the left umbilical vein of the fetus, and the paraumbilical vein, which connects the left branch of the portal vein with the subcutaneous veins in the region of the umbilicus.

7. Ligamentum Teres Hepatis (Round Ligament of the Liver)
—lies in the free margin of the falciform ligament and ascends from the umbilicus to the inferior (visceral) surface of the liver, lying in the fissure that forms the left boundary of the quadrate lobe of the liver.
—is formed after birth from the remnant of the left umbilical vein, which carries oxygenated blood from the placenta to the left branch of the portal vein in the fetus. (The right umbilical vein obliterates during the embryonic period.)

8. Coronary Ligament
—is a peritoneal reflection from the diaphragmatic surface of the liver onto the diaphragm and encloses a triangular area of the right lobe, the **bare area** of the liver.
—forms the **right** and **left triangular ligaments** by its right and left extensions, respectively.

9. Ligamentum Venosum
—is the fibrous remnant of the ductus venosus.
—lies in the fissure on the inferior surface of the liver, forming the left boundary of the caudate lobe of the liver.

D. Peritoneal Folds
—are peritoneal reflections with free edges.

1. Umbilical Folds
—are five folds of peritoneum below the umbilicus, including the median, medial, and lateral umbilical folds (see p 153).

2. Rectouterine Fold
—extends from the cervix to the uterus, along the side of the rectum, to the posterior pelvic wall.

3. Ileocecal Fold
—extends from the terminal ileum to the cecum.

III. Peritoneal Cavity (see Figures 5.4 and 5.5)

—is a potential space between the parietal and visceral peritoneum.

—contains a film of fluid that lubricates the surface of the peritoneum and facilitates free movements of the viscerae.

—is a completely closed sac in the male but communicates with the exterior through the openings of the uterine tubes in the female.

—is divided into the lesser and greater sacs.

A. Lesser Sac (Omental Bursa)

—is a closed sac, except for its communication with the greater sac through the epiploic foramen.

—is an irregular space that lies behind the liver, lesser omentum, stomach, and upper anterior part of the greater omentum.

—presents three recesses:

1. **Superior recess,** which lies behind the stomach, lesser omentum, and liver.

2. **Inferior recess,** which lies behind the stomach, extending into the layers of the greater omentum.

3. **Splenic recess,** which extends to the left at the hilus of the spleen.

B. Greater Sac

—extends across the entire breadth of the abdomen and from the diaphragm to the pelvic floor.

—presents recesses and Morison's pouch.

1. **Subphrenic Recess**

   —is a peritoneal pocket located anterior and superior to the liver, beneath the diaphragm; it is divided into right and left subphrenic recesses by the falciform ligament.

2. **Subhepatic Recess**

   —is a peritoneal pocket between the liver and the transverse colon.

3. **Hepatorenal Recess**

   —is a deep recess of the peritoneal cavity located between the liver anteriorly and the kidney and suprarenal gland posteriorly.

4. **Morison's Pouch**

   —is a peritoneal pouch (recess) located below the liver and to the right of the right kidney.

   —communicates with the subphrenic recess, the lesser sac (via the epiploic foramen), and the right paracolic gutter (which lies to the right of the ascending colon) and, hence, the pelvic cavity.

C. Epiploic (Winslow's) Foramen

—is a natural opening between the lesser and greater sacs.

—is bounded superiorly by peritoneum on the caudate lobe of the liver, inferiorly by peritoneum on the first part of the duodenum, anteriorly by the free edge of the lesser omentum, and posteriorly by peritoneum covering the inferior vena cava.

Gastrointestinal Viscera

I. Stomach (Figure 5.6)
—rests, in the supine position, on the stomach bed, which is formed by the pancreas, spleen, left kidney, left suprarenal gland, transverse colon and its mesocolon, and diaphragm.
—is covered entirely by peritoneum and is located in the left hypochondriac and epigastric regions of the abdomen.
—has greater and lesser curvatures, anterior and posterior walls, cardiac and pyloric openings, and cardiac and angular notches.
—is divided into four regions: the **cardia, fundus, body,** and **pylorus**. The pylorus is divided into the pyloric antrum and canal. The pyloric orifice is surrounded by a thickened ring of gastric circular muscle constituting the **pyloric sphincter**.
—receives blood from the right and left gastric, right and left gastroepiploic, and short gastric arteries.
—has longitudinal folds of the mucous membrane, the **rugae,** when the organ is contracted.
—has the **gastric canal,** which is the grooved channel formed by the rugae, and directs fluids toward the pylorus.
—produces acid and pepsin in its fundus and body, and produces the hormone gastrin in its pyloric antrum.

II. Small Intestine (Figure 5.7)
—extends from the pyloric opening to the ileocecal junction.
—is the location of complete digestion and absorption of most of the digestion products and water, electrolytes, and minerals such as calcium and iron.
—consists of the **duodenum, jejunum,** and **ileum**.

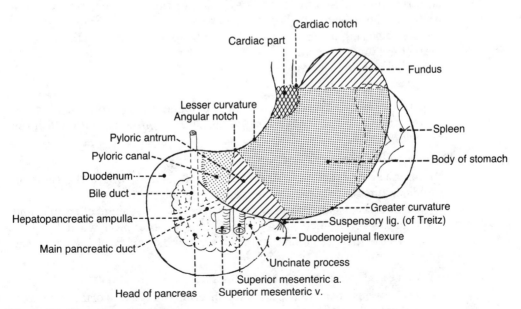

Figure 5.6. Stomach and duodenum.

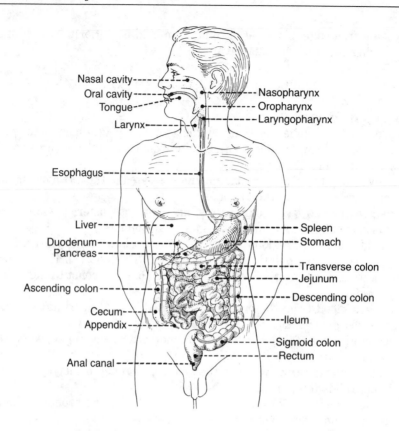

Nasal cavity
Oral cavity
Tongue
Larynx
Nasopharynx
Oropharynx
Laryngopharynx
Esophagus
Liver
Duodenum
Pancreas
Spleen
Stomach
Transverse colon
Jejunum
Ascending colon
Descending colon
Cecum
Appendix
Ileum
Sigmoid colon
Rectum
Anal canal

Figure 5.7. Diagram of the digestive system.

A. Duodenum

—is a C-shaped tube surrounding the head of the pancreas, and is the shortest (25 cm [10 inches] long) but widest part of the small intestine.

—is retroperitoneal except for the beginning of the first part, which is connected to the liver by the hepatoduodenal part of the lesser omentum.

—receives blood supply from both the celiac (foregut) artery and the superior mesenteric (midgut) artery.

—is divided into four parts:

1. Superior (First) Portion

—has a mobile or free section at the beginning of the first part, termed the **duodenal cap** (because of its appearance on x-ray), into which the pylorus invaginates.

2. Descending (Second) Portion

—contains the junction of the foregut and midgut, where the common bile and main pancreatic ducts open.

—contains the **greater papilla,** on which terminal openings of the bile and main pancreatic ducts are located, and the **lesser papilla,** which lies 2 cm above the greater papilla and marks the site of entry of the accessory pancreatic duct.

3. Transverse (Inferior, or Third) Portion

—is the longest part and crosses the inferior vena cava, aorta, and vertebral column to the left.

4. Ascending (Fourth) Portion

—ascends to the left of the aorta to the level of the second lumbar vertebra and terminates at the duodenojejunal junction, which is fixed in position by the **suspensory ligament (of Treitz)**, a surgical landmark. This fibromuscular band is attached to the right crus of the diaphragm.

B. Jejunum

—makes up the proximal two-fifths of the small intestine (the ileum makes up the distal three-fifths).

—is emptier, larger in diameter, and thicker-walled than the ileum.

—has the **plicae circulares** (circular folds), which are tall and closely packed.

—contains *no* **Peyer's patches** (aggregations of lymphoid tissue).

—has translucent areas called **"windows"** between the blood vessels of its mesentery.

—has less prominent **arterial arcades** (anastomotic loops) in its mesentery than does the ileum.

—has longer **vasa recta** (straight arteries, or arteriae rectae) than those of the ileum.

C. Ileum

—is longer than the jejunum and occupies the false pelvis in the right lower quadrant of the abdomen.

—is characterized by the presence of Peyer's patches (lower portion), shorter plicae circulares and vasa recta, and more mesenteric fat and arterial arcades when compared with the jejunum.

Meckel's Diverticulum

—is an evagination of the terminal part of the ileum located on the antimesenteric side of the ileum; it occurs in about 2% of the population.

—may contain gastric and pancreatic tissues in its wall.

—represents persistent portions of the embryonic yolk stalk (vitelline, or the omphalomesenteric duct) that is present in some adults.

—is clinically important because bleeding may occur from an ulcer in its wall.

III. Large Intestine (see Figures 5.7 and 5.12)

—extends from the ileocecal junction to the anus and is about 1.5 m (5 feet) long.

—consists of the cecum, appendix, colon, rectum, and anal canal.

—functions to convert the liquid contents of the ileum into semisolid feces by absorbing fluid and electrolytes.

A. Colon

—consists of the **ascending, transverse, descending,** and **sigmoid segments**.

—its ascending and descending segments are retroperitoneal, and its transverse and sigmoid segments are surrounded by peritoneum (they have their own mesenteries called the transverse mesocolon and the sigmoid mesocolon, respectively).

—its ascending and transverse segments are supplied by the superior mesenteric artery and the vagus nerve, whereas its descending and sigmoid segments are supplied by the inferior mesenteric artery and the pelvic splanchnic nerves.

—is characterized by the presence of the following:

1. **Teniae coli:** three narrow bands of the outer longitudinal muscular coats.
2. **Sacculations or haustra:** produced by the teniae, which are slightly shorter than the gut.
3. **Epiploic appendages:** peritoneum-covered sacs of fat, attached in rows along the teniae.

B. Cecum

—is the blind pouch of the large intestine, lies in the right iliac fossa, and is usually surrounded by peritoneum.

C. Appendix

—is a narrow, hollow, muscular tube containing large aggregations of lymphoid tissue in its wall.
—is suspended from the terminal ileum by a small mesentery, the mesoappendix, which contains the appendicular vessels.
—may undergo inflammation that causes spasm and distension, resulting in pain that is referred to the epigastrium.

D. Rectum and Anal Canal

—is the part of the large intestine that extends from the sigmoid colon to the anal canal, which ends at the anus.
—are described as pelvic organs (see pp 212–213).

IV. Accessory Organs of Digestive System

A. Liver (Figures 5.8 and 5.9)

—is the largest visceral organ and the largest gland in the body.
—plays an important role in bile secretion, detoxification, blood-clotting mechanisms, and storage of glycogen, vitamin, iron, and copper. In the fetus, it is important in the manufacture of red blood cells.
—is surrounded by the peritoneum and is attached to the diaphragm by the **coronary** and **falciform ligaments** and the right and left **triangular ligaments**.
—has a **bare area** on the diaphragmatic surface, which is limited by layers of the coronary ligament but is devoid of peritoneum.
—is attached to the anterior abdominal wall and the diaphragm by the falciform ligament, which appears to divide the **right lobe** from the **left lobe** on the diaphragmatic surface. (These lobes serve as landmarks but do not correspond to the structural units or hepatic segments.)

1. Lobes of the Liver

—divide into right and left lobes by the fossae for the gallbladder and the inferior vena cava.
—consist of a **right lobe,** which is divided into **anterior** and **posterior** segments, each in turn is subdivided into superior and inferior areas or segments (see Figure 5.9).
—consist of a **left lobe,** which is divided into medial and lateral segments, each of which is subdivided into superior and inferior areas (segments). Thus, the segments of the left lobe include the medial superior (**caudate lobe**), medial inferior (**quadrate lobe**), lateral superior, and lateral inferior segments (see Figure 5.9).

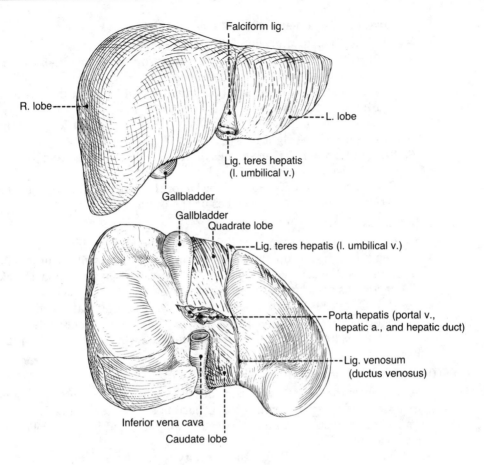

Falciform lig.

R. lobe

L. lobe

Lig. teres hepatis
(l. umbilical v.)

Gallbladder

Gallbladder

Quadrate lobe

Lig. teres hepatis (l. umbilical v.)

Porta hepatis (portal v.,
hepatic a., and hepatic duct)

Lig. venosum
(ductus venosus)

Inferior vena cava

Caudate lobe

Figure 5.8. Anterior and visceral surfaces of the liver.

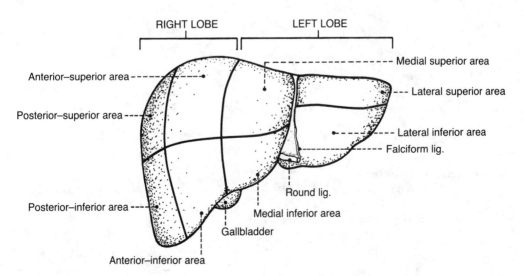

RIGHT LOBE

LEFT LOBE

Anterior–superior area

Medial superior area

Lateral superior area

Posterior–superior area

Lateral inferior area

Falciform lig.

Round lig.

Posterior–inferior area

Medial inferior area

Gallbladder

Anterior–inferior area

Figure 5.9. Divisions of the liver, based on hepatic drainage and blood supply.

2. **Fissures and Ligaments of the Liver**

—include an **H**-shaped group of fissures:

a. Fissure for the **round ligament (ligamentum teres hepatis)**, located between the left lobe and the quadrate lobe.

b. Fissure for the **ligamentum venosum,** located between the left lobe and the caudate lobe.

c. Fossa for the **gallbladder,** located between the quadrate lobe and the major part of the right lobe.

d. Fissure for the **inferior vena cava,** located between the caudate lobe and the major part of the right lobe.

e. **Porta hepatis** (the crossbar of the H) for the hepatic ducts, the hepatic artery proper, and the branches of the portal vein.

B. **Gallbladder** (Figure 5.10)

—is a pear-shaped sac lying on the inferior surface of the liver in a fossa between the right and quadrate lobes and has a capacity of about 30 ml.

—has a **fundus** (the rounded blind end); a **body** (the major part); and a **neck** (the narrow part), which gives rise to the cystic duct with spiral valves.

—receives bile, stores it, and concentrates it by absorbing water and salts.

—contracts to expel bile as a result of stimulation by the hormone cholecystokinin, which is produced by the duodenal mucosa when food arrives in the duodenum.

C. **Pancreas** (see Figure 5.10)

—lies largely in the floor of the lesser sac in the epigastric and left hypochondriac regions, where it forms a major portion of the stomach bed.

—is a retroperitoneal organ except for a small portion of its tail, which lies in the lienorenal ligament.

—its head lies within the C-shaped concavity of the duodenum.

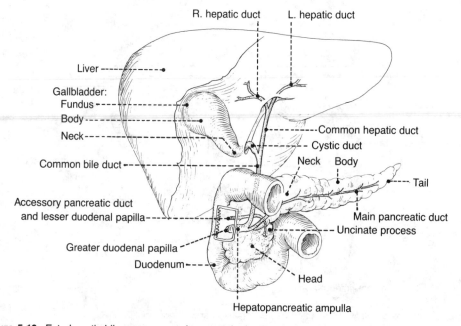

Figure 5.10. Extrahepatic bile passages and pancreatic ducts.

—its **uncinate process** is a projection of the lower part of its head to the left behind the superior mesenteric vessels.

—receives blood from branches of the splenic artery and from the superior and inferior pancreaticoduodenal arteries.

—is both an exocrine gland, which produces digestive enzymes, and an endocrine gland, which secretes two hormones, **insulin** and **glucagon**.

—its **main duct** (Wirsung's duct) begins in the tail, runs to the right along the entire pancreas, and joins the bile duct to form the hepatopancreatic ampulla (Vater's ampulla) before entering the second part of the duodenum at the greater papilla.

—its **accessory duct** (Santorini's duct) begins in the lower portion of the head and empties at the lesser duodenal papilla about 2 cm above the greater papilla.

—may have tumors in its head, which obstruct the bile flow, resulting in jaundice.

D. Duct System for Bile Passage (see Figure 5.10)

1. Right and Left Hepatic Ducts

—are formed by union of the intrahepatic ductules from each lobe of the liver.

—drain bile from the corresponding halves of the liver.

2. Common Hepatic Duct

—is formed by union of the right and left hepatic ducts.

—is accompanied by the proper hepatic artery and the portal vein.

3. Cystic Duct

—has spiral folds (valves) to keep it constantly open, and thus bile can pass upward into the gallbladder when the bile duct is closed.

—runs alongside the hepatic duct before joining the common hepatic duct.

—is a common site of impaction of gallstones.

4. Bile Duct (Ductus Choledochus)

—is formed by union of the common hepatic duct and the cystic duct.

—lies with the hepatic artery and portal vein in the right free margin of the lesser omentum.

—descends behind the first part of the duodenum and runs through the head of the pancreas.

—joins the main pancreatic duct to form the hepatopancreatic duct or hepatopancreatic ampulla, which enter the second part of the duodenum at the greater papilla.

—is located lateral to the hepatic artery proper and anterior to the portal vein.

5. Hepatopancreatic Duct or Hepatopancreatic Ampulla (Vater's Ampulla)

—is formed by union of the bile duct and the main pancreatic duct and enters the second part of the duodenum at the greater papilla. This entrance represents the junction of the embryonic foregut and midgut.

V. Spleen (Figure 5.11)

—is a large lymphatic organ lying against the diaphragm and ribs 9 to 11 in the left hypochondriac region.

—is developed in the dorsal mesogastrium and supported by the lienogastric and lienorenal ligaments.

—is composed of **white pulp,** which consists of lymphatic nodules and diffuse lymphatic tissue, and **red pulp,** which consists of venous sinusoids connected by splenic cords.

—is hematopoietic in early life and later functions in blood cell destruction.

—filters blood (removes particulate matter and cellular residue from the blood), stores red corpuscles, and produces lymphocytes and antibodies.

—is supplied by the splenic artery and is drained by the splenic vein.

—may be removed surgically with minimal effect on body function.

VI. Celiac and Mesenteric Arteries

A. Celiac Trunk (see Figure 5.11)

—arises from the front of the abdominal aorta immediately below the aortic hiatus of the diaphragm, between the right and left crura.

—divides into the left gastric, splenic, and common hepatic arteries.

1. Left Gastric Artery

—is the smallest branch of the celiac trunk.

—runs upward and to the left toward the cardia, giving off esophageal and hepatic branches, and then turns to the right and runs along the lesser curvature within the lesser omentum to anastomose with the right gastric artery.

2. Splenic Artery

—is the largest branch of the celiac trunk.

—runs a highly tortuous course along the superior border of the pancreas and enters the lienorenal ligament.

—gives rise to:

a. A number of pancreatic branches including the **dorsal pancreatic artery**.

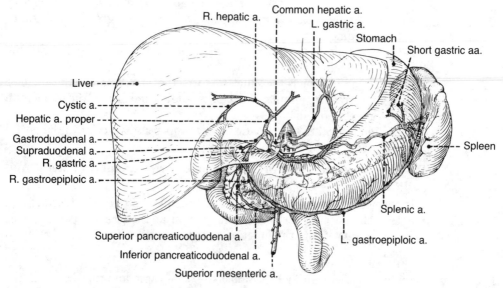

Figure 5.11. Branches of the celiac trunk.

   b. A few **short gastric arteries,** which pass through the lienogastric ligament to reach the fundus of the stomach.

   c. The **left gastroepiploic artery,** which reaches the greater omentum through the lienogastric ligament and runs along the greater curvature of the stomach to distribute to the stomach and greater omentum.

3. **Common Hepatic Artery**

—runs to the right along the upper border of the pancreas and divides into the **proper hepatic artery,** the **gastroduodenal artery,** and probably the **right gastric artery.**

   a. **Proper Hepatic Artery**

   —ascends in the free edge of the lesser omentum and divides, near the porta hepatis, into the left and right hepatic arteries; the right hepatic artery gives off a cystic artery to the gallbladder.

   —gives off, near its beginning, the right gastric artery.

   b. **Right Gastric Artery**

   —runs to the pylorus and then along the lesser curvature of the stomach and anastomoses with the left gastric artery.

   c. **Gastroduodenal Artery**

   —descends behind the first part of the duodenum, giving off the **supraduodenal artery** to its superior aspect and a few retroduodenal arteries to its inferior aspect.

   —divides into two major branches:

   (1) The **right gastroepiploic artery** runs to the left along the greater curvature of the stomach, supplying the stomach and the greater omentum.

   (2) The **superior pancreaticoduodenal artery** passes between the duodenum and the head of the pancreas and further divides into the anterior–superior pancreaticoduodenal artery and the posterior–superior pancreaticoduodenal artery.

B. **Superior Mesenteric Artery** (Figure 5.12)

—arises from the aorta behind the neck of the pancreas.

—descends across the uncinate process of the pancreas and the third part of the duodenum and then enters the root of the mesentery behind the transverse colon to run to the right iliac fossa.

—gives off the following branches:

1. **Inferior Pancreaticoduodenal Artery**

—passes to the right and divides into the anterior–inferior pancreaticoduodenal artery and the posterior–inferior pancreaticoduodenal artery, which anastomose with the corresponding branches of the superior pancreaticoduodenal arteries.

2. **Middle Colic Artery**

—enters the transverse mesocolon and divides into the right branch, which anastomoses with the right colic artery, and the left branch, which anastomoses with the ascending branch of the left colic artery. The branches of the mesenteric arteries form an anastomotic channel, the **marginal artery,** along the large intestine.

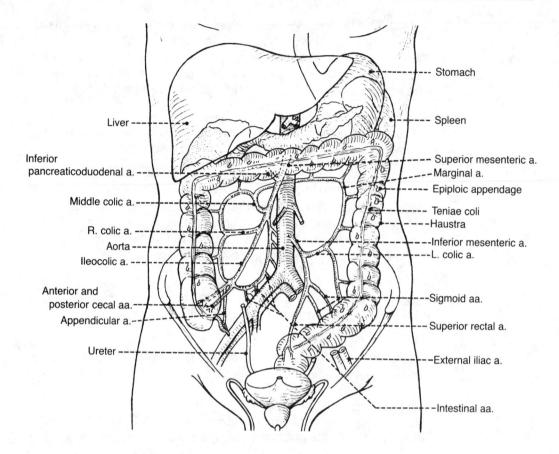

Figure 5.12. Branches of the superior and inferior mesenteric arteries.

3. Ileocolic Artery

—descends behind the peritoneum toward the right iliac fossa and ends by dividing into an ascending branch, which anastomoses with the right colic artery, anterior and posterior cecal branches, an appendicular artery, and an ileal branch.

4. Right Colic Artery

—arises from the superior mesenteric artery or the ileocolic artery.
—runs to the right behind the peritoneum and divides into ascending and descending branches, distributing to the ascending colon.

5. Intestinal Arteries

—are 12 to 15 in number and supply the jejunum and ileum.
—branch and anastomose to form a series of arcades in the mesentery.

C. Inferior Mesenteric Artery (see Figure 5.12)

—passes to the left behind the peritoneum and distributes to the descending and sigmoid colons and the upper portion of the rectum.
—gives rise to:

1. Left Colic Artery

—runs to the left behind the peritoneum toward the descending colon and divides into ascending and descending branches.

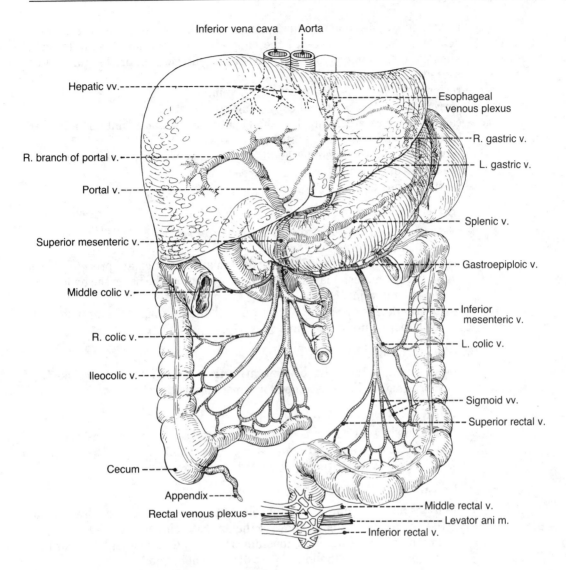

Inferior vena cava Aorta

Hepatic vv.

R. branch of portal v.

Portal v.

Superior mesenteric v.

Middle colic v.

R. colic v.

Ileocolic v.

Cecum

Appendix

Rectal venous plexus

Esophageal venous plexus

R. gastric v.

L. gastric v.

Splenic v.

Gastroepiploic v.

Inferior mesenteric v.

L. colic v.

Sigmoid vv.

Superior rectal v.

Middle rectal v.

Levator ani m.

Inferior rectal v.

Figure 5.13. Portal venous system.

2. Sigmoid Arteries

—are two to three in number, run toward the sigmoid colon in its mesentery, and divide into ascending and descending branches.

3. Superior Rectal Artery

—is the termination of the inferior mesenteric artery, descends into the pelvis, divides into two branches that follow the sides of the rectum, and anastomoses with the middle and inferior rectal arteries. (The middle and inferior rectal arteries arise from the internal iliac and internal pudendal arteries, respectively.)

VII. Portal Venous System (Figure 5.13)

A. Portal Vein

—drains the abdominal part of the gut, spleen, pancreas, and gallbladder and is 8 cm (3.2 inches) long.

—is formed by the union of the **splenic vein** and the **superior mesenteric vein** posterior to the neck of the pancreas. The inferior mesenteric vein joins either the splenic or the superior mesenteric vein or the junction of these two veins.

—receives the left gastric (or coronary) vein.

—carries deoxygenated blood containing nutrients.

—carries twice as much blood as the hepatic artery and maintains a higher blood pressure than in the inferior vena cava.

—ascends behind the bile duct and hepatic artery within the free margin of the lesser omentum.

—its tributaries are:

1. **Superior Mesenteric Vein**

   —accompanies the superior mesenteric artery on its right side in the root of the mesentery.

   —crosses the third part of the duodenum and the uncinate process of the pancreas and terminates posterior to the neck of the pancreas by joining the splenic vein, thereby forming the portal vein.

   —its tributaries are some of the veins that accompany the branches of the superior mesenteric artery.

2. **Splenic Vein**

   —is formed by the union of tributaries from the spleen and receives the short gastric, left gastroepiploic, and pancreatic veins.

3. **Inferior Mesenteric Vein**

   —is formed by the union of the superior rectal and sigmoid veins, receives the left colic vein, and empties into the splenic vein or the superior mesenteric vein or the junction of these two veins.

4. **Left Gastric (Coronary) Vein**

   —drains normally into the portal vein, but the esophageal vein drains into the systemic venous system via the azygos vein.

   —its esophageal tributaries anastomose with the esophageal veins of the azygos system at the lower part of the esophagus.

5. **Paraumbilical Veins**

   —are found in the falciform ligament and are virtually closed; however, they dilate in portal hypertension.

   —connect the left branch of the portal vein with the small subcutaneous veins in the region of the umbilicus, which are radicles of the superior epigastric, inferior epigastric, thoracoepigastric, and superficial epigastric veins.

B. Important Portal–Caval (Systemic) Anastomoses

—occur between:

1. The left gastric vein and the esophageal vein of the azygos system.
2. The superior rectal vein and the middle and inferior rectal veins.
3. The paraumbilical vein and radicles of the epigastric (superficial and inferior) veins.
4. The retroperitoneal veins draining the colon and twigs of the renal, suprarenal, and gonadal veins.

C. Portal Hypertension

—results from thrombosis of the portal vein or liver cirrhosis.

—causes a dilatation of veins in the lower part of the esophagus, forming **esophageal varices**. Their rupture results in vomiting of blood (hematemesis).

—results in **caput medusae** (dilated veins radiating from the umbilicus), which occurs because the paraumbilical vein enclosed in the free margin of the falciform ligament anastomoses with branches of the epigastric (superficial and inferior) veins around the umbilicus.

—may result in **hemorrhoids**, which are caused by enlargement of veins around the anal canal.

—can be reduced by diverting blood from the portal to the caval system; this is done by anastomosing the splenic vein to the renal vein or by creating a communication between the portal vein and the inferior vena cava.

VIII. Clinical Considerations

A. Gastric Ulcer

—erodes the mucosa and penetrates the gastric wall to various depths.

—may perforate into the lesser sac and erode the pancreas and the splenic artery, causing fatal hemorrhage.

B. Duodenal Ulcer

—penetrates the duodenal wall of the first part of the duodenum, erodes the gastroduodenal artery, and is commonly located in the duodenal cap.

C. Liver Cirrhosis

—is a condition in which liver cells are progressively destroyed and replaced by fibrous tissue that surrounds the intrahepatic blood vessels and biliary radicles, impeding the circulation of blood through the liver.

—causes portal hypertension, resulting in esophageal varices, hemorrhoids, and caput medusae.

D. Gallstones

—are formed by solidification of bile constituents.

—are composed chiefly of cholesterol crystals.

—may become lodged in the fundus of the gallbladder, which may ulcerate through the wall of the gallbladder into the transverse colon or into the duodenum. In the former case, they are passed naturally to the rectum, but in the latter case they may be held up at the ileocolic junction, producing an intestinal obstruction.

—may become lodged in the bile duct, obstructing the bile flow to the duodenum and leading to **jaundice**.

—may become lodged in the hepatopancreatic ampulla, blocking both the biliary and the pancreatic duct systems. In this case, bile may enter the pancreatic duct system, causing aseptic or noninfectious **pancreatitis**.

E. Megacolon (Hirschsprung's Disease)

—is caused by the absence of enteric ganglia in the lower part of the colon, which leads to the dilatation of the colon proximal to the inactive segment.

—is of congenital origin and is usually found during infancy and childhood; its symptoms are constipation, abdominal distension, and vomiting.

Retroperitoneal Viscera, Diaphragm, and Posterior Abdominal Wall

I. Kidney, Ureter, and Suprarenal Gland

A. Kidney (Figures 5.14 and 5.15)

—is retroperitoneal and extends from the level of lumbar vertebrae 1 to 4 in the erect position. The right kidney lies a little lower than the left, owing to the large size of the right lobe of the liver, and usually is related to the twelfth rib posteriorly. The left kidney is related to the eleventh and twelfth ribs posteriorly.

—is invested by a firm fibrous capsule.

—is surrounded by a mass of fat and fibrous fascia (the renal fascia), which divides the fat into two regions: The **perirenal fat** lies in the perinephric space between the capsule of the kidney and the renal fascia, and the **pararenal fat** lies external to the renal fascia.

—has an indentation, the **hilus,** on its medial border, through which the ureter, renal vessels, and nerves enter or leave the organ.

—consists of the medulla and the cortex, containing 1 to 2 million **nephrons,** which are the anatomical and functional units of the kidney. The nephron consists of a glomerular capsule, a proximal convoluted tubule, Henle's loop, and a distal convoluted tubule.

1. Cortex

—forms the outer part of the kidney and also projects into the medullary region between the renal pyramids as **renal columns**.

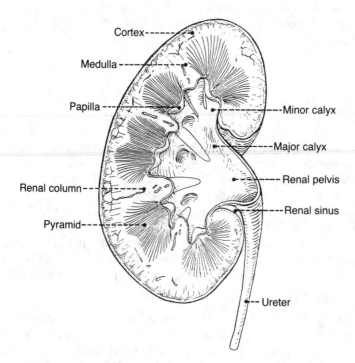

Figure 5.14. Frontal section of the kidney.

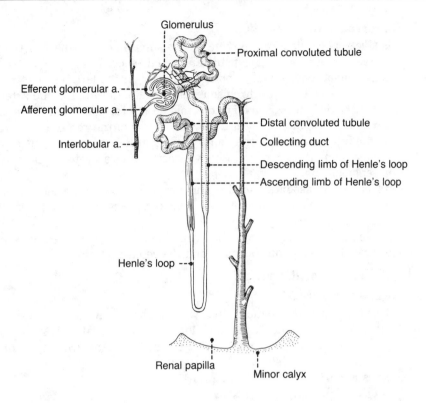

Glomerulus

Proximal convoluted tubule

Efferent glomerular a.

Afferent glomerular a.

Distal convoluted tubule

Interlobular a.

Collecting duct

Descending limb of Henle's loop

Ascending limb of Henle's loop

Henle's loop

Renal papilla

Minor calyx

Figure 5.15. Renal tubules.

—contains renal corpuscles and proximal and distal convoluted tubules. (The **renal corpuscle** consists of a **glomerular capsule**, which is the invaginated blind end of the nephron.)

2. **Medulla**

—forms the inner part of the kidney and consists of 8 to 12 **renal pyramids,** which contain straight tubules (Henle's loops) and collecting tubules. An apex of the renal pyramid, the **renal papilla,** fits into the cup-shaped **minor calyx** on which the **collecting tubules** open.

3. **Minor Calyces**

—receive urine from the collecting tubules and empty into two or three **major calyces,** which in turn empty into an upper dilated portion of the ureter, the **renal pelvis**.

B. **Ureter** (Figure 5.16; see Figure 5.14)

—is a muscular tube that extends from the kidney to the urinary bladder.

—is retroperitoneal, descends on the psoas muscle, is crossed by the gonadal vessels, and crosses the bifurcation of the common iliac artery.

—may be obstructed where it joins the renal pelvis (the ureteropelvic junction), where it crosses the pelvic brim over the distal end of the common iliac artery, or where it enters the wall of the urinary bladder.

—receives blood from the aorta and the following arteries: the renal, gonadal, common and internal iliac, umbilical, superior and inferior vesical, and middle rectal.

C. Suprarenal (Adrenal) Gland

—is a retroperitoneal organ lying on the superomedial aspect of the kidney, and is surrounded by a capsule and renal fascia.

—is pyramidal on the right and is semilunar on the left.

—receives arteries from three sources: the superior suprarenal artery from the **inferior phrenic artery;** the middle suprarenal from the abdominal **aorta;** and the inferior suprarenal **artery** from the **renal artery.**

—its **medulla** is derived from embryonic neural crest cells, receives preganglionic sympathetic nerve fibers directly, and secretes epinephrine and norepinephrine.

—its **cortex** is essential to life and produces steroid hormones including mineralocorticoids (aldosterone), glucocorticoids (such as cortisone), and sex hormones.

—its venous drainage is via the **suprarenal vein,** which empties into the inferior vena cava on the right and the renal vein on the left.

II. Blood Vessels and Lymphatics

A. Aorta (see Figure 5.16)

—passes through the aortic hiatus in the diaphragm at the level of the twelfth thoracic vertebra, descends anterior to the vertebral bodies, and ends by bifurcating into the right and left common iliac arteries anterior to the fourth lumbar vertebra.

—gives rise to:

1. Inferior Phrenic Arteries

—arise from the aorta immediately below the aortic hiatus and give off **superior suprarenal arteries.**

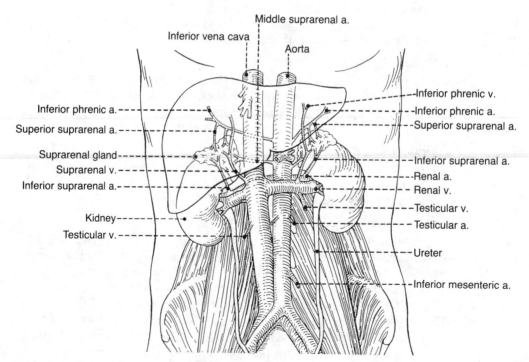

Figure 5.16. Abdominal aorta and its branches.

—diverge across the crura of the diaphragm, the left artery passing posterior to the esophagus and the right passing posterior to the inferior vena cava.

2. Middle Suprarenal Arteries

—arise from the aorta and run laterally on the crura of the diaphragm just superior to the renal arteries.

3. Renal Arteries

—arise from the aorta inferior to the origin of the superior mesenteric artery. The right artery is longer and a little lower than the left and passes posterior to the inferior vena cava; the left artery passes posterior to the left renal vein.

—give rise to the **inferior suprarenal arteries**.

4. Testicular or Ovarian Arteries

—descend retroperitoneally and run laterally on the psoas major muscle and across the ureter. The testicular artery accompanies the ductus deferens into the scrotum, where it supplies the spermatic cord, epididymis, and testis. The ovarian artery enters the suspensory ligament of the ovary, supplies the ovary, and anastomoses with the ovarian branch of the uterine artery.

5. Lumbar Arteries

—have four or five pairs and arise from the back of the aorta.

—run posterior to the sympathetic trunk, the inferior vena cava (on the right side), the psoas major muscle, the lumbar plexus, and the quadratus lumborum.

—divide into smaller anterior branches (to supply adjacent muscles) and larger posterior branches, which accompany the dorsal primary rami of the corresponding spinal nerves and divide into spinal and muscular branches.

6. Celiac and Superior and Inferior Mesenteric Arteries (see pp 166–169)

B. Inferior Vena Cava (see Figure 5.16)

—is formed on the right side of the fifth lumbar vertebra by the union of the two common iliac veins, below the bifurcation of the aorta.

—is longer than the abdominal aorta and ascends at the right side of the aorta.

—passes through its opening (opening for the inferior vena cava) in the central tendon of the diaphragm at the level of the eighth thoracic vertebra and enters the right atrium of the heart.

—receives the right gonadal, suprarenal, and inferior phrenic veins. On the left side, these veins usually drain into the left renal veins.

—also receives the three (left, middle, and right) hepatic veins. The middle and left hepatic veins frequently unite for about 1 cm before entering the vena cava.

C. Cisterna Chyli

—is the lower dilated end of the thoracic duct and lies posterior to and just to the right of the aorta, usually between two crura of the diaphragm.

—is formed by the intestinal and lumbar lymph trunks.

III. Lumbar Plexus (Figure 5.17)

—is formed by the union of the ventral rami of the first three lumbar nerves and a part of the fourth lumbar nerve.

—lies anterior to the transverse processes of the lumbar vertebrae within the substance of the psoas muscle.

A. Subcostal Nerve (T12)

—runs behind the lateral lumbocostal arch and in front of the quadratus lumborum; penetrates the transversus abdominis muscle to run between it and the internal oblique muscle; and supplies the external oblique, internal oblique, transversus abdominis, rectus abdominis, and pyramidalis muscles.

B. Iliohypogastric Nerve (L1)

—emerges from the lateral border of the psoas muscle, runs in front of the quadratus lumborum muscle, and pierces the transversus abdominis muscle near the iliac crest to run between this muscle and the internal oblique muscle.

—pierces the internal abdominal oblique muscle and then continues medially deep to the external abdominal oblique muscle.

—supplies the internal oblique and transversus abdominis muscles and divides into an anterior cutaneous branch, which supplies the skin above the

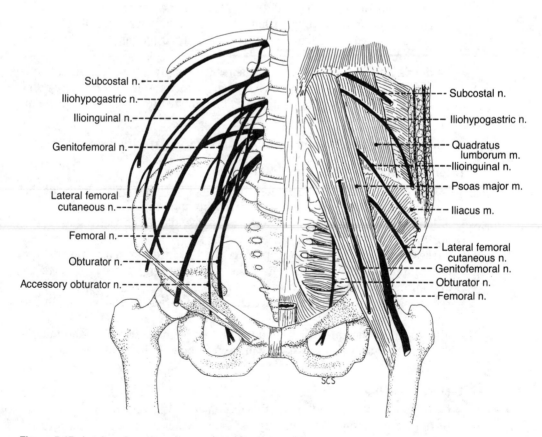

Figure 5.17. Lumbar plexus.

pubis, and a lateral cutaneous branch, which supplies the skin of the gluteal region.

C. Iliolingual Nerve (L1)

—runs in front of the quadratus lumborum muscle; pierces the transversus abdominis muscle and then the internal oblique muscle to run between the internal and external oblique aponeuroses.

—accompanies the spermatic cord (or the round ligament of the uterus), continues through the inguinal canal, and emerges through the superficial inguinal ring.

—supplies the internal oblique and transversus abdominis muscles and gives off cutaneous branches to the thigh and anterior scrotal or labial branches.

D. Genitofemoral Nerve (L1 and L2)

—emerges on the front of the psoas muscle and descends on its anterior surface.

—divides into a **genital branch,** which enters the inguinal canal through the deep inguinal ring to supply the cremaster muscle and the scrotum (or labium majus), and a **femoral branch,** which supplies the skin of the femoral triangle.

E. Lateral Femoral Cutaneous Nerve (L2 and L3)

—emerges from the lateral side of the psoas muscle, runs in front of the iliacus and behind the inguinal ligament, and supplies the skin of the anterior and lateral aspects of the thigh.

F. Femoral Nerve (L2 to L4)

—emerges from the lateral border of the psoas major muscle and descends in the groove between the psoas and iliacus muscles.

—enters the femoral triangle deep to the inguinal ligament and lateral to the femoral vessels outside the femoral sheath, and divides into numerous branches.

—supplies the skin of the thigh and leg, the muscles of the front of the thigh, and the hip and knee joints.

—supplies the quadriceps femoris, pectineal, and sartorius muscles and gives off the anterior femoral cutaneous nerve and the saphenous nerve.

G. Obturator Nerve (L2 to L4)

—arises from the second, third, and fourth lumbar nerves and descends along the medial border of the psoas muscle. It runs forward on the lateral wall of the pelvis and enters the thigh through the obturator foramen.

—divides into an **anterior branch** and a **posterior branch** and supplies the adductor group of muscles, the hip and knee joints, and the skin of the medial side of the thigh.

H. Accessory Obturator Nerve (L3 and L4)

—is present in about 9% of the population.

—descends medial to the psoas muscle, passes over the superior pubic ramus, and supplies the hip joint and the pectineal muscle.

IV. Autonomic Nervous System (Figures 5.18 and 5.19)

—is composed of motor, or efferent, nerves through which cardiac muscle, smooth muscle, and glands are innervated.

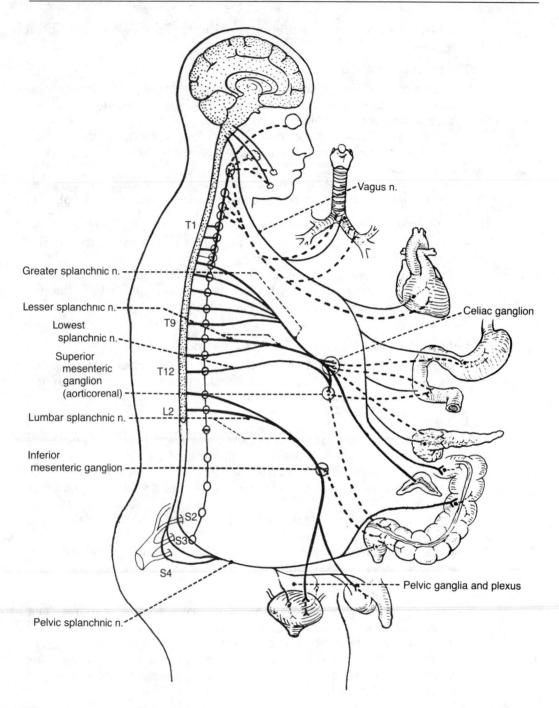

Figure 5.18. Autonomic nervous system.

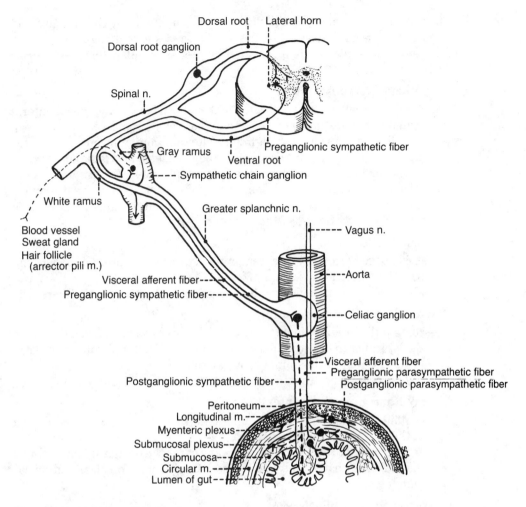

Figure 5.19. Nerve supply to the viscera.

—also includes general visceral afferent (GVA) fibers that run along with general visceral efferent (GVE) fibers.

—consists of sympathetic (or thoracolumbar outflow) and parasympathetic (or craniosacral outflow) systems.

—involves two neurons: preganglionic and postganglionic.

A. Parasympathetic Nervous System

—is also known as the craniosacral division.

—promotes quiet and orderly processes of the body, thereby conserving energy.

—is not as widely distributed over the entire body as sympathetic fibers; the body wall and extremities have no parasympathetic supply.

—preganglionic fibers running in cranial nerves (CN) III, VII, and IX pass to cranial autonomic ganglia (i.e., the ciliary, submandibular, and otic ganglia), where they synapse with postganglionic neurons.

—preganglionic fibers in CN X and in pelvic splanchnic nerves (originating from S2–S4) pass to terminal ganglia, where they synapse.

—parasympathetic fibers in the vagus nerve supply all of the thoracic and abdominal viscerae except the descending and sigmoid colons and other pelvic viscerae, which are innervated by the pelvic splanchnic nerves (S2–S4).

B. Sympathetic Nervous System

—preganglionic cell bodies are located in the lateral horn or intermediolateral cell column of the spinal cord segments between T1 and L2.

—preganglionic fibers pass through the white rami communicantes and enter the sympathetic chain ganglion, where they synapse. Postganglionic fibers join each spinal nerve by way of the gray rami communicantes and supply the blood vessels, hair follicles (arrector pili muscles), and sweat glands.

—enables the body to cope with crises or emergencies and thus often is referred to as the fight-or-flight division.

1. Sympathetic Trunk

—is composed primarily of ascending and descending preganglionic sympathetic fibers and visceral afferent fibers.

—also contains cell bodies of the postganglionic sympathetic fibers.

2. White Rami Communicantes

—contain the preganglionic sympathetic fibers with cell bodies located in the lateral horn of the spinal cord, and visceral afferent fibers with cell bodies located in the dorsal root ganglia.

—are limited to the spinal cord segments between T1 and L2.

3. Gray Rami Communicantes

—contain the postganglionic sympathetic fibers with cell bodies located in the sympathetic chain ganglia.

—are connected to every spinal nerve and carry the postganglionic sympathetic fibers that supply the blood vessels, hair follicles, and sweat glands.

4. Greater Splanchnic Nerve

—arises from the thoracic ganglia of T5 through T9 and enters the abdomen, passing through the aortic hiatus.

—enters the celiac ganglion, which is formed by the cell bodies of the postganglionic fibers.

—contains the sympathetic preganglionic fibers with cell bodies located in the lateral horn of the spinal cord, and also contains the visceral afferent fibers with cell bodies located in the dorsal root ganglia.

5. Lesser Splanchnic Nerve

—arises from the thoracic ganglia of T10 and T11 and enters the aorticorenal or superior mesenteric ganglion.

—contains the preganglionic sympathetic and visceral afferent fibers.

6. Least (Lowest) Splanchnic Nerve

—arises from the thoracic ganglia of T12 and enters the renal plexus.

—contains the preganglionic sympathetic and visceral afferent fibers.

7. Collateral (Prevertebral) Ganglia

—include the celiac, superior mesenteric, aorticorenal, and inferior mesenteric ganglia, which usually are located near the origin of the respective arteries.

C. **Autonomic Plexuses** (see Figure 6.15)

1. **Celiac Plexus**
   —is formed by splanchnic nerves and branches from the vagus nerves.
   —also contains the celiac ganglia, which receive the greater splanchnic nerves.
   —lies on the front of the crura of the diaphragm and on the abdominal aorta at the origins of the celiac trunk and the superior mesenteric and renal arteries.
   —extends its branches along the branches of the celiac trunk and forms the subsidiary plexuses, which are named according to the arteries along which they pass, such as gastric, splenic, hepatic, suprarenal, and renal plexuses.

2. **Aortic Plexus**
   —extends from the celiac plexus along the front of the aorta.
   —extends its branches along the arteries and forms plexuses that are named accordingly—**superior mesenteric, testicular** (or ovarian), and **inferior mesenteric**.
   —continues down along the aorta and forms the superior hypogastric plexus just below the bifurcation of the aorta.

3. **Superior and Inferior Hypogastric Plexuses** (see pp 217–218)

V. Diaphragm and Its Openings

A. **Diaphragm** (Figure 5.20)
   —arises from the xiphoid process (sternal part), lower six costal cartilages (costal part), medial and lateral lumbocostal arches (lumbar part), and lumbar vertebrae L1 to L3 for the right crus, or vertebrae L1 and L2 for the left crus.
   —inserts into the central tendon.
   —is innervated by the phrenic nerve (central part) and intercostal nerves (peripheral part).
   —receives blood from the musculophrenic, pericardiacophrenic, superior phrenic, and inferior phrenic arteries.
   —descends when it contracts, causing an increase in thoracic volume with a decreased thoracic pressure.
   —ascends when it relaxes, causing a decrease in thoracic volume with an increased thoracic pressure.

1. **Right Crus**
   —is larger and longer than the left crus.
   —originates from the upper three lumbar vertebrae (the left crus originates from the upper two).
   —splits to enclose the esophagus.

2. **Medial Arcuate Ligament (Medial Lumbocostal Arch)**
   —extends from the body of the first lumbar vertebra to the transverse process of the first lumbar vertebra and passes over the psoas muscle and the sympathetic trunk.

3. **Lateral Arcuate Ligament (Lateral Lumbocostal Arch)**
   —extends from the transverse process of the first lumbar vertebra to the twelfth rib and passes over the quadratus lumborum muscle.

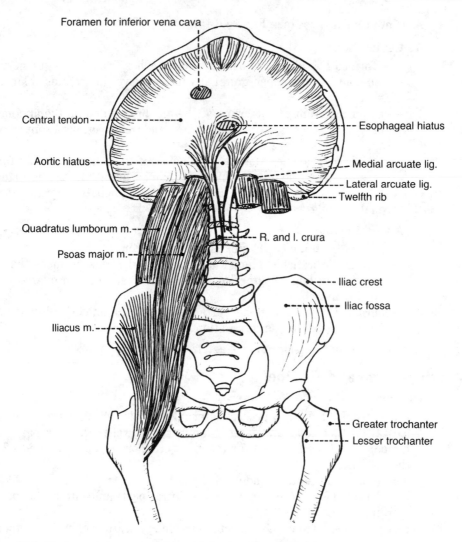

Foramen for inferior vena cava

Central tendon

Aortic hiatus

Quadratus lumborum m.

Psoas major m.

Iliacus m.

Esophageal hiatus

Medial arcuate lig.

Lateral arcuate lig.

Twelfth rib

R. and l. crura

Iliac crest

Iliac fossa

Greater trochanter

Lesser trochanter

Figure 5.20. Diaphragm and muscles of the posterior abdominal wall.

B. Hiatus (see Figure 5.20)

1. Vena Cava Hiatus (Vena Caval Foramen)

—lies in the diaphragm (central tendon) at the level of the eighth thoracic vertebra and transmits the inferior vena cava and the right phrenic nerve.

2. Esophageal Hiatus

—lies in the muscular part of the diaphragm (right crus) at the level of the tenth thoracic vertebra and transmits the esophagus and anterior and posterior trunks of the vagus nerves.

3. Aortic Hiatus

—lies behind or between two crura at the level of the twelfth thoracic vertebra and transmits the aorta, thoracic duct, greater splanchnic nerve, and azygos vein.

VI. Muscles of the Posterior Abdominal Wall

Muscle	Origin	Insertion	Nerve	Action
Quadratus lumborum	Transverse processes of LV3–LV5; iliolumbar ligament; iliac crest	Lower border of last rib; transverse processes of LV1–LV3	Subcostal n.; L1–L3	Depresses twelfth rib; flexes trunk laterally
Psoas major	Transverse processes, intervertebral disks and bodies of TV12–LV5	Lesser trochanter	L2–L3	Flexes thigh and trunk
Psoas minor	Bodies and intervertebral disks of TV12–LV1	Pectineal line; iliopectineal eminence	L1	Aids in flexing of trunk

LV = lumbar vertebra; TV = thoracic vertebra.

Review Test

DIRECTIONS: Each of the numbered items or incomplete statements in this section is followed by answers or by completions of the statement. Select the **one** lettered answer or completion that is **best** in each case.

1. Each statement below concerning the portal venous system is true EXCEPT

(A) blood pressure in the portal vein is higher than blood pressure in the inferior vena cava.
(B) the lower end of the esophagus is the area with the highest risk of varices or venous dilatations due to portal hypertension.
(C) the valves that permit distension of the portal vein can result in abnormally high blood pressures and injury to the portal vein.
(D) portal hypertension can cause caput medusae and hemorrhoids.
(E) the volume of blood delivered to the liver by the portal vein is at least twice the volume delivered by the hepatic artery.

2. Each of the following statements comparing the ileum to the jejunum is true EXCEPT

(A) the ileum has fewer plicae circulares and they are less prominent.
(B) the ileum has fewer mesenteric arterial arcades.
(C) less digestion and absorption of nutrients occur in the ileum.
(D) the ileum has shorter vasa recta.
(E) the mesentery of the ileum contains more fat.

3. Which of the following fetal vessels becomes the round ligament of the liver after birth?

(A) Ductus venosus
(B) Inferior epigastric vein
(C) Umbilical artery
(D) Ductus arteriosus
(E) Umbilical vein

4. Each of the following veins belongs to the portal system EXCEPT the

(A) right colic vein.
(B) superior rectal vein.
(C) splenic vein.
(D) right suprarenal vein.
(E) left gastroepiploic vein.

5. Which of the following statements concerning the structures on the anterior abdominal wall is true?

(A) The conjoined tendon includes fibers of the aponeurosis of the transversus abdominis and internal oblique abdominis muscles.
(B) The lacunar ligament is a triangular expansion of the lateral end of the inguinal ligament.
(C) The inguinal ligament is formed by the aponeuroses of the external and internal oblique abdominis muscles.
(D) The inferior epigastric vessels course between the superficial inguinal ring and the medial umbilical ligament.
(E) Scarpa's fascia continues over the penis as the deep fascia of the penis.

6. Which of the following conditions would most likely cause obstructive jaundice?

(A) Aneurysm of the splenic artery
(B) Perforated ulcer of the stomach
(C) Damage to the pancreas during splenectomy
(D) Cancer of the head of the pancreas
(E) Cancer of the body of the pancreas

7. Each artery below follows the mesentery to reach the organ it supplies EXCEPT the

(A) hepatic artery.
(B) mesenteric artery.
(C) middle colic artery.
(D) sigmoid branch of the left colic artery.
(E) dorsal pancreatic artery.

8. Each structure below is related to embryonic or fetal blood vessels EXCEPT the

(A) lateral umbilical fold.
(B) medial umbilical fold.
(C) ligamentum venosum.
(D) ligamentum teres hepatis.
(E) ligamentum arteriosum.

9. Which of the following statements concerning the anterior abdominal wall is true?

(A) Between the costal margin and the umbilicus, the posterior layer of the sheath of the rectus abdominis muscle is formed by the aponeuroses of the internal oblique and transversus abdominis muscles.
(B) The deep inguinal ring is in the aponeurosis of the transversus abdominis muscle.
(C) The external spermatic fascia is continuous with the transversalis fascia.
(D) The internal spermatic fascia is derived from the internal oblique aponeurosis.
(E) The anterior wall of the inguinal canal is formed primarily by the inguinal and lacunar ligaments.

10. Each of the following veins is a direct tributary of the superior mesenteric vein EXCEPT the

(A) middle colic vein.
(B) right colic vein.
(C) inferior pancreaticoduodenal vein.
(D) ileocolic vein.
(E) left colic vein.

11. Each of the following structures enters or leaves the liver through the porta hepatis EXCEPT

(A) the right hepatic artery.
(B) branches of the anterior gastric nerve.
(C) the common hepatic artery.
(D) the left hepatic duct.
(E) the portal vein.

12. Which of the following structures carries the preganglionic parasympathetic fibers of the liver?

(A) Sympathetic trunks
(B) Lumbar splanchnic nerves
(C) Pelvic splanchnic nerves
(D) Vagus nerves
(E) Greater splanchnic nerves

13. Which selection below describes the respective locations of the cell bodies of efferent and afferent nerve fibers in visceral branches of the sympathetic trunk?

(A) Collateral ganglia; ventral horn of the spinal cord
(B) Intermediolateral cell column of the spinal cord; dorsal root ganglia
(C) Dorsal root ganglia; ventral horn of the spinal cord
(D) Sympathetic chain ganglia; terminal ganglia
(E) Intermediolateral cell column of the spinal cord; sympathetic chain ganglia

14. Which of the following structures is a retroperitoneal organ?

(A) Stomach
(B) Transverse colon
(C) Jejunum
(D) Descending colon
(E) Spleen

15. The nerve fibers that innervate the suprarenal medulla to secrete noradrenaline are

(A) preganglionic sympathetic fibers.
(B) postganglionic sympathetic fibers.
(C) somatic motor fibers.
(D) postganglionic parasympathetic fibers.
(E) preganglionic parasympathetic fibers.

16. Which of the following structures is part of, or is formed by, the internal oblique muscle?

(A) Lacunar ligament
(B) Inguinal ligament
(C) Cremaster muscle
(D) External spermatic fascia
(E) Internal spermatic fascia

17. Which of the following arteries is least likely to send branches to the pancreas?

(A) Splenic artery
(B) Gastroduodenal artery
(C) Superior mesenteric artery
(D) Middle colic artery
(E) None of the above

18. Which of the following structures is formed by remnants of the embryonic urachus?

(A) Medial umbilical fold
(B) Round ligament of the uterus
(C) Inguinal ligament
(D) Median umbilical fold
(E) Lateral umbilical fold

19. Each statement below concerning the suprarenal glands is true EXCEPT

(A) each suprarenal gland is associated with more arteries than veins.
(B) thoracic splanchnic nerves carry postganglionic sympathetic fibers to the medulla of each gland.
(C) each gland is surrounded by an extension of the renal fascia.
(D) part of the left gland lies posterior to the lesser peritoneal sac.
(E) all or most of each gland lies on the diaphragm.

20. Which of the following pairs of veins (right and left) terminates in the same vein?

(A) Ovarian veins
(B) Testicular veins
(C) Inferior phrenic veins
(D) Suprarenal veins
(E) Third lumbar veins

21. Which of the following structures crosses anterior to the inferior vena cava?

(A) Right sympathetic trunk
(B) Third right lumbar vein
(C) Third part of the duodenum
(D) Right renal artery
(E) Cisterna chyli

22. Which of the following structures is posterior to the aorta?

(A) Third left lumbar vein
(B) Left renal vein
(C) Fourth part of the duodenum
(D) Root of the mesentery
(E) Uncinate process of the pancreas

23. The white rami communicantes contain

(A) preganglionic sympathetic fibers with cell bodies in the anterior horn of the spinal cord.
(B) preganglionic sympathetic fibers with cell bodies in the lateral horn of the spinal cord.
(C) postganglionic sympathetic fibers with cell bodies in the sympathetic chain ganglia.
(D) postganglionic sympathetic fibers that innervate the blood vessels and sweat glands.
(E) somatic afferent fibers with cell bodies in the dorsal root ganglia.

24. Which of the following statements concerning the inferior epigastric artery is true?

(A) It lies medial to a direct inguinal hernia.
(B) It lies lateral and posterior to an indirect inguinal hernia.
(C) It is a branch of the internal iliac artery.
(D) It is a route of collateral circulation in coarctation of the aorta.
(E) It anastomoses with the musculophrenic artery.

25. Each of the following statements concerning structures that open on the greater duodenal papilla is true EXCEPT

(A) the bile duct and the main pancreatic duct have separate openings into the duodenum.
(B) the common bile duct traverses the head of the pancreas.
(C) the greater duodenal papilla is in the posterior medial wall of the second part of the duodenum.
(D) the common bile duct lies in the free border of the lesser omentum.
(E) the bile duct lies anterior to the portal vein in the right free edge of the lesser omentum.

26. Which of the following statements concerning the inguinal ligament is true?

(A) It is formed by the inferior free edge of the internal abdominal oblique muscle.
(B) It extends between the anterior–inferior iliac spine and the ischial tuberosity.
(C) It forms the roof of the inguinal canal.
(D) It forms the floor of the inguinal canal.
(E) It forms the lateral boundary of the inguinal triangle.

27. Each structure below has a mesentery EXCEPT the

(A) first inch of the first part of the duodenum.
(B) transverse colon.
(C) sigmoid colon.
(D) appendix.
(E) ascending colon.

28. The peritoneal cavity contains

(A) the stomach.
(B) the pancreas.
(C) the spleen.
(D) the kidneys.
(E) none of the above.

29. Each of the following arteries sends branches to the ureter EXCEPT the

(A) renal artery.
(B) gonadal artery.
(C) middle rectal artery.
(D) inferior phrenic artery.
(E) common iliac artery.

30. Which structure below is least likely to contain pain fibers from the abdominal viscera?

(A) Splanchnic nerves
(B) Roots of the brachial plexus
(C) Sympathetic nerves to the viscera
(D) Vagus nerve fibers
(E) White rami communicantes

DIRECTIONS: Each group of items in this section consists of lettered options followed by a set of numbered items. For each item, select the **one** lettered option that is most closely associated with it. Each lettered option may be selected once, more than once, or not at all.

Questions 31–34

Match each statement below with the vein it describes.

(A) Hepatic vein
(B) Portal vein
(C) Superior mesenteric vein
(D) Coronary vein
(E) Inferior mesenteric vein

31. Lies in front of the uncinate process of the pancreas

32. Drains blood directly from the lesser curvature of the stomach

33. Usually empties into the splenic vein

34. Lies immediately anterior to the epiploic foramen

Questions 35–39

Match each statement below with the appropriate abdominal structure or component.

(A) Linea alba
(B) Linea semilunaris
(C) Linea semicircularis
(D) Transversalis fascia
(E) Conjoint tendon (falx inguinalis)

35. Defines the lateral margin of the rectus abdominis muscle

36. Forms the posterior layer of the sheath of the rectus abdominis below the arcuate line

37. The tendinous medial line extending from the xiphoid process to the pubic symphysis

38. Composed of the aponeuroses of the internal oblique and transversus abdominis muscles

39. The crescentic line marking the termination of the posterior layer of the sheath of the rectus abdominis muscle

Questions 40–44

Match each statement below with the most appropriate ligament.

(A) Lienorenal ligament
(B) Lienogastric ligament
(C) Gastrophrenic ligament
(D) Falciform ligament
(E) Hepatoduodenal ligament

40. Contains a small portion of the tail of the pancreas

41. Contains the bile duct

42. Contains a paraumbilical vein

43. Contains short gastric arteries

44. Contains splenic vessels

Questions 45–49

Match each statement below with the most appropriate artery.

(A) Right gastric artery
(B) Left gastroepiploic artery
(C) Splenic artery
(D) Gastroduodenal artery
(E) Cystic artery

45. Located within the lienogastric ligament

46. Gives rise to the superior pancreaticoduodenal artery

47. A direct branch of the celiac artery

48. Runs along the lesser curvature of the stomach

49. Runs along the superior border of the pancreas

Questions 50–53

Match each structure below with the appropriate lettered component of this computed tomogram of the upper abdomen.

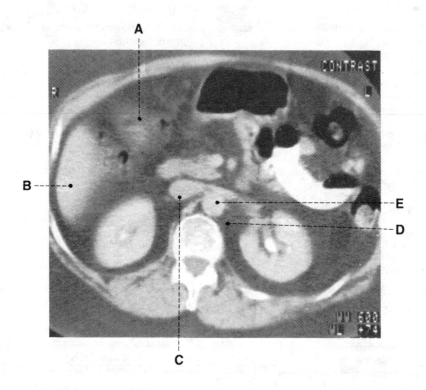

50. Inferior vena cava

51. Aorta

52. Crus of the diaphragm

53. Gallbladder

DIRECTIONS: Each question below contains four suggested answers of which **one or more** is correct. Choose answer

> **A** if **1, 2, and 3** are correct
> **B** if **1 and 3** are correct
> **C** if **2 and 4** are correct
> **D** if **4** is correct
> **E** if **1, 2, 3, and 4** are correct

54. The portal–caval anastomosis exists between the

(1) hepatic veins and inferior vena cava.
(2) superior rectal vein and middle rectal vein.
(3) middle rectal vein and inferior rectal vein.
(4) paraumbilical vein and superficial epigastric vein.

55. Which of the following arteries are sources of suprarenal arteries?

(1) Aorta
(2) Renal artery
(3) Inferior phrenic artery
(4) Superior mesenteric artery

56. Which of the following veins drain into the left renal vein?

(1) Suprarenal vein
(2) Hepatic vein
(3) Testicular (ovarian) vein
(4) Lumbar veins

57. Which of the following statements concerning the portal vein are true?

(1) It is formed posterior to the neck of the pancreas by the union of the splenic and superior mesenteric veins.
(2) It ascends anterior to the bile duct and the hepatic artery proper.
(3) It passes anterior to the epiploic foramen in the free edge of the lesser omentum.
(4) It has no tributaries superior to its beginning.

58. A tumor in the uncinate process of the pancreas may compress the

(1) common bile duct.
(2) superior mesenteric vein.
(3) portal vein.
(4) duodenojejunal junction.

59. Which of the following statements concerning the descending colon are true?

(1) It has teniae coli and epiploic appendices.
(2) It is a retroperitoneal organ.
(3) It is innervated by the pelvic splanchnic nerves.
(4) The inferior mesenteric artery provides most of its blood supply.

60. Which of the following statements concerning the free edge of the hepatoduodenal ligament of the lesser omentum are true?

(1) It forms a boundary of the epiploic foramen.
(2) It contains the common bile duct, hepatic artery, and portal vein.
(3) It is part of the lesser omentum.
(4) It is attached to the second part of the duodenum.

61. Which of the following statements concerning the inguinal canal are true?

(1) It extends from the anterior–superior iliac spine to the pubic tubercle.
(2) It begins at the deep inguinal ring in the transversalis fascia.
(3) Its anterior wall is primarily formed by the external oblique muscle, including aponeurosis and fascia.
(4) It transmits the round ligament of the uterus or spermatic cord.

62. Which of the following statements concerning a direct inguinal hernia are true?

(1) It enters the inguinal canal through its posterior wall.
(2) It lies medial to the inferior epigastric artery.
(3) It has a peritoneal covering.
(4) It is acquired.

Answers and Explanations

1–C. The portal vein and its tributaries have no valves, or, if present, they are insignificant.

2–B. The ileum has more mesenteric arterial arcades, more mesenteric fat, and shorter vasa recta than the jejunum. The plicae circulares (circular folds) in the upper part of the ileum are less prominent than those in the jejunum (the lower part of the ileum has no plicae circulares). More digestion and absorption of nutrients occurs in the jejunum than in the ileum.

3–E. The left umbilical vein becomes the round ligament of the liver after birth.

4–D. The suprarenal vein belongs to the systemic (or caval) venous system and drains into the inferior vena cava on the right and into the renal vein on the left.

5–A. The lacunar ligament is a triangular expansion of the medial end of the inguinal ligament. The inguinal ligament is the folded lower border of the aponeurosis of the external oblique abdominis muscle. The inferior epigastric vessels run from the deep inguinal ring toward the umbilicus, forming the lateral umbilical fold. The superficial fascia of the penis is continuous with Scarpa's fascia; the deep fascia of the penis (Buck's fascia) is continuous with the deep fascia covering the muscles of the anterior abdominal wall.

6–D. The bile duct traverses the head of the pancreas. Cancer in the head of the pancreas obstructs the bile duct, resulting in jaundice.

7–E. The pancreas is a retroperitoneal organ, except for a small portion of its tail; thus, the dorsal pancreatic artery arising from the splenic artery runs behind the peritoneum.

8–A. The lateral umbilical fold (ligament) contains the inferior epigastric artery and vein, which are adult blood vessels.

9–A. The deep inguinal ring is in the transversalis fascia. The external spermatic fascia is continuous with the aponeurosis of the external oblique abdominis muscle. The internal spermatic fascia is derived from the transversalis fascia. The anterior wall of the inguinal canal is formed by the aponeuroses of the external and internal oblique abdominal muscles.

10–E. The lower left colic vein is a tributary of the inferior mesenteric vein.

11–C. The common hepatic artery gives off the gastroduodenal artery and the hepatic artery proper, which divides into the right and left hepatic arteries.

12–D. Vagus nerves carry preganglionic parasympathetic fibers to the liver.

13–B. The cell bodies of the visceral efferent fibers are located in the intermediolateral cell column (or lateral horn) of the spinal cord; the cell bodies of visceral afferent fibers are located in the dorsal root ganglia.

14–D. The descending colon is a retroperitoneal organ.

15–A. The suprarenal medulla is the only organ that receives preganglionic sympathetic fibers.

16–C. The cremaster muscle and fascia arise from the internal oblique abdominis muscle.

17–D. The middle colic artery does not supply the pancreas. The splenic artery gives off a number of pancreatic branches, including the dorsal pancreatic artery. The superior pancreaticoduodenal artery arises from the gastroduodenal artery and the inferior pancreaticoduodenal artery arises from the superior mesenteric artery.

18–D. The median umbilical fold contains the urachus, which is the fibrous remnant of the embryonic allantois.

19–B. The suprarenal medulla is the only organ that receives the preganglionic sympathetic fibers. The chromaffin cells in the suprarenal medulla may be considered modified postganglionic sympathetic neurons.

20–E. The right and left third lumbar veins drain into the inferior vena cava.

21–C. The third part of the duodenum crosses anterior to the inferior vena cava.

22–A. The third left lumbar vein runs posterior to the abdominal aorta.

23–B. The white rami communicantes contain preganglionic sympathetic and visceral afferent fibers, which have cell bodies in the lateral horn of the spinal cord and the dorsal root ganglia, respectively.

24–D. The inferior epigastric artery is a branch of the external iliac artery. It lies lateral to the direct inguinal hernia and medial to the indirect inguinal hernia. In coarctation of the aorta, the epigastric artery provides a route of collateral circulation by anastomosing with the superior epigastric branch of the internal thoracic artery.

25–A. The common bile duct traverses the head of the pancreas and is joined by the main pancreatic duct to form the hepatopancreatic duct and/or ampulla, which opens into the greater papilla in the posterior medial wall of the second part of the duodenum. The portal vein lies posterior to the bile duct and the hepatic artery proper in the free margin of the lesser omentum.

26–D. The inguinal ligament extends from the anterior–superior iliac spine to the pubic tubercle. It forms the floor of the inguinal canal and the inferior boundary of the inguinal triangle.

27–E. Structures described as having a mesentery are suspended from the body wall by a double layer extension of the peritoneum. The ascending and descending colons are retroperitoneal and, therefore, do not have a mesentery. The first inch of the first part of the duodenum is surrounded by peritoneum (hepatoduodenal ligament). The transverse colon, sigmoid colon, and appendix have their own mesenteries: the transverse mesocolon, sigmoid mesocolon, and mesoappendix, respectively.

28–E. The peritoneal cavity contains nothing but a film of fluid that lubricates the surface of the peritoneum.

29–D. The ureter receives blood from the aorta, the renal, gonadal, common iliac, internal iliac, umbilical, superior and inferior vesical, and middle rectal arteries.

30–D. The vagus nerve contains sensory fibers associated with reflexes; it does not contain pain fibers.

31–C. The superior mesenteric vessels lie in front of the uncinate process of the pancreas.

32–D. The coronary (left gastric) vein drains blood from the lesser curvature of the stomach into the portal vein.

33–E. The inferior mesenteric vein usually empties into the splenic vein; however, it may empty into the superior mesenteric vein or the junction of the superior mesenteric and splenic veins.

34–B. The portal vein lies immediately anterior to the epiploic foramen in the free margin of the lesser omentum.

35–B. The linea semilunaris is a curved line along the lateral margin of the rectus abdominis muscle.

36–D. The transversalis fascia forms the posterior layer of the rectus sheath below the arcuate line.

37–A. The linea alba is the tendinous medial line extending from the xiphoid process to the symphysis pubis.

38–E. The aponeuroses of the internal oblique and transversus abdominis muscles form the conjoint tendon (falx inguinalis).

39–C. The linea semicircularis is the crescentic line marking the termination of the posterior layer of the rectus sheath.

40–A. The lienorenal ligament contains the splenic vessels and a small portion of the tail of the pancreas.

41–E. The hepatoduodenal ligament is part of the lesser omentum and contains the bile duct, the hepatic artery, and the portal vein.

42–D. The falciform ligament contains a paraumbilical vein that connects the left branch of the portal vein with subcutaneous veins around the umbilicus.

43–B. The lienogastric (gastrosplenic) ligament contains several short gastric and left gastroepiploic vessels.

44–A. The lienorenal ligament contains the splenic vessels.

45–B. The lienogastric ligament contains the left gastroepiploic and short gastric vessels.

46–D. The gastroduodenal artery is divided into the superior pancreaticoduodenal and right gastroepiploic arteries.

47–C. The splenic, left gastric, and common hepatic arteries are the direct branches of the celiac trunk.

48–A. The right gastric artery arises either from the hepatic artery proper or from the common hepatic artery and runs between layers of the lesser omentum along the lesser curvature of the stomach.

49–C. The splenic artery follows a tortuous course along the superior border of the pancreas.

50–C.
51–E.
52–D.
53–A.

54–C. The hepatic and superficial epigastric veins, inferior vena cava, and middle and inferior rectal veins are systemic (caval) veins. The superior rectal and paraumbilical veins are part of the portal venous system.

55–A. The suprarenal gland receives arteries from three sources: The superior suprarenal artery arises from the inferior phrenic artery, the middle suprarenal artery arises from the abdominal aorta, and the inferior suprarenal artery arises from the renal artery.

56–B. The suprarenal, inferior phrenic, and testicular or ovarian veins drain into the renal vein on the left and into the inferior vena cava on the right.

57–B. The portal vein is formed posterior to the neck of the pancreas by the union of the splenic and superior mesenteric veins. It ascends posterior to the bile duct and the hepatic artery proper and passes anterior to the epiploic foramen in the free edge of the lesser omentum. Tributaries of the portal vein include the left and right gastric veins and the posterior–superior pancreaticoduodenal vein.

58–C. The common bile duct traverses the head of the pancreas. The superior mesenteric vein anteriorly crosses the uncinate process of the pancreas near the duodenojejunal junction and joins the splenic vein posterior to the neck of the pancreas to form the portal vein.

59–E. The descending colon is retroperitoneal and has external characteristics that include teniae coli, epiploic appendices, and haustra. It receives most of its blood supply from the inferior mesenteric artery. Parasympathetic fibers from the pelvic splanchnic nerves, which arise from the sacral segments of the spinal cord (S2–S4), innervate the descending colon.

60–A. The hepatoduodenal ligament is ated to the beginning of the first part of the duodenum.

61–E. The inguinal canal extends from the anterior–superior iliac spine to the pubic tubercle. It begins at the deep inguinal ring in the transversalis fascia and ends at the superficial inguinal ring in the external oblique abdominis aponeurosis. It transmits the round ligament of the uterus or spermatic cord.

62–E. The direct inguinal hernia occurs through the posterior wall of the inguinal canal and, hence, lies medial to the inferior epigastric vessels. It is acquired (develops after birth), and its sac is formed by a peritoneum.

6
Perineum and Pelvis

Perineal Region

I. Perineum
—its floor is skin and fascia, and its roof is formed by the pelvic diaphragm with its fascial covering.

—is divided into an anterior urogenital triangle and a posterior anal triangle by a line connecting the two ischial tuberosities.

—is a diamond-shaped space that has the same boundaries as the inferior aperture of the pelvis:

A. Anterior: pubic symphysis.

B. Anterolateral: ischiopubic rami.

C. Lateral: ischial tuberosities.

D. Posterolateral: sacrotuberous ligaments.

E. Posterior: tip of the coccyx.

II. Urogenital Triangle (Figures 6.1 and 6.2)

A. Superficial Perineal Space (Pouch)

—lies between the inferior fascia of the urogenital diaphragm (the perineal membrane) and the membranous layer of the superficial perineal fascia (Colles' fascia).

—contains the superficial transverse perineal muscle, the ischiocavernosus muscle and crura of the penis or clitoris, the bulbospongiosus muscle and the bulb of the penis or the vestibule, the central tendon of the perineum, branches of the internal pudendal vessels, and the perineal nerve and its branches.

1. Colles' Fascia

—is the deep membranous layer of the superficial perineal fascia.

—is continuous with the dartos tunic in the male, passes over the penis as the superficial fascia of the penis, and is continuous with Scarpa's fascia.

—forms the lower boundary of the superficial perineal pouch.

2. Perineal Membrane

—is the inferior fascia of the urogenital diaphragm.

—forms the inferior boundary of the deep perineal pouch and the superior boundary of the superficial pouch.

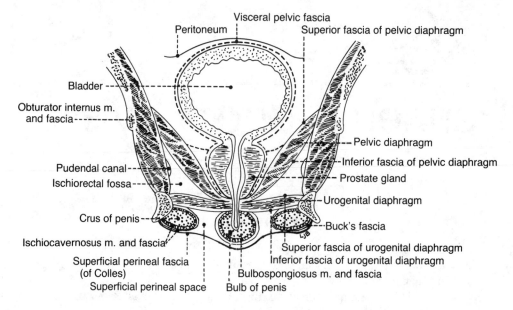

Figure 6.1. Frontal section of the male perineum and pelvis.

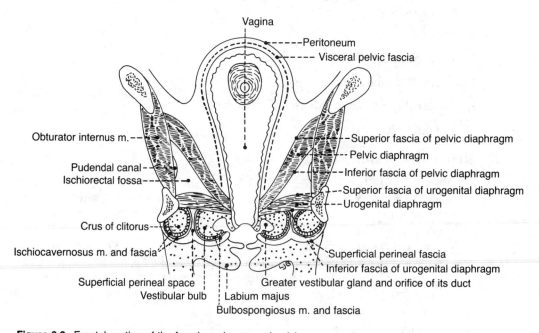

Figure 6.2. Frontal section of the female perineum and pelvis.

3. Muscles (Figures 6.3 and 6.4)

a. Ischiocavernosus Muscle

—arises from the inner surface of the ischial tuberosity and the ischiopubic ramus.

—inserts into the corpus cavernosum (the crus of the penis or clitoris).

—is innervated by the perineal branch of the pudendal nerve.

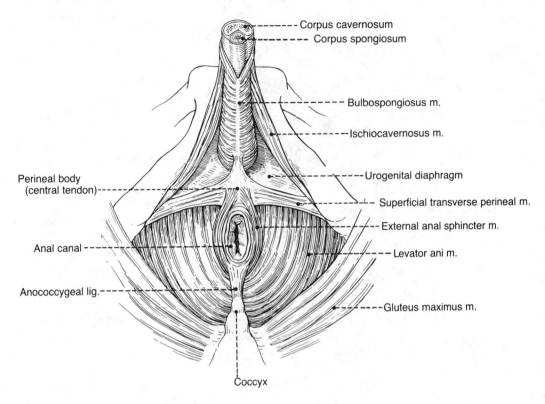

Figure 6.3. Muscles of the male perineum.

—maintains erection of the penis by compressing the crus and the deep dorsal vein of the penis, thereby retarding the venous return.

b. **Bulbospongiosus Muscle**

—arises from the perineal body and the fibrous raphe of the bulb of the penis.

—inserts into the corpus spongiosum.

—is innervated by the perineal branch of the pudendal nerve.

—its contraction (along with contraction of the ischiocavernosus) constricts the corpus spongiosum, thereby expelling the last drops of urine, or expelling the final semen in ejaculation.

—in the male, compresses the bulb, impeding venous return from the penis and thereby maintaining erection.

—in the female, compresses the erectile tissue of the bulb of the vestibule and constricts the vaginal orifice.

c. **Superficial Transverse Perineal Muscle**

—arises from the ischial ramus and tuberosity.

—inserts into the central tendon.

—is supplied by the perineal branch of the pudendal nerve.

—stabilizes the central tendon.

4. **Perineal Body (Central Tendon of the Perineum)**

—is a fibromuscular mass located in the center of the perineum between the anal canal and the vagina (or the bulb of the penis).

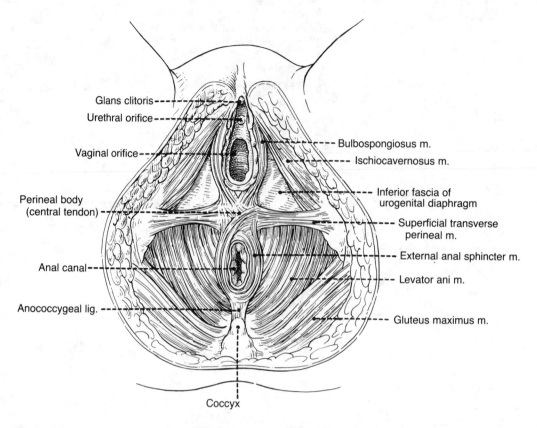

Figure 6.4. Muscles of the female perineum.

—serves as a site of attachment for the superficial and deep transverse perineal, bulbospongiosus, and levator ani muscles, and the external anal sphincter.

5. Greater Vestibular (Bartholin's) Gland

—lies in the superficial perineal pouch, under cover of and/or behind the bulb of the vestibule.

—is homologous to the bulbourethral gland in the male, is compressed during coitus, and secretes mucus that lubricates the vagina.

—its duct opens into the vestibule between the labium minus and the hymen.

B. Deep Perineal Space (Pouch)

—lies between the superior and inferior fasciae of the urogenital diaphragm.

—contains the deep transverse perineal muscle, sphincter urethrae, membranous part of the urethra, bulbourethral glands, branches of the internal pudendal vessels, and pudendal nerve.

1. Muscles

a. Deep Transverse Perineal Muscle

—arises from the inner surface of the ischial ramus.

—inserts into the medial tendinous raphe and the wall of the vagina (in the female) and the perineal body.

—is supplied by the perineal branches of the pudendal nerve.

—stabilizes the perineal body and supports the prostate gland or the vagina.

b. Sphincter Urethrae Muscle

—arises from the inferior pubic ramus.

—inserts into the median raphe and perineal body.

—is supplied by the perineal branch of the pudendal nerve.

—encircles the urethra, in the female and in the male, and its inferior fibers are attached to the anterolateral wall of the vagina.

2. Urogenital Diaphragm

—consists of the deep transverse perineal muscle and the sphincter urethrae, and is invested by the superior and inferior fasciae.

—stretches between the two pubic rami and ischial rami, and its inferior fascia gives attachment to the bulb of the penis.

—is pierced by the urethra and the vagina in the female and by the membranous urethra in the male.

—does not reach the pubic symphysis anteriorly.

3. Bulbourethral (Cowper's) Gland

—lies among the fibers of the sphincter urethrae in the deep perineal pouch, on the posterolateral sides of the membranous portion of the urethra.

—its ducts pass through the inferior fascia of the urogenital diaphragm to open into the bulbous portion of the spongy (penile) urethra.

III. Anal Triangle

A. Ischiorectal Fossa (see Figures 6.1 and 6.2)

—is located in the anal triangle and contains fat, the inferior rectal nerve and vessels, the internal pudendal vessels and the pudendal nerve, and a perineal branch of the posterior femoral cutaneous nerve.

—has the following **boundaries:**

1. **Anterior:** the posterior borders of the superficial and deep transverse perineal muscles.
2. **Posterior:** the gluteus maximus muscle and the sacrotuberous ligaments.
3. **Superomedial:** the sphincter ani externus and levator ani muscles.
4. **Lateral:** the obturator fascia covering the obturator internus muscle.
5. **Floor:** the skin over the anal triangle.

B. Muscles (Figure 6.5)

1. Obturator Internus Muscle

—arises from the inner surface of the obturator membrane.

—its tendon passes around the lesser sciatic notch to insert into the medial surface of the greater trochanter of the femur.

—is innervated by the nerve to the obturator internus.

—laterally rotates the thigh.

2. Sphincter Ani Externus Muscle

—arises from the tip of the coccyx and the anococcygeal ligament.

—inserts into the central tendon of the perineum.

—is supplied by the inferior rectal nerve.

—closes the anus.

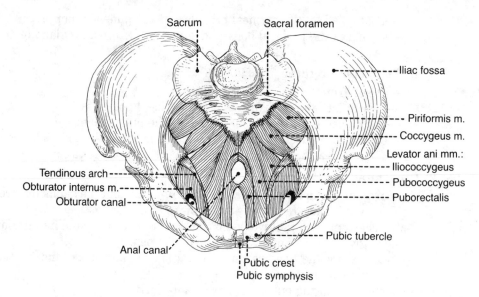

Figure 6.5. Muscles of the perineum and pelvis.

3. Levator Ani Muscle

—arises from the body of the pubis, the arcus tendineus, and the ischial spine.

—inserts into the coccyx and the anococcygeal raphe.

—is supplied by the branches of the anterior rami of the third and fourth sacral nerves and the perineal branch of the pudendal nerve.

—supports and raises the pelvic floor.

—consists of the puborectalis, pubococcygeus, and iliococcygeus.

4. Coccygeus Muscle

—arises from the ischial spine and the sacrospinous ligament.

—inserts into the coccyx and the lower part of the sacrum.

—is supplied by the branches of the third and fourth sacral nerves.

—supports and raises the pelvic floor.

IV. Male External Genitalia and Associated Structures (Figures 6.6 and 6.7)

A. Fascia and Ligaments

1. Fundiform Ligament of the Penis

—arises from the linea alba and the membranous layer of the superficial fascia of the abdomen.

—splits into left and right parts, encircles the body of the penis, and blends with the superficial penile fascia.

—enters the septum of the scrotum.

2. Suspensory Ligament of the Penis (or Clitoris)

—arises from the pubic symphysis and the arcuate pubic ligament and inserts into the deep fascia of the penis (or clitoris).

—lies deep to the fundiform ligaments.

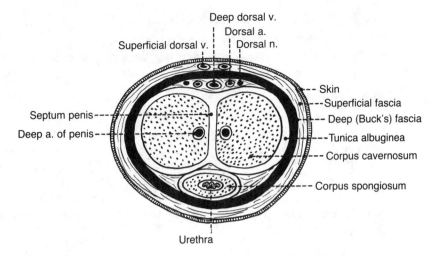

Figure 6.6. Cross section of the penis.

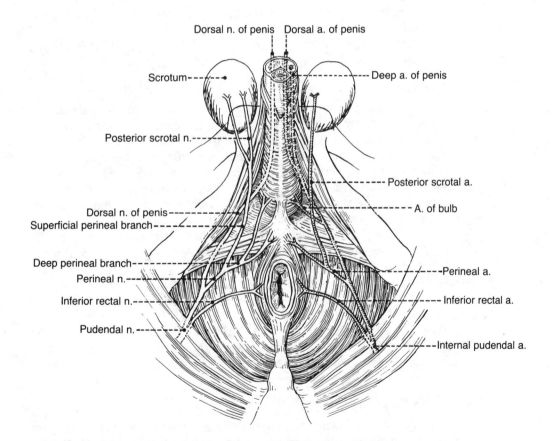

Figure 6.7. Internal pudendal artery and pudendal nerve and branches.

3. Deep Fascia of the Penis (Buck's Fascia)

—is a continuation of the deep perineal fascia.

—is continuous with the fascia covering the external oblique abdominis muscle and the rectus sheath.

4. Tunica Albuginea

—is a fibrous connective-tissue layer that envelops both the corpora cavernosa and the corpus spongiosum.

—is very dense around the corpora cavernosa, thereby greatly impeding venous return and resulting in the extreme turgidity of these structures when the erectile tissue becomes engorged with blood.

—is more elastic surrounding the corpus spongiosum, which therefore does *not* become excessively turgid on erection and permits passage of the ejaculate.

5. Tunica Vaginalis

—is a double serous membrane, a peritoneal sac on the end of the processus vaginalis that covers the front and sides of the testis and epididymis.

—is a closed sac, derived from the abdominal peritoneum, forming the innermost layer of the scrotum.

—consists of a parietal layer adjacent to the internal spermatic fascia and a visceral layer adherent to the testis and epididymis.

B. Scrotum

—is the cutaneous pouch consisting of skin and the underlying dartos.

—its skin is relatively thin with little or no fat, which is important in maintaining a temperature lower than in the rest of the body.

—contains the testis and its covering and the epididymis (see p 207).

—its **dartos tunic** is continuous with the superficial penile fascia and the superficial perineal fascia; is formed by fusion of the superficial and deep layers of superficial fascia; and consists largely of smooth muscle fibers, contains no fat, and functions in temperature regulation.

—is contracted and wrinkled when cold (or sexually stimulated), bringing the testis into close contact with the body to conserve heat; is relaxed when warm and hence is flaccid and distended to dissipate heat.

—receives blood from the external pudendal arteries and the posterior scrotal branches of the internal pudendal arteries. It also receives branches of the testicular and cremaster arteries.

—is innervated by the anterior scrotal branch of the ilioinguinal nerve, the genital branch of the genitofemoral nerve, the posterior scrotal branch of the perineal branch of the pudendal nerve, and the perineal branch of the posterior femoral cutaneous nerve.

C. Penis (see Figure 6.6)

—consists of three masses of vascular erectile tissue, the paired corpora cavernosa and the midline corpus spongiosum, which are bounded by tunica albuginea.

—has a root, which includes two crura and the bulb of the penis, and the body, which contains the single corpus spongiosum and the paired corpora cavernosa.

—the glans is formed by the terminal part of the corpus spongiosum and is covered by a double fold of skin, the **prepuce**. The frenulum of the prepuce is a median ventral fold that is attached to the external urethral orifice.

V. Female External Genitalia (see Figure 6.4)

A. Labia Majora

—are two longitudinal folds of skin that run downward and backward from the mons pubis.

—are joined anteriorly by the anterior labial commissure.

—their outer surfaces are covered with pigmented skin containing sebaceous and sweat glands and, after puberty, are covered with hairs.

—are homologous to the scrotum of the male.

—contain the terminations of the round ligaments of the uterus.

B. Labia Minora

—unlike the labia majora, are hairless and contain no fat.

—are divided into two parts, upper (lateral) and lower (medial).

1. The **lateral parts,** above the clitoris, fuse to form the **prepuce** of the clitoris.

2. The **medial parts,** below the clitoris, fuse to form the **frenulum** of the clitoris.

C. Vestibule of the Vagina

—is the space or cleft between the labia minora and has the openings of the urethra, the vagina, and the ducts of the greater vestibular glands in its floor.

D. Clitoris

—is homologous to the penis in the male; consists of erectile tissue and is enlarged as a result of engorgement with blood.

—consists of two crura, two corpora cavernosa, and a glans but has no corpus spongiosum. The glans is derived from the corpora cavernosa and is covered by a sensitive epithelium.

VI. Nerve Supply of the Perineal Region (see Figure 6.7)

Pudendal Nerve (S2–S4)

—passes through the greater sciatic foramen between the piriformis and coccygeus muscles.

—crosses the ischial spine and enters the perineum with the internal pudendal artery through the lesser sciatic foramen.

—enters the pudendal canal, gives off the inferior rectal nerve and the perineal nerve, and terminates as the dorsal nerve of the penis (or clitoris).

A. Inferior Rectal Nerve

—arises within the pudendal canal, divides into several branches, crosses the ischiorectal fossa, and supplies the sphincter ani externus and the skin around the anus.

B. Perineal Nerve

—arises within the pudendal canal and divides into the deep branch, which supplies all of the perineal muscles and the superficial (posterior scrotal or labial) branch, which in turn divides into two branches to supply the scrotum or labia majora.

C. Dorsal Nerve of the Penis (or Clitoris)

—pierces the perineal membrane, runs between the two layers of the suspensory ligament, and runs deep to the deep fascia on the dorsum of the penis or clitoris to supply the skin, prepuce, and glans.

VII. Blood Supply of the Perineal Region (see Figure 6.7)

A. Internal Pudendal Artery

—arises from the internal iliac artery.

—leaves the pelvis by way of the greater sciatic foramen below the piriformis and coccygeus muscles and immediately enters the perineum through the lesser sciatic foramen by hooking around the ischial spine.

—is accompanied by the pudendal nerve during its course.

—passes along the lateral wall of the ischiorectal fossa in the pudendal canal.

1. Inferior Rectal Artery

—arises within the pudendal canal, pierces the wall of the pudendal canal, and breaks into several branches, which cross the ischiorectal fossa to muscles and skin around the anal canal.

2. Perineal Artery

—supplies the superficial perineal muscles and gives rise to a transverse perineal branch and a posterior scrotal branch.

3. Artery of the Bulb

—arises within the deep perineal space, pierces the perineal membrane, and supplies the bulb of the penis or vestibule and the bulbourethral gland or the greater vestibular gland.

4. Urethral Artery

—pierces the perineal membrane, enters the corpus spongiosum penis or clitoris, and continues to the glans penis or clitoris.

5. Deep Artery of the Penis or Clitoris

—is one of the two terminal branches of the internal pudendal artery.

—pierces the perineal membrane and runs through the center of the corpus cavernosum penis or clitoris.

6. Dorsal Artery of the Penis or Clitoris

—is a paired artery on each side of the deep dorsal vein and runs deep to the deep fascia (Buck's fascia) and superficial to the tunica albuginea to supply the glans and the prepuce.

—pierces the perineal membrane and passes through the suspensory ligament of the penis or clitoris.

B. External Pudendal Artery

—arises from the femoral artery, emerges through the saphenous ring, and passes medially over the spermatic cord or the round ligament of the uterus to supply the skin above the pubis, penis, and scrotum or labium majus.

C. Veins of the Penis

1. Deep Dorsal Vein of the Penis

—is an unpaired vein that begins in the sulcus behind the glans and lies in the dorsal midline deep to the deep fascia and superficial to the tunica albuginea.

—leaves the perineum through the gap between the **arcuate pubic ligament** and the **transverse perineal ligament**.

—passes through the suspensory ligament of the penis below the arcuate pubic ligament and drains into the prostatic and pelvic venous plexuses.

2. Superficial Dorsal Vein of the Penis

—runs toward the pubic symphysis between the superficial and deep fasciae on the dorsum of the penis and divides into the right and left branches, which end in the external (superficial) pudendal veins. The external pudendal vein drains into the greater saphenous vein.

VIII. Clinical Considerations

A. Extravasated Urine

—may result from rupture of the spongy urethra below the urogenital diaphragm and may pass into the superficial perineal space.

—cannot spread laterally into the thigh because the inferior fascia of the urogenital diaphragm (the perineal membrane) and the superficial fascia of the perineum are firmly attached to the ischiopubic rami and are connected with the deep fascia of the thigh (fascia lata).

—cannot spread posteriorly into the anal region because the perineal membrane and Colles' fascia are continuous with each other around the superficial transverse perineal muscles.

—spreads inferiorly into the scrotum, anteriorly around the penis, and superiorly into the abdominal wall.

B. Episiotomy

—is a surgical incision through the posterolateral vaginal wall, just lateral to the perineal body, to enlarge the birth canal and thus prevent uncontrolled tearing during parturition.

C. Hydrocele

—is an accumulation of fluid in the cavity of the tunica vaginalis of the testis or along the spermatic cord.

D. Varicocele

—is a condition when enlargement (varicosity) of the veins of the spermatic cord form an appearance of a "bag of worms," accompanied by a constant pulling and dragging, and frequently causing oligospermia.

—is more common on the left side, probably due to a malignant tumor of the left kidney, which blocks the exit of the testicular vein.

Pelvis

I. Bony Pelvis (Figure 6.8)

A. Pelvis

—is the basin-shaped ring of bone formed by four bones: the two hip bones, the sacrum, and the coccyx. (The hip, or coxal bone, consists of the ilium, ischium, and pubis.)

—is divided by the **pelvic brim** or iliopectineal line into the **pelvis major** (false pelvis) above and the **pelvis minor** (true pelvis) below.

—its outlet is closed by the coccygeus and levator ani muscles, which form the floor of the pelvis.

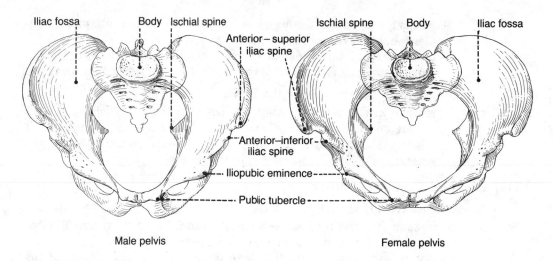

Figure 6.8. Male and female pelvic bones.

—is tilted in the normal anatomical position; thus:
1. The anterior–superior iliac spine and the pubic tubercles are in the same vertical plane.
2. The coccyx is in the same horizontal plane as the upper margin of the symphysis pubis.
3. The axis of the pelvic cavity running through the central point of the inlet and the outlet almost parallels the curvature of the sacrum.

B. Upper Pelvic Aperture (Pelvic Inlet or Pelvic Brim)

—is the superior rim of the pelvic cavity; is bounded posteriorly by the promontory of the sacrum, laterally by the arcuate line of the ilium (the iliopectineal line), and anteriorly by the pubic crest and the superior margin of the symphysis pubis.

—has transverse, oblique, and anteroposterior (conjugate) diameters.

C. Lower Pelvic Aperture (Pelvic Outlet)

—is a diamond-shaped aperture; is bounded posteriorly by the sacrum and coccyx, laterally by the ischial tuberosities and sacrotuberous ligaments, and anteriorly by the pubic symphysis, arcuate ligament, and rami of the pubis and ischium.

—is closed by the pelvic and urogenital diaphragms.

D. Pelvis Major (False Pelvis)

—is the expanded portion of the pelvis above the pelvic brim.

E. Pelvis Minor (True Pelvis)

—is the cavity of the pelvis below the pelvic brim (or superior aperture) and above the pelvic outlet (or inferior aperture).

—its outlet is closed by the coccygeus and levator ani muscles and the perineal fascia, which form the floor of the pelvis.

F. Differences Between the Female and Male Pelvis

—the female pelvis usually has smaller, lighter, and thinner bones than the male.

—the inlet is transversely oval in the female but is heart-shaped in the male.

—the outlet is larger in the female than in the male because of the everted ischial tuberosities in the female.

—the cavity is wider and shallower in the female than in the male.

—the subpubic angle or pubic arch is larger and the greater sciatic notch is wider in the female than in the male.

—the female sacrum is shorter and wider than the male sacrum.

—the obturator foramen is oval or triangular in the female but round in the male.

II. Pelvic Diaphragm (see Figure 6.5)

—forms the pelvic floor and supports all of the pelvic viscerae.

—is formed by the **levator ani** and **coccygeus muscles** and their fascial coverings.

—lies posterior and deep to the urogenital diaphragm as well as medial and deep to the ischiorectal fossa.

—upon contraction, raises the entire pelvic floor.

—flexes the anorectal canal during defecation and helps the voluntary control of micturition.

—helps to direct the fetal head toward the birth canal at parturition.

III. Ligaments

A. Broad Ligament of the Uterus (Figure 6.9)

—consists of two layers of peritoneum and extends from the lateral margin of the uterus to the lateral pelvic wall.

—contains the uterine tube, uterine vessels, round ligament of the uterus, ovarian ligament, ureter, uterovaginal nerve plexus, and lymphatic vessels.

—does not contain the ovary but gives attachment to the ovary through the mesovarium.

—its posterior layer curves from the isthmus of the uterus, as the rectouterine fold, to the posterior wall of the pelvis alongside the rectum.

1. Mesovarium

—is a fold of peritoneum that connects the anterior surface of the ovary with the posterior layer of the broad ligament.

2. Mesosalpinx

—is a fold of the broad ligament that suspends the uterine tube.

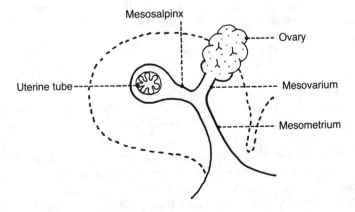

Figure 6.9. Sagittal section of the broad ligament.

3. Mesometrium

—is a major part of the broad ligament below the mesosalpinx and meso-varium.

B. Round Ligament of the Uterus

—is attached to the uterus in front of and below the attachment of the uterine tube and represents the remains of the lower part of the gubernaculum.

—runs within the layers of the broad ligament, contains smooth muscle fibers, and holds the fundus of the uterus forward, keeping the uterus anteverted and anteflexed.

—enters the inguinal canal at the deep inguinal ring, emerges from the superficial inguinal ring, and becomes lost in the subcutaneous tissue of the labium majus.

C. Ovarian Ligament

—is a fibromuscular cord that extends from the uterine end of the ovary to the side of the uterus below the uterine tube through the broad ligament.

—runs within the layers of the broad ligament.

D. Suspensory Ligament of the Ovary

—is a band of peritoneum that extends upward from the ovary to the pelvic wall and transmits the ovarian vessels, nerves, and lymphatics.

E. Lateral or Transverse Cervical (Cardinal or Mackenrodt's) Ligaments of the Uterus

—are fibromuscular condensations of pelvic fascia from the cervix and the lateral fornices of the vagina to the pelvic walls.

—extend laterally below the base of the broad ligament.

—contain smooth muscle fibers and support the uterus.

F. Pubocervical Ligaments

—are firm bands of connective tissue that extend from the posterior surface of the pubis to the cervix of the uterus.

G. Sacrocervical Ligaments

—are firm fibromuscular bands of pelvic fascia that extend from the lower end of the sacrum to the cervix and the upper end of the vagina.

H. Puboprostatic (or Pubovesical) Ligaments

—are condensations of the pelvic fascia that extend from the prostate gland (or the neck of the bladder in the female) to the pelvic bone.

IV. Ureter and Urinary Bladder

A. Ureter

—is a muscular tube that transmits urine by peristaltic waves.

—has three constrictions along its course: at its origin where the pelvis of the ureter joins the ureter; where it crosses the pelvic brim; and at its junction with the bladder.

—crosses the pelvic brim in front of the bifurcation of the common iliac artery and descends retroperitoneally on the lateral pelvic wall.

—lies 1 to 2 cm lateral to the cervix of the uterus in the female, whereas it passes posterior and medial to the ductus deferens and lies in front of the seminal vesicle in the male.

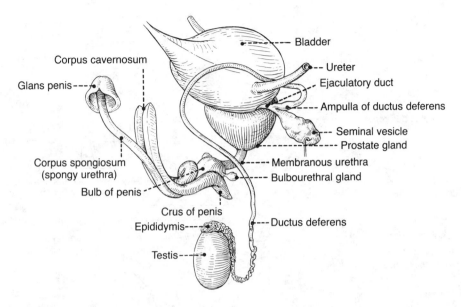

Figure 6.10. Male reproductive organs.

—in the female, is accompanied in its course by the uterine artery, which runs above and anterior to it.

—is sometimes injured by a clamp; may be ligated and sectioned by mistake in the process of a hysterectomy.

B. Urinary Bladder

—is situated below the peritoneum and slightly lower in the female than in the male.

—its trigone is bounded by the two orifices of the ureters and the internal urethral orifice.

—its musculature (bundles of smooth muscle fibers) as a whole is known as the **detrusor urinae muscle**.

—extends upward above the pelvic brim as it fills and may reach as high as the umbilicus if fully distended.

—receives blood from the superior and inferior vesical arteries (and from the vaginal artery in the female).

—its venous blood is drained by the prostatic (or vesical) plexus of veins, which empties into the internal iliac vein.

—is supplied by nerve fibers from the vesical and prostatic plexuses, which are extensions from the inferior hypogastric plexuses. The parasympathetic nerve originating from the cord segment, S2 to S4, causes the musculature of the bladder wall to contract, relaxes the internal sphincter, and promotes emptying.

V. Male Genital Organs (Figures 6.10 and 6.11)

A. Testis

—develops retroperitoneally and descends into the scrotum retroperitoneally.

—is covered by the **tunica albuginea,** which lies beneath the visceral layer of the tunica vaginalis.

—produces spermatozoa and secretes sex hormones.

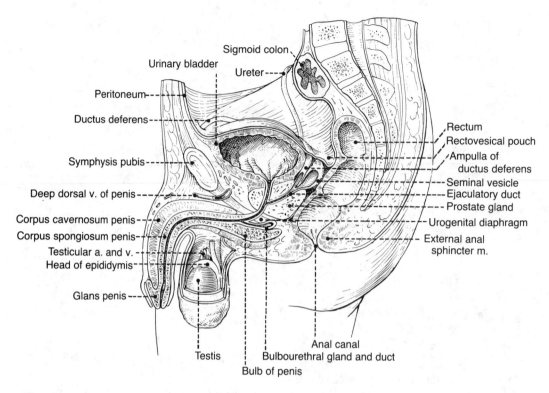

Figure 6.11. Sagittal section of the male pelvis.

—is supplied by the testicular artery from the abdominal aorta and is drained by veins of the pampiniform plexus muscle.

—its lymph vessels ascend with the testicular vessels and drain into the lumbar (aortic) nodes; lymphatic vessels in the scrotum drain into the superficial inguinal nodes.

B. Epididymis

—consists of the head, body, and tail. The tail contains a convoluted duct about 6 m (20 feet) long.

C. Ductus Deferens

—is a thick-walled tube with a relatively small lumen.

—enters the pelvis at the deep inguinal ring at the lateral side of the inferior epigastric artery.

—crosses the medial side of the umbilical artery, ureter, and obturator nerve and vessels during its course.

—its dilated part is termed the **ampulla**.

—receives innervation mainly from sympathetics of the hypogastric plexus.

D. Ejaculatory Duct

—is formed by the union of the ductus deferens with the duct of the seminal vesicle.

—opens into the prostatic urethra on the **seminal colliculus** just lateral to the blind prostatic utricle.

E. Seminal Vesicles

—are enclosed by dense endopelvic fascia and are lobulated glandular structures, which are diverticula of the ductus deferens.
—lie lateral to the ampullae of the ductus deferens against the fundus (base) of the bladder.
—produce an alkaline constituent of the seminal fluid.
—their lower ends become narrow and form ducts that join the ampullae of the ductus deferens to form the ejaculatory ducts.
—normally do not store spermatozoa. (Spermatozoa are stored in the ampulla of the ductus deferens and in the epididymis.)

F. Prostate Gland

—consists chiefly of glandular tissue, mixed with smooth muscle and fibrous tissue.
—has five lobes: the **anterior lobe** (or **isthmus**), which lies in front of the urethra and is devoid of glandular substance; the **middle** (or **median lobe**), which lies between the urethra and the ejaculatory ducts; the **posterior lobe,** which lies behind the urethra and below the ejaculatory ducts and contains glandular tissue; and the **right** and **left lateral lobes,** which are situated on either side of the urethra and form the main mass of the gland.
—secretes a fluid that causes the characteristic odor of semen and, together with the secretion from the seminal vesicles, forms the seminal fluid.
—its ducts open into the **prostatic sinus,** which is a groove on either side of the **urethral crest.**
—receives the ejaculatory duct, which opens into the urethra on the seminal colliculus just lateral to the blind prostatic utricle.

G. Urethral Crest

—is located on the posterior wall of the prostatic urethra and has numerous openings for the prostatic ducts on either side.
—has an ovoid-shaped enlargement called the **seminal colliculus.**
—at the summit of the colliculus is the **prostatic utricle,** which is an invagination (a blind pouch) about 5 mm deep.

H. Prostatic Sinus

—is a groove between the urethral crest and the wall of the prostatic urethra and receives the ducts of the prostate gland.

I. Seminal Colliculus

—is a small elevated portion of the urethral crest upon which the two ejaculatory ducts and the prostatic utricle open.

J. Prostatic Utricle

—is a minute pouch on the summit of the seminal colliculus.
—is analogous to the uterus and vagina in the female.

VI. Female Genital Organs (Figures 6.12 and 6.13)

A. Ovary

—lies on the posterior aspect of the broad ligament on the side wall of the pelvic minor and is bounded by the external and internal iliac vessels.
—is not covered by the peritoneum, and thus the ovum or oocyte is expelled into the peritoneal cavity and then into the uterine tube.

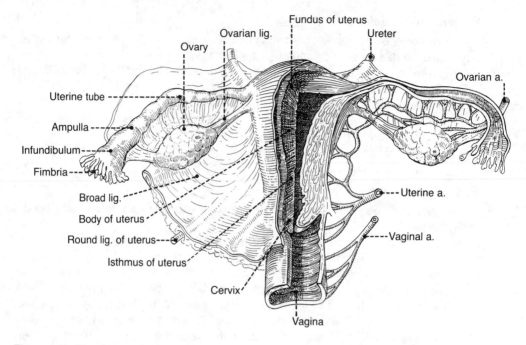

Figure 6.12. Female reproductive organs.

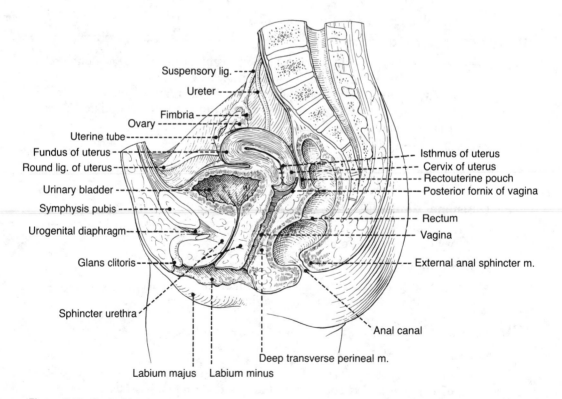

Figure 6.13. Sagittal section of the female pelvis.

—is not enclosed in the broad ligament, but its anterior surface is attached to the posterior surface of the broad ligament by the mesovarium.

—its surface is covered by germinal (columnar) epithelium, which is modified from the developmental peritoneal covering of the ovary.

—is supplied primarily by the ovarian artery, which is contained in the suspensory ligament and anastomoses with branches of the uterine artery.

—is drained by the ovarian vein, which joins the inferior vena cava on the right and the left renal vein on the left.

B. Uterine Tube

—extends from the uterus to the uterine end of the ovary and connects the uterine cavity to the peritoneal cavity.

—is subdivided into four parts: the **uterine part,** the **isthmus,** the **ampulla** (the longest and widest part), and the **infundibulum** (the funnel-shaped termination formed of **fimbriae**).

—conveys the fertilized or unfertilized oocytes to the uterus by its ciliary action and muscular contraction, which takes 3 to 4 days.

—transports spermatozoa in the opposite direction; fertilization takes place within the tube, usually in the **infundibulum** or **ampulla.**

C. Uterus

—is the organ of gestation, in which the fertilized oocyte normally becomes embedded and the developing organism grows until its birth.

—is normally **anteverted** with respect to the vagina and **anteflexed** at the junction of its cervix and body.

—is supported by the pelvic diaphragm; the urogenital diaphragm; the round, broad, lateral, or transverse cervical (cardinal) ligaments; and the pubocervical and sacrocervical ligaments.

—is supplied primarily by the uterine artery and secondarily by the ovarian artery.

—its anterior surface rests upon the posterosuperior surface of the bladder.

—is divided into four parts for the purpose of description:

1. Fundus

—is the rounded part of the uterus located superior and anterior to the plane of the entrance of the uterine tube.

2. Body

—is the main part of the uterus located inferior to the fundus and superior to the isthmus. The uterine cavity is triangular in the coronal section and is continuous with the lumina of the uterine tube and with the **internal os.**

3. Isthmus

—is the constricted part located between the body and cervix of the uterus. It corresponds to the internal os.

4. Cervix

—is the inferior narrow part of the uterus that projects into the vagina and divides into the following regions:

a. Internal os: the junction of the cervical canal with the uterine body.

b. Cervical canal: the cavity of the cervix between the internal and external ostia.

c. External os: the opening of the cervical canal into the vagina.

D. Vagina

—extends between the vestibule and the cervix of the uterus.

—is located at the lower end of the birth canal.

—serves as the excretory channel for the products of menstruation; also serves to receive the penis during coitus.

—has a subdivision called the **fornix,** which is the recess between the cervix and the wall of the vagina.

—its opening into the vestibule is partially closed by a membranous crescentic fold, the **hymen.**

—is supported by the levator ani muscles; the transverse cervical, pubocervical, and sacrocervical ligaments (upper part); the urogenital diaphragm (middle part); and the perineal body (lower part).

—receives blood from the vaginal branches of the uterine artery and of the internal iliac artery.

—has lymphatic drainage in two directions: The lymphatics from the upper three-fourths drain into the internal iliac nodes; those from the lower one-fourth, below the hymen, drain downward to the perineum and thus into the superficial inguinal nodes.

VII. Rectum and Anal Canal

A. Rectum

—is the part of the large intestine that extends from the sigmoid colon to the anal canal and follows the curvature of the sacrum and coccyx.

—has a peritoneal covering on its anterior, right, and left sides for the proximal third; only on its front for the middle third; and no covering for the distal third.

—its mucous membrane and the circular muscle layer form three folds called transverse folds of the rectum (Houston's valves).

—receives blood from the superior, middle, and inferior rectal arteries and the middle sacral artery.

—its venous blood returns to the portal venous system via the superior rectal vein and to the caval (systemic) system via the middle and inferior rectal veins.

—receives parasympathetic nerve fibers by way of the pelvic splanchnic nerve.

B. Anal Canal

—lies below the pelvic diaphragm and ends at the anus.

—is divided into an upper two-thirds (visceral portion), which belongs to the intestine, and a lower one-third (somatic portion), which belongs to the perineum with respect to the mucosa, blood supply, and nerve supply.

—the **anal columns** are 5 to 10 longitudinal folds of mucosa in its upper half (each column contains a small artery and a small vein).

—has crescent-shaped mucosal folds (**anal valves**) that connect the lower ends of the anal columns.

—has a series of pouch-like recesses at the lower end of the anal column (the **anal sinuses**) in which the anal glands open.

—contains the **internal anal sphincter,** which is a thickening of the circular smooth muscle in the lower part of the rectum that is separated from the **external anal sphincter** (skeletal muscle) by the intermuscular groove called **Hilton's white line.**

—has the **pectinate line,** which is a serrated line following the anal valves and crossing the bases of the anal columns.

1. The epithelium is **columnar** or **cuboidal** above the pectinate line and **stratified squamous** below it.
2. Venous drainage above the pectinate line goes into the **portal venous system** mainly via the superior rectal vein, and venous drainage below the pectinate line goes into the **caval system** via the middle and inferior rectal veins.
3. The lymphatic vessels drain into the **internal iliac nodes** above the line and into the **superficial inguinal nodes** below it.
4. The sensory innervation above the line is through fibers from the pelvic plexus and thus is of the **visceral** type, but the sensory innervation below it is by **somatic** nerve fibers of the pudendal nerve (which are very sensitive).
5. **Internal hemorrhoids** occur above the pectinate line, and **external hemorrhoids** occur below it. (A hemorrhoid is a varicosity of a venous plexus around the rectum and anal canal.)

VIII. Clinical Considerations

A. Uterine Prolapse

—is the protrusion of the cervix of the uterus into the vagina close to the vestibule.

—causes a bearing-down sensation in the womb and an increased frequency of and burning sensation on urination.

—occurs as a result of advancing age and is characterized by increased relaxation and loss of tonus of the muscular and fascial structures that constitute the support of the pelvic viscera.

—may be surgically corrected; however, during prolapse surgery, the ureter may be mistaken for the uterine artery and erroneously ligated in surgical removal of the uterus. (The uterine artery crosses cranially and anterior to the ureter.)

B. Hysterectomy

—is surgical removal of the uterus, performed either through the abdominal wall or through the vagina.

C. Prostatectomy

—is surgical removal of a part of the hypertrophied prostate gland, which occurs most in the median lobe, obstructing the internal urethral orifice and thus leading to nocturia (excessive urination at night), dysuria (difficulty or pain in urination), and urgency (sudden desire to urinate).

IX. Blood Vessels (Figure 6.14)

A. Internal Iliac Artery

—arises from the bifurcation of the common iliac artery, in front of the sacroiliac joint, and is crossed in front by the ureter at the pelvic brim.

—is commonly divided into a **posterior division,** which gives off the iliolumbar, lateral sacral, and superior gluteal arteries; and an **anterior division,** which gives off the inferior gluteal, internal pudendal, umbilical, obturator, inferior vesical, middle rectal, and uterine arteries.

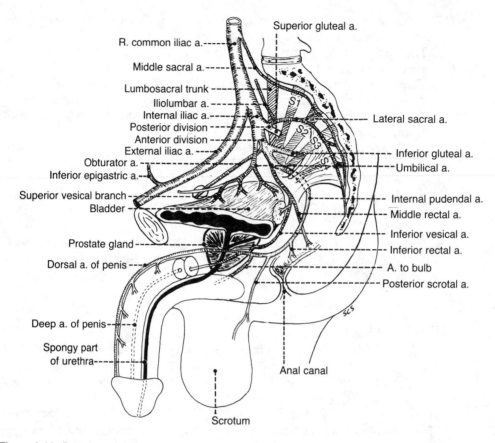

Figure 6.14. Branches of the internal iliac artery.

1. Iliolumbar Artery

—runs superolaterally to the iliac fossa, passing deep to the psoas major muscle.

—divides into an iliac branch supplying the iliacus muscle and the ilium, and a lumbar branch supplying the psoas major and the quadratus lumborum muscles.

2. Lateral Sacral Artery

—consists usually of a **superior artery** and an **inferior artery,** which may arise from a common trunk.

a. Superior Artery

—enters the first or second pelvic sacral foramen and supplies the contents of the sacral canal and the muscles on the dorsum of the sacrum.

—anastomoses with branches of the superior gluteal artery.

b. Inferior Artery

—descends in front of the piriformis muscle and the sacral nerves and then on the pelvic surface of the sacrum between the sacral sympathetic trunk and the sacral foramina.

—anastomoses with the median sacral artery.

3. **Superior Gluteal Artery**

—usually runs between the lumbosacral trunk and the first sacral nerve.
—leaves the pelvis through the greater sciatic foramen above the piriformis muscle to supply muscles in the buttocks.

4. **Inferior Gluteal Artery**

—runs between the first and second or between the second and third sacral nerves.
—leaves the pelvis through the greater sciatic foramen, inferior to the piriformis.

5. **Internal Pudendal Artery**

—leaves the pelvis through the greater sciatic foramen, passing between the piriformis and coccygeus muscles, and enters the perineum through the lesser sciatic foramen (see p 202).

6. **Umbilical Artery**

—runs forward along the lateral pelvic wall and along the side of the bladder.
—its proximal part gives off the **superior vesical artery** to the superior part of the bladder and the **artery of the ductus deferens,** which supplies the ductus deferens, the seminal vesicles, the lower part of the ureter, and the bladder.
—its distal part is obliterated and continues forward as the medial umbilical ligament.

7. **Obturator Artery**

—usually arises from the internal iliac artery, but in about 20% to 30% of the population it arises from the inferior epigastric artery. It then passes close to or across the femoral canal to reach the obturator foramen and hence is susceptible to damage during hernia operations.
—runs through the upper part of the obturator foramen, divides into the anterior and posterior branches, and supplies the muscles of the thigh.
—its posterior branch gives rise to an acetabular branch, which enters the joint through the acetabular notch and reaches the head of the femur by way of the ligamentum capitis femoris.

8. **Inferior Vesical Artery**

—corresponds to the vaginal artery in the female.
—supplies the fundus of the bladder, prostate gland, seminal vesicles, ductus deferens, and lower part of the ureter.

9. **Middle Rectal Artery**

—runs medially to the middle portion of the rectum.
—also supplies the prostate gland, seminal vesicles, ureter, and vagina.

10. **Uterine Artery**

—is homologous to the artery of the ductus deferens in the male.
—arises from the internal iliac artery or in common with the vaginal or middle rectal artery.
—runs medially in the base of the broad ligament to reach the junction of the cervix and the body of the uterus and runs in front of and above the ureter near the lateral fornix of the vagina.

—divides into a large superior branch, supplying the body and fundus of the uterus, and a smaller vaginal branch, supplying the cervix and vagina.

—takes a tortuous course along the lateral margin of the uterus and ends by anastomosing with the ovarian artery.

11. Vaginal Artery

—arises from the uterine or internal iliac artery.

—gives rise to numerous branches to the anterior and posterior wall of the vagina and makes longitudinal anastomoses in the median plane to form the **anterior** and **posterior azygos arteries** of the vagina.

B. Median Sacral Artery

—is an unpaired artery, arising from the posterior aspect of the abdominal aorta just before its bifurcation.

—descends in front of the sacrum and ends in the **coccygeal body,** which is a small cellular and vascular mass located in front of the tip of the coccyx.

C. Superior Rectal Artery

—is the direct continuation of the inferior mesenteric artery.

D. Ovarian Artery

—arises from the abdominal aorta, crosses the proximal end of the external iliac artery to enter the pelvic minor, and reaches the ovary through the suspensory ligament of the ovary.

X. Sacral Plexus

—is formed by the fourth and fifth lumbar ventral rami (the lumbosacral trunk) and the first four sacral ventral rami.

—lies largely on the internal surface of the piriformis muscle in the pelvis.

A. Superior Gluteal Nerve (L4–S1)

—leaves the pelvis through the greater sciatic foramen, above the piriformis.

—supplies the gluteus medius, gluteus minimis, and tensor fascia lata muscles.

B. Inferior Gluteal Nerve (L5–S2)

—leaves the pelvis through the greater sciatic foramen, below the piriformis.

—supplies the gluteus maximus muscle.

C. Sciatic Nerve (L4–S3)

—is the largest nerve in the body and is composed of peroneal and tibial parts.

—leaves the pelvis through the greater sciatic foramen below the piriformis.

—enters the thigh in the hollow between the ischial tuberosity and the greater trochanter of the femur.

D. Nerve to the Obturator Internus Muscle (L5–S2)

—leaves the pelvis through the greater sciatic foramen below the piriformis.

—enters the perineum through the lesser sciatic foramen.

—supplies the obturator internus and superior gemellus muscles.

E. Nerve to the Quadratus Femoris Muscle (L5–S1)

—leaves the pelvis through the greater sciatic foramen, below the piriformis.

—descends deep to the gemelli and obturator internus muscles and ends in the deep surface of the quadratus femoris, supplying the quadratus femoris and the inferior gemellus muscles.

F. Posterior Femoral Cutaneous Nerve (S1–S3)

—leaves the pelvis through the greater sciatic foramen below the piriformis.
—lies alongside the sciatic nerve and descends on the back of the knee.
—gives off several inferior cluneal nerves and perineal branches.

G. Pudendal Nerve (S2–S4)

—leaves the pelvis through the greater sciatic nerve below the piriformis.
—enters the perineum through the lesser sciatic foramen and the pudendal canal in the lateral wall of the ischiorectal fossa.

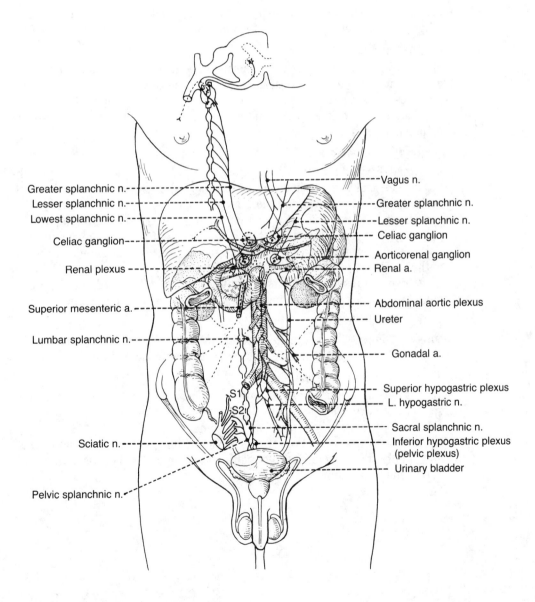

Figure 6.15. Autonomic ganglia and plexuses.

H. **Branches Distributed to the Pelvis**

—include the nerve to the piriformis muscle (S1–S2), the nerves to the levator ani and coccygeus muscles (S3–S4), the nerve to the sphincter ani externus muscle, and the pelvic splanchnic nerves (S2–S4).

XI. Autonomic Nerves (Figure 6.15)

A. **Superior Hypogastric Plexus**

—is the continuation of the aortic plexus below the aortic bifurcation and receives the lower two lumbar splanchnic nerves.

—lies behind the peritoneum, descends in front of the fifth lumbar vertebra, and ends by bifurcation into the **right** and **left hypogastric nerves** in front of the sacrum.

—contains preganglionic and postganglionic sympathetic fibers and visceral afferent fibers but apparently does not contain parasympathetic fibers.

B. **Hypogastric Nerve**

—is the lateral extension of the superior hypogastric plexus and lies in the extraperitoneal connective tissue lateral to the rectum.

—provides branches to the sigmoid colon and the descending colon.

—is joined by the pelvic splanchnic nerves to form the inferior hypogastric or pelvic plexus.

C. **Inferior Hypogastric Plexus (or Pelvic Plexus)**

—is formed by the union of hypogastric, pelvic splanchnic, and sacral splanchnic nerves and lies against the posterolateral pelvic wall, lateral to the rectum, vagina, and base of the bladder.

—contains pelvic ganglia, in which both sympathetic and parasympathetic preganglionic fibers synapse. Hence, it consists of preganglionic and postganglionic sympathetic fibers, preganglionic and postganglionic parasympathetic fibers, and visceral afferent fibers.

—gives rise to subsidiary plexuses including the middle rectal plexus, uterovaginal plexus, vesical plexus, differential plexus, and prostatic plexus.

D. **Sacral Splanchnic Nerves**

—consist primarily of preganglionic sympathetic fibers that come off the chain and synapse in the inferior hypogastric (pelvic) plexus.

E. **Pelvic Splanchnic Nerves**

—are the only splanchnic nerves that carry parasympathetic fibers. (All other splanchnic nerves are sympathetic.)

—arise from the sacral segment of the spinal cord (S2–S4).

—contribute to the formation of the pelvic (or inferior hypogastric) plexus, and supply the descending colon, sigmoid colon, and other viscerae in the pelvis and perineum.

Review Test

DIRECTIONS: Each of the numbered items or incomplete statements in this section is followed by answers or by completions of the statement. Select the **one** lettered answer or completion that is **best** in each case.

1. The processus vaginalis is an extension of the

(A) external spermatic fascia.
(B) internal spermatic fascia.
(C) extraperitoneal connective tissue.
(D) peritoneum.
(E) vaginal mucosa.

2. Sensation in which of the following groups of nerves must be blocked to completely anesthetize the skin of the urogenital triangle?

(A) Pudendal and ilioinguinal nerves
(B) Pudendal nerve and perineal branches of the posterior cutaneous nerve of the thigh
(C) Pudendal nerve, perineal branches of the posterior cutaneous nerve of the thigh, and the ilioinguinal nerve
(D) Perineal branches of the posterior cutaneous nerve of the thigh and the ilioinguinal nerve
(E) Pudendal, ilioinguinal, and genitofemoral nerves

3. Carcinoma of the uterus can spread directly to the labia majora in lymphatics that follow the

(A) pubic arcuate ligament.
(B) suspensory ligament of the ovary.
(C) cardinal ligament.
(D) suspensory ligament of the clitoris.
(E) round ligament of the uterus.

4. Tenderness and swelling of the left testicle may be produced by thrombosis in the

(A) internal pudendal vein.
(B) left renal vein.
(C) internal iliac vein.
(D) inferior epigastric vein.
(E) external pudendal vein.

5. Which of the following structures leaves the pelvis above the piriformis muscle?

(A) Sciatic nerve
(B) Internal pudendal artery
(C) Superior gluteal nerve
(D) Inferior gluteal artery
(E) Posterior femoral cutaneous nerve

6. Each structure below forms part of the boundary of the perineum EXCEPT the

(A) pubic arcuate ligament.
(B) tip of the coccyx.
(C) ischial tuberosities.
(D) sacrospinous ligament.
(E) sacrotuberous ligament.

7. Each of the following statements concerning the scrotum is true EXCEPT

(A) it is innervated anteriorly by the ilioinguinal nerve.
(B) it receives its blood supply from the testicular artery.
(C) it is partitioned into two sacs by a septum of superficial fascia.
(D) it has lymphatic drainage primarily into superficial inguinal lymph nodes.
(E) it has a dartos layer of fascia and muscle that is continuous with Colles' layer of superficial fascia in the perineum.

8. Each of the following structures helps support the uterus EXCEPT the

(A) sacrogenital folds.
(B) round ligament of the uterus.
(C) lateral cervical ligaments.
(D) pelvic diaphragm.
(E) rectouterine ligaments.

9. Each of the following structures is part of the boundary of the inferior pelvic aperture EXCEPT the

(A) sacrospinous ligament.
(B) ischial tuberosities.
(C) inferior rami of the pubis.
(D) rami of the ischium.
(E) arcuate ligament of the pubis.

10. Which of the following ligaments normally is found in the inguinal canal?

(A) Suspensory ligament of the ovary
(B) Ovarian ligament
(C) Broad ligament
(D) Round ligament of the uterus
(E) Pubovesical ligament

11. All of the following arteries send branches to the labia majora EXCEPT the

(A) internal iliac artery.
(B) obturator artery.
(C) uterine artery.
(D) internal pudendal artery.
(E) external pudendal artery.

12. Which of the following statements concerning the perineal membrane is true?

(A) It covers the superior surface of the urogenital diaphragm.
(B) It covers the inferior surface of the levator ani muscle.
(C) It is part of the deep layer of the superficial fascia covering the body.
(D) It covers the inferior surface of the urogenital diaphragm.
(E) It covers the superior surface of the pelvic diaphragm.

13. All of the following nerves from the lumbosacral plexus leave the abdominal or pelvic cavity EXCEPT the

(A) ilioinguinal nerve.
(B) genitofemoral nerve.
(C) lumbosacral trunk.
(D) femoral nerve.
(E) lateral femoral cutaneous nerve.

14. Pelvic splanchnic nerves carry preganglionic efferent fibers that synapse in

(A) terminal ganglia on or near the viscera.
(B) sympathetic chain ganglia.
(C) collateral ganglia.
(D) dorsal root ganglia.
(E) the ganglion impar.

15. As the uterine artery passes from the anterior division of the internal iliac artery to the uterus, it crosses a structure that is sometimes mistakenly ligated during pelvic surgery. This structure is the

(A) ovarian artery.
(B) ovarian ligament.
(C) mesovarium.
(D) ureter.
(E) vaginal artery.

16. Which of the following statements concerning pelvic vasculature is true?

(A) The inferior gluteal artery is a branch of the superior gluteal artery.
(B) Branches of the uterine artery usually anastomose with branches of the ovarian artery.
(C) The prostatic plexus of veins connects directly or indirectly with veins on the external part of the body.
(D) The inferior vesical arteries are branches of the obturator artery.
(E) The internal pudendal artery is a branch of the posterior division of the internal iliac artery.

17. Pelvic splanchnic nerves primarily consist of

(A) postganglionic parasympathetic fibers.
(B) postganglionic sympathetic fibers.
(C) preganglionic sympathetic fibers.
(D) preganglionic parasympathetic fibers.
(E) all of the above.

18. Which of the following statements concerning the ischiorectal fossa is true?

(A) It is bounded anteriorly by the transverse ligament of the urogenital diaphragm.
(B) It is partially bounded posteriorly by the gluteus maximus muscle.
(C) The pudendal canal runs along its medial wall.
(D) The levator ani muscle separates it from the urogenital triangle.
(E) It contains a perineal branch of the fifth lumbar nerve.

19. Which of the following structures constitutes the superior boundary of the superficial perineal space?

(A) Pelvic diaphragm
(B) Colles' fascia
(C) Superficial layer of the superficial fascia
(D) Deep layer of the superficial fascia
(E) Perineal membrane

20. Which group of structures below is found in the urogenital diaphragm of males?

(A) Deep transverse perineal muscles; bulbourethral glands; the membranous urethra
(B) Deep transverse perineal muscles; bulbospongiosus muscles; part of the spongy urethra
(C) Arteries to the bulb; ischiocavernosus muscles; bulbourethral glands
(D) Superficial transverse perineal muscles; prostatic urethra; sphincter urethrae muscles
(E) Sphincter urethrae muscles; bulbourethral glands; great vestibular glands

21. Each structure below lies in the superficial perineal space EXCEPT the

(A) ischiocavernosus muscles.
(B) transversus perinei superficialis muscles.
(C) greater vestibular glands in females.
(D) bulbourethral gland in males.
(E) bulb of the vestibule.

22. Which of the following statements concerning the external anal sphincter is true?

(A) It is primarily innervated and controlled by autonomic nerves.
(B) It contains mostly smooth muscles.
(C) It has deep, superficial, and subcutaneous components.
(D) It has lateral fibers that interdigitate with fibers of the obturator internus muscle.
(E) It extends superiorly as far as the lower end of the sigmoid colon.

23. Which of the following statements concerning ducts from the prostate gland is true?

(A) They open into the membranous part of the urethra.
(B) They open onto the seminal colliculus.
(C) They open onto the cavernous urethra.
(D) They open on either side of the urethral crest.
(E) They open into the prostatic utricle.

24. Which of the following statements concerning the duct of the seminal vesicle is true?

(A) It joins the duct of a bulbourethral gland to form the ejaculatory duct.
(B) It opens into the membranous urethra.
(C) It widens just before it opens into the urinary bladder to form the ejaculatory duct.
(D) It widens to form the ampulla of the ductus deferens.
(E) It unites with the ductus deferens to form a single ejaculatory duct.

25. Which of the following statements concerning the round ligament of the uterus is true?

(A) It does not contain smooth muscle fibers.
(B) It inserts into the pecten pubis.
(C) It traverses the inguinal canal.
(D) It contains the uterine artery.
(E) It runs medial to the inferior epigastric vessels.

26. Which of the following muscles forms the urogenital diaphragm?

(A) Sphincter urethrae muscle
(B) Coccygeus muscle
(C) Superficial transverse perineal muscle
(D) Levator ani muscle
(E) Obturator internus muscle

27. Superficial inguinal nodes receive lymph from each organ or structure below EXCEPT the

(A) lower part of the anal canal.
(B) labium majus.
(C) clitoris.
(D) testis.
(E) scrotum.

28. Which of the following structures passes through a gap between the arcuate pubic ligament and the transverse perineal ligament?

(A) Dorsal nerve of the penis
(B) Deep dorsal vein of the penis
(C) Superficial dorsal vein
(D) Dorsal artery of the penis
(E) Deep artery of the penis

29. Which of the following pairs of muscles is attached to the perineal body?

(A) Ischiocavernosus and sphincter urethrae muscles
(B) Deep transverse perineal and obturator internus muscles
(C) Bulbospongiosus and superficial transverse perineal muscles
(D) Superficial sphincter of anus and sphincter urethrae muscles
(E) Bulbospongiosus and ischiocavernosus muscles

30. Each structure below forms part of the broad ligament EXCEPT the

(A) ovarian ligament.
(B) mesometrium.
(C) mesosalpinx.
(D) peritoneum and underlying connective tissues covering the oviducts.
(E) peritoneum and underlying connective tissues on the posterior surface of the uterus.

31. Each of the following statements concerning the prostate gland is true EXCEPT

(A) the prostatic utricle opens at the apex of the seminal colliculus.
(B) the middle lobe of the prostate gland is posterior to the urethra.
(C) the uvula of the male bladder often is accentuated by the hypertrophy of the middle lobe of the prostate.
(D) the lateral lobes form the main mass of the prostate gland.
(E) the prostatic utricle is the terminal end of the duct of the prostate gland.

32. Which of the following structures is located between the inferior fascia of the urogenital diaphragm and the superficial perineal fascia?

(A) Prostate gland
(B) Bulbourethral gland
(C) Membranous part of the male urethra
(D) Superficial transverse perineal muscle
(E) None of the above

33. Which of the following ducts opens into the prostatic urethra on the seminal colliculus?

(A) Duct of the seminal vesicles
(B) Duct of the prostate gland
(C) Ejaculatory duct
(D) Duct of the bulbourethral gland
(E) All of the above

34. Which of the following structures lies in the broad ligament for all or part of its course?

(A) Ovarian ligament
(B) Uterine artery
(C) Round ligament of the uterus
(D) Uterine tube
(E) All of the above

35. Each of the following structures crosses the pelvic brim EXCEPT the

(A) ovarian artery.
(B) ureter.
(C) round ligament of the uterus.
(D) uterine artery.
(E) lumbosacral trunk.

DIRECTIONS: Each group of items in this section consists of lettered options followed by a set of numbered items. For each item, select the **one** lettered option that is most closely associated with it. Each lettered option may be selected once, more than once, or not at all.

Questions 36–40

Match each description below with the most appropriate muscle.

(A) Bulbospongiosus muscle
(B) Ischiocavernosus muscle
(C) Sphincter urethrae muscle
(D) Levator ani muscle
(E) Obturator internus muscle

36. Helps form a major uterine support

37. Covers, or is close to, the major vestibular glands

38. Lies on the surface of the crus

39. Forms the lateral wall of the ischiorectal fossa

40. Embedded in this muscle (in males) is an accessory reproductive gland

Questions 41–45

Match each description below with the most appropriate structure.

(A) Prostate gland
(B) Seminal vesicle
(C) Great vestibular gland
(D) Bulbourethral gland
(E) Anal gland

41. Lies on the posterolateral aspect of the bladder

42. Lies lateral to the membranous urethra

43. Lies in the superficial perineal space

44. Has numerous ducts that open into the urethra

45. Has ducts that empty into the bulb of the penis

Questions 46–50

Match each statement on the right with the most appropriate organ or structure.

(A) Ovary
(B) Uterus
(C) Uterine tube
(D) Vagina
(E) Clitoris

46. Ovum fertilization occurs here

47. Is supported by the cardinal ligaments

48. Opens into the peritoneal cavity

49. Is bounded by the external and internal iliac vessels

50. Is attached to the pubic symphysis by the suspensory ligament

DIRECTIONS: Each question below contains four suggested answers of which **one or more** is correct. Choose answer

A if **1, 2, and** 3 are correct
B if **1 and 3** are correct
C if **2 and 4** are correct
D if **4** is correct
E if **1, 2, 3, and** 4 are correct

51. The superficial inguinal lymph nodes drain the

(1) perineum.
(2) lower end of the anal canal.
(3) external genitalia.
(4) lower part of the anterior abdominal wall.

52. Which of the following actions occur during ejaculation?

(1) The urethral sphincters at the neck of the bladder close.
(2) The prostate gland, seminal vesicles, and bulbourethral glands contract.
(3) Smooth muscles in the efferent ducts and the vas deferens contract.
(4) Semen is propelled into the urethra.

53. Which of the following statements concerning the bulbourethral gland are true?

(1) It lies in the deep perineal space.
(2) It is embedded in the sphincter urethrae muscles.
(3) Its duct opens into the bulbous portion of the penile urethra.
(4) It lies on the posterolateral side of the membranous portion of the urethra.

54. Which of the following statements concerning the sphincter urethrae are true?

(1) It is striated muscle.
(2) It is innervated by the perineal nerve.
(3) Its innervation originates in spinal cord segments S2–S4.
(4) It is enclosed in the pelvic fascia.

55. Which of the following statements concerning the ejaculatory duct are true?

(1) It is formed by the union of the excretory ducts of the bulbourethral gland and ductus deferens.
(2) It passes through the prostate gland.
(3) It opens into the membranous urethra.
(4) It is formed by the union of the ampulla of the ductus deferens and the excretory duct of the seminal vesicle.

56. The normal position of the uterus is

(1) anteflexed.
(2) retroflexed.
(3) anteverted.
(4) retroverted.

57. Which of the following statements concerning perineal structures are true?

(1) The dorsal artery of the penis lies superficial to Buck's fascia.
(2) Buck's fascia intervenes between the bulb of the penis and Colles' fascia.
(3) The superficial perineal pouch lies between the pelvic fascia and the perineal membrane.
(4) The deep artery of the penis and the dorsal artery of the penis are the terminal branches of the internal pudendal artery.

58. Which of the following structures are part of the wall of the pelvic inlet?

(1) Promontory of the sacrum
(2) Ischiopubic rami
(3) Pectineal line
(4) Iliac crest

59. Which of the following statements concerning the pudendal nerve are true?

(1) It passes through the lesser sciatic foramen.
(2) It innervates the scrotum.
(3) It gives off branches that innervate the labia majora.
(4) It can be blocked by injecting an anesthetic near the inferior margin of the ischial spine.

60. Which of the following statements concerning the levator ani muscle are true?

(1) The puborectal sling (puborectalis) relaxes during defecation.
(2) The iliococcygeus muscle usually is the most developed muscle in the levator ani.
(3) The pubococcygeus may be torn or damaged during parturition, allowing the descent of the pelvic viscera.
(4) It forms the lateral wall of the ischiorectal fossa.

Answers and Explanations

1–D. The processus vaginalis is an extension of the peritoneum.

2–C. The skin of the urogenital triangle is innervated by the pudendal nerve, perineal branches of the posterior femoral cutaneous nerve, and the anterior scrotal or labial branches of the ilioinguinal nerve.

3–E. The round ligament of the uterus runs laterally from the uterus through the deep inguinal ring, inguinal canal, and superficial inguinal ring and becomes lost in the subcutaneous tissues of the labium majus. Thus, carcinoma of the uterus can spread directly to the labium majus by traveling in lymphatics that follow the ligament.

4–B. A tender swollen left testis may be produced by thrombosis in the left renal vein because the left testicular vein drains into the left renal vein.

5–C. The superior gluteal nerve and vessels leave the pelvis above the piriformis muscle. The sciatic nerve, internal pudendal vessels, inferior gluteal vessels and nerve, and posterior femoral cutaneous nerve leave the pelvis below the piriformis muscle.

6–D. The pubic arcuate ligament, tip of the coccyx, ischial tuberosities, and the sacrotuberous ligament all form part of the boundary of the perineum. The sacrospinous ligament forms a boundary of the lesser sciatic foramen.

7–B. The scrotum receives blood from the posterior scrotal branches of the internal pudendal arteries and the anterior scrotal branches of the external pudendal arteries. The lymph vessels from the scrotum drain into the superficial inguinal nodes, whereas the lymph vessels from the testis drain into the upper lumbar nodes.

8–A. The uterus is supported by the pelvic diaphragm, the round ligament of the uterus, and the lateral cervical (cardinal), rectouterine, transverse cervical, pubocervical, and sacrocervical ligaments.

9–A. The inferior pelvic aperture (pelvic outlet) is formed by the ischial tuberosities, inferior pubic rami, rami of the ischium, arcuate ligament of the pubis, and pubic symphysis.

10–D. The round ligament of the uterus is found in the inguinal canal along its course.

11–C. The uterine artery remains within the pelvic cavity. It does not leave the pelvic canal.

12–D. The perineal membrane is a connective tissue covering the inferior surface of the urogenital diaphragm.

13–C. The lumbosacral trunk is formed by part of the ventral ramus of the fourth lumbar nerve and the ventral ramus of the fifth lumbar nerve. This trunk contributes to the formation of the sacral plexus by joining the ventral ramus of the first sacral nerve in the pelvic cavity and does not leave the pelvic cavity.

14–A. The pelvic splanchnic nerves carry preganglionic efferent fibers that synapse in the ganglia of the inferior hypogastric plexus and in minute ganglia in the muscular walls of the pelvic organs.

15–D. The ureter runs under the uterine artery near the cervix; thus, the ureter is sometimes mistakenly ligated during pelvic surgery.

16–B. Branches of the uterine artery usually anastomose with branches of the ovarian artery. The prostatic plexus of veins is connected with the external pudendal vein and the superficial and deep dorsal veins of the penis. The inferior vesical artery is a branch of the internal iliac artery. The internal pudendal and inferior gluteal arteries are branches of the anterior division of the internal iliac artery.

17–D. The pelvic splanchnic nerves consist primarily of preganglionic parasympathetic neurons.

18–B. The ischiorectal fossa is bounded anteriorly by the superficial transverse perineal and profundus muscles, and posteriorly by the gluteus maximus muscle and the sacrotuberous ligament. It contains the inferior rectal nerve and vessels and the pudendal canal, which runs along the inner surface of its lateral wall.

19–E. The superior (deep) boundary of the superficial perineal space is the perineal membrane (inferior fascia of the urogenital diaphragm). Colles' fascia is the deep membranous layer of the superficial perineal fascia.

20–A. The urogenital diaphragm (deep perineal space) contains the transversus perineal profundus, the sphincter urethrae, the membranous urethra, and the bulbourethral (Cowper's) gland.

21–D. The superficial perineal space contains the ischiocavernosus muscle, the superficial transverse perineal muscle, the greater vestibular gland, and the bulb of the vestibule, which is covered with the bulbospongiosus.

22–C. The external anal sphincter has deep, superficial, and subcutaneous components.

23–D. Ducts from the prostate gland open on either side of the urethral crest.

24–E. A duct from a seminal vesicle joins the ductus deferens to form an ejaculatory duct.

25–C. The round ligament of the uterus enters the inguinal canal through the deep inguinal ring, emerges through the superficial inguinal ring, and becomes lost in the subcutaneous tissue of the labium majus. It contains smooth muscle fibers and runs lateral to the inferior epigastric vessels.

26–A. The urogenital diaphragm consists of the sphincter urethrae and deep transverse perineal muscles.

27–D. Lymphatic vessels from the testis and epididymis ascend along the testicular vessels in the spermatic cord through the inguinal canal and continue upward in the abdomen to drain into the upper lumbar nodes.

28–B. The deep dorsal vein of the penis enters the pelvis through a gap between the arcuate pubic ligament and the transverse perineal ligament.

29–C. The perineal body (central tendon of the perineum) is a fibromuscular node at the center of the perineum. It provides attachment for the bulbospongiosus, the superficial and deep transverse perineal, and the sphincter and ani externus muscles.

30–A. An ovarian ligament is not considered part of the broad ligament.

31–E. The prostatic utricle is a minute pouch on the summit of the seminal colliculus.

32–D. The superficial transversus perineal muscle is located in the superficial perineal space between the inferior fascia of the urogenital diaphragm and the superficial perineal (Colles') fascia. The bulbourethral (Cowper's) gland and the membranous urethra are found in the deep perineal pouch.

33–C. The ejaculatory duct opens into the prostatic urethra on the seminal colliculus. The ducts of the seminal vesicles and the ductus deferens form the ejaculatory duct. The ducts of the prostate gland open into the prostatic sinus, which is a groove on each side of the urethral crest. The duct of the bulbourethral gland opens into the lumen of the bulbous portion of the penile urethra.

34–E. The broad ligament contains the ovarian ligament, uterine artery, round ligament of the uterus, and uterine tube.

35–D. The ovarian artery, ureter, round ligament of the uterus, and the lumbosacral trunk all cross the pelvic brim. The uterine artery does not cross the pelvic brim. It arises from the internal iliac artery and then runs medially in the base of the broad ligament to the junction of the cervix and body of the uterus.

36–D. The levator ani and coccygeus muscles form the pelvic diaphragm, which is a major uterine support.

37–A. The bulbospongiosus covers or is in close proximity to the major vestibular glands.

38–B. The ischiocavernosus lies on the surface of the crus of the penis or clitoris.

39–E. The obturator internus muscle forms the lateral wall of the ischiorectal fossa.

40–C. The bulbourethral gland is embedded in the sphincter urethrae muscle.

41–B. The seminal vesicles are lobulated glandular structures and lie lateral to the ampullae of the ductus deferens, against the posterolateral aspect of the bladder.

42–D. The bulbourethral glands lie lateral to the membranous urethra in the deep perineal pouch.

43–C. The greater vestibular glands are found in the superficial perineal space (pouch).

44–A. The prostate gland opens its ducts into the prostatic sinus, which is a groove on each side of the urethral crest.

45–D. Ducts of the bulbourethral glands open into the bulb of the penis.

46–C. Ova are fertilized in the uterine tube, usually in the infundibulum or ampulla.

47–B. The uterus is supported by the cardinal ligaments.

48–C. The uterine tube opens into the peritoneal cavity.

49–A. The ovary is bounded by the external and internal iliac vessels. It is not enclosed by the peritoneum (broad ligament), but is attached along its posterior surface by the mesovarium.

50–E. The clitoris is attached to the pubic symphysis by the suspensory ligament.

51–E. The superficial inguinal lymph nodes drain the perineum, lower end of the anal canal, external genitalia, and lower part of the anterior abdominal wall.

52–E. Ejaculation occurs with the contraction of smooth muscle of the epididymal ducts and ductus deferens. During ejaculation, a sphincter at the neck of the bladder contracts (preventing sperm from entering the bladder and preventing urine from leaving it); the seminal vesicles, prostate gland, and bulbourethral glands contract to pump their secretions into the urethra; and semen is propelled through the ducts and out the external urethral opening.

53–E. The bulbourethral (Cowper's) gland is embedded in the substance of the sphincter urethrae muscle, on the posterolateral side of the membranous portion of the urethra in the deep perineal space. Its duct opens into the bulbous portion of the spongy (penile) urethra.

54–A. The sphincter urethrae muscle consists of skeletal (striated) muscle and forms a part of the urogenital diaphragm, which is covered by the superior and inferior fasciae of the urogenital diaphragm. It is innervated by the perineal branch of the pudendal nerve that originated from spinal cord segments S2–S4.

55–C. The ejaculatory duct is formed by the ampulla of the ductus deferens and the duct of the seminal vesicle, passes through the prostate gland, and opens into the prostatic urethra on the seminal colliculus.

56–B. The normal position of the uterus is anteflexed and anteverted.

57–C. The dorsal artery of the penis lies deep to Buck's fascia. The superficial perineal pouch lies between the membranous layer of Colles' (superficial perineal) fascia and the perineal membrane.

58–B. The pelvic inlet (pelvic brim) is bounded by the promontory of the sacrum, the arcuate line of the ilium, the pectineal line, the pubic crest, and the superior margin of the pubic symphysis.

59–E. The pudendal nerve leaves the pelvis through the greater sciatic foramen and enters the perineum through the lesser sciatic foramen near the inferior margin of the ischial spine. It provides sensory innervation to the scrotum or labium majus.

60–A. The lateral wall of the ischiorectal fossa is formed by the obturator internus muscle.

7

Back

<hr>

Back

I. Superficial Back

A. Superficial Muscles (Figure 7.1)

Muscle	Origin	Insertion	Nerve	Action
Trapezius	External occipital protuberance, superior nuchal line, ligamentum nuchae, spines of C7–T12	Spine of scapula, acromion, and lateral third of clavicle	Spinal accessory n., C3–C4	Adducts, rotates, elevates, and depresses scapula
Levator scapulae	Transverse processes of C1–C4	Medial border of scapula	Nerves to levator scapulae, C3–C4; dorsal scapular n.	Elevates scapula
Rhomboid minor	Spines of C7–T1	Root of spine of scapula	Dorsal scapular n., C5	Adducts scapula
Rhomboid major	Spines of T2–T5	Medial border of scapula	Dorsal scapular n.	Adducts scapula
Latissimus dorsi	Spines of T5–T12, thoracodorsal fascia, iliac crest, ribs 9–12	Floor of bicipital groove of humerus	Thoracodorsal n.	Adducts, extends, and rotates arm medially
Serratus posterior–superior	Ligamentum nuchae, supraspinal ligament, and spines of C7–T3	Upper border of ribs 2–5	Intercostal n., T1–T4	Elevates ribs
Serratus posterior–inferior	Supraspinous ligament and spines of T11–L3	Lower border of ribs 9–12	Intercostal n., T9–12	Depresses ribs

B. Triangles and Fascia

1. **Triangle of Auscultation** (see Figure 7.1)

   —is bounded by the upper border of the latissimus dorsi, the lateral border of the trapezius, and the medial border of the scapula.

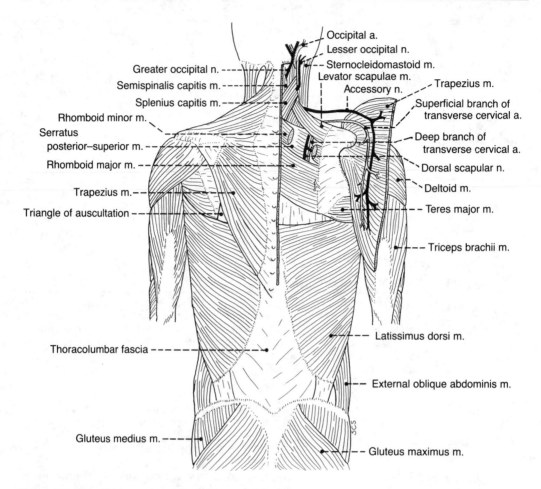

Figure 7.1. Superficial muscles of the back.

—its floor is formed by the rhomboid major muscle.
—is the site where breathing sounds can be heard most clearly.

  2. **Lumbar Triangle**
      —is formed by the iliac crest, latissimus dorsi muscle, and posterior free border of the external oblique abdominis muscle.

  3. **Thoracolumbar (Lumbodorsal) Fascia**
      —invests the deep muscles of the back.
      —its anterior layer lies anterior to the erector spinae muscle and attaches to the vertebral transverse process.
      —its posterior layer lies posterior to the erector spinae and attaches to the spinous processes.

II. Deep Back

A. Deep or Intrinsic Muscles

  1. **Muscles of the Superficial Layer: Spinotransverse Group**
      —consist of the splenius capitis and the splenius cervicis.
      —originate from the spinous processes and insert into the transverse processes. (The splenius capitis inserts on the mastoid process and the superior nuchal line.)

—are innervated by the dorsal primary rami of the middle and lower cervical spinal nerves.

—act to rotate the head and neck toward the same side and to extend the head and the trunk.

2. Muscles of the Intermediate Layer: Sacrospinalis Group

—originate from the sacrum, ilium, ribs, and vertebral spines.

—are innervated by the dorsal primary rami of the spinal nerves.

—act to flex, extend, and rotate the vertebral column and the head.

—consist of the erector spinae (sacrospinales), which are divided into three columns: iliocostalis, longissimus, and spinalis.

a. Iliocostalis (Lateral Column)

—inserts on the ribs and cervical transverse processes.

b. Longissimus (Medial Column)

—inserts into the ribs, transverse processes, and mastoid process.

c. Spinalis (Most Medial Column)

—arises from and inserts into the spinous processes.

3. Muscles of the Deep Layer: Transversospinalis Group

—originate from the transverse processes and insert into the spinous processes. (The semispinalis capitis inserts into the skull. The long rotators run from the transverse processes to spinous processes two vertebrae above, and the short rotators run from the transverse processes to the spinous processes of adjacent vertebrae.)

—are innervated by the dorsal primary rami of the spinal nerves.

—act to rotate and extend the head and neck and trunk.

—consist of:

a. Semispinalis capitis, cervicis, and thoracis

b. Multifidus (deep to semispinalis)

c. Rotators

B. Segmental Muscles

—are innervated by the dorsal primary rami of the spinal nerves.

—consist of:

1. Interspinales

—run between adjacent spinous processes and aid in extension of the vertebral column.

2. Intertransversarii

—run between adjacent transverse processes and aid in lateral flexion of the vertebral column.

III. Suboccipital Area (Figure 7.2)

A. Suboccipital Triangle

—is bound medially by the rectus capitis posterior major muscle, laterally by the oblique capitis superior muscle, and inferiorly by the oblique capitis inferior muscle.

—its roof is formed by the semispinalis capitis and longissimus capitis muscles.

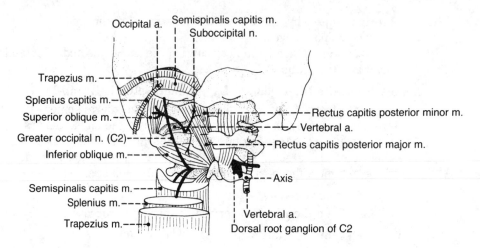

Figure 7.2. Suboccipital triangle.

—its floor is formed by the posterior arch of the atlas and posterior atlanto-occipital membrane.
—contains the vertebral artery and suboccipital nerve and vessels.

B. Suboccipital Nerve

—is derived from the dorsal ramus of C1 and emerges between the vertebral artery above and the posterior arch of the atlas below.
—supplies the muscles of the suboccipital triangle and semispinalis capitis.

C. Suboccipital Muscles

Muscle	Origin	Insertion	Nerve	Action
Rectus capitis posterior major	Spine of axis	Lateral portion of inferior nuchal line	Suboccipital n.	Extends, rotates, and flexes head laterally
Rectus capitis posterior minor	Posterior tubercle of atlas	Occipital bone below inferior nuchal line	Suboccipital n.	Extends and flexes head laterally
Obliquus capitis superior	Transverse process of atlas	Occipital bone above inferior nuchal line	Suboccipital n.	Extends, rotates, and flexes head laterally
Obliquus capitis inferior	Spine of axis	Transverse process of atlas	Suboccipital n.	Extends head and rotates it laterally

D. Joints

1. Atlanto-Occipital Joint

—is an ellipsoidal synovial joint.
—occurs between the superior articular facets of the atlas and the occipital condyles.
—is involved primarily in flexion, extension, and lateral bending of the head.

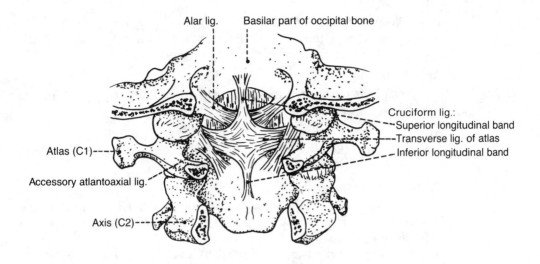

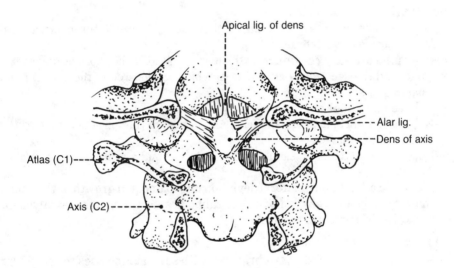

Figure 7.3. Ligaments of the atlas and the axis.

2. Atlantoaxial Joints

—are synovial joints.

—are three in number: **two planes,** which are between the superior and inferior articular facets of the atlas and axis, and **one pivot,** which is the median joint between the dens of the axis and the anterior arch of the atlas.

—are involved in rotation of the head.

E. Components of the Occipitoaxial Ligament (Figure 7.3)

1. Cruciform Ligament

a. Transverse Ligament

—runs between the lateral masses of the atlas, arching over the dens of the axis.

> **b. Longitudinal Ligament**
> —extends from the dens of the axis to the anterior aspect of the foramen magnum and to the body of the axis.

> **2. Apical Ligament**
> —extends from the apex of the dens to the anterior aspect of the foramen magnum (of the occipital bone).

> **3. Alar Ligament**
> —extends from the apex of the dens to the tubercle of the medial side of the occipital condyle.

> **4. Tectorial Membrane**
> —is an upward extension of the posterior longitudinal ligament from the body of the axis to the basilar part of the occipital bone anterior to the foramen magnum.
> —covers the posterior surface of the dens and the apical, alar, and cruciform ligaments.

IV. Vertebral Column

—consists of 33 vertebrae (7 cervical, 12 thoracic, 5 lumbar, 5 fused sacral, and 4 fused coccygeal vertebrae).

—presents the **primary curvatures,** which are located in the thoracic and sacral regions, and the **secondary curvatures,** which are located in the cervical and lumbar regions.

—may have **abnormal curvatures:**
—**Kyphosis** (hunchback): an abnormal exaggerated thoracic curvature of the vertebral column.
—**Lordosis** (swayback or saddle back): an abnormal accentuation of lumbar curvature.
—**Scoliosis:** a condition of lateral deviation due to unequal growth of the spinal column, pathologic erosion of vertebral bodies, or asymmetric paralysis or weakness of vertebral muscles.

A. Typical Vertebra (Figure 7.4)

—consists of a body and a vertebral arch with several processes for muscular and articular attachments.

> **1. Body**
> —is a short cylinder, functions to support weight, and is separated and also bound together by the intervertebral disks.
> —also articulates with the heads of the corresponding and subjacent ribs.

> **2. Vertebral Arch**
> —consists of paired **pedicles** laterally and paired **laminae** posteriorly.
> —gives rise to **seven processes:** one spinous, two transverse, and four articular.
> —failure of the vertebral arches to fuse results in **spina bifida,** which is classified as follows:
> **a. Spina bifida occulta:** bony defect only.
> **b. Meningocele:** protrusion of the meninges through the unfused arch of the vertebra.
> **c. Meningomyelocele:** protrusion of the spinal cord as well as the meninges.

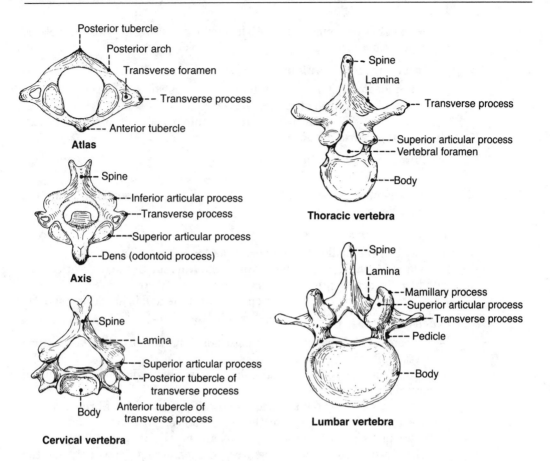

Figure 7.4. Typical cervical, thoracic, and lumbar vertebrae.

3. Processes

a. Spinous Process

—projects posteriorly from the vertebral arch.

b. Transverse Processes

—project on each side from the junction of the pedicle and the lamina and articulate with the tubercles of ribs 1 to 10, in the thoracic region.

c. Costal (Facets) Processes

—arise on the sides of a vertebral body anterior to the pedicle and articulate with the heads of the ribs.

d. Mamillary Processes

—are tubercles on the superior articular processes of the lumbar vertebrae.

4. Foramina

a. Vertebral Foramina

—are formed by the vertebral bodies and vertebral arches (pedicles and laminae). A series of vertebral foramina forms the **vertebral canal,** which contains the spinal cord, its meningeal coverings, and the associated vessels as well as the nerve roots (caudal to L2).

—transmit the spinal cord with its meningeal coverings and associated vessels.

b. Intervertebral Foramina
—are located between the inferior and superior surfaces of the pedicles of the adjacent vertebrae.
—transmit the spinal nerves and accompanying vessels as they exit the vertebral canal.

c. Transverse Foramina
—are present in each transverse process of the cervical vertebra.
—transmit the vertebral artery (except for C7), vertebral vein, and autonomic nerves.

B. Intervertebral Disk
—lies between the bodies of two vertebrae from the axis to the sacrum.
—consists of a central mucoid substance (the **nucleus pulposus**) with a surrounding fibrocartilaginous lamina (the **anulus fibrosus**).
—is important in movements between the vertebrae and in absorbing shocks.

1. Nucleus Pulposus
—is a remnant of the embryonic notochord and is situated in the central portion of the intervertebral disk.
—is composed of reticular and collagenous fibers embedded in mucoid material.
—may protrude or extrude (herniate) through the anulus fibrosus, thereby compressing the roots of the spinal nerve.
—functions as a shock-absorbing mechanism.
—is involved in equalizing pressure and exchanging fluids between the disk and the capillaries of the vertebrae.

2. Anulus Fibrosus
—consists of concentric layers of fibrous tissue and fibrocartilage.
—binds the vertebral column together, retains the nucleus, and permits a small amount of movement.
—functions as a shock-absorbing mechanism.

C. Regional Characteristics of Vertebrae (see Figure 7.4)

1. First Cervical Vertebra (Atlas)
—supports the skull.
—has no body and no spinous process but consists of anterior and posterior arches and paired transverse processes.
—articulates superiorly with the occipital condyles of the skull to form the atlanto-occipital joints and inferiorly with the axis to form the atlantoaxial joints.

2. Second Cervical Vertebra (Axis)
—has the smallest transverse process and is characterized by the presence of the dens (odontoid process).
Dens of the Axis
—projects superiorly from the body of the axis.
—articulates with the anterior arch of the atlas and thus forms the pivot around which the atlas rotates.

—is supported by the cruciform, apical, and alar ligaments and the tectorial membrane.

3. **Seventh Cervical Vertebra**
   —is called the vertebra prominens because of its long spinous process.
   —is nearly horizontal (usually is not bifid) and forms a visible protrusion.
   —gives attachment to the ligamentum nuchae.

4. **Fifth Lumbar Vertebra**
   —has the largest body of the vertebrae. (Lumbar vertebrae are distinguished by their large bodies.)
   —is characterized by a strong, massive transverse process and has the mamillary and accessory processes.

5. **Sacrum** (Figure 7.5)
   —is a large, triangular, wedge-shaped bone composed of five fused sacral vertebrae.
   —has four pairs of foramina for the exit of the ventral and dorsal primary rami of the first four sacral nerves.
   —forms the posterior part of the pelvis and provides the strength and stability of the pelvis.
   —exhibits its **promontory,** which is the prominent anterior edge of the first sacral vertebra.
   —is characterized by the following structures:
   a. The **ala** is formed by the fused transverse processes and fused costal processes.
   b. The **median sacral crest** is formed by the fused spinous processes.
   c. The **sacral hiatus** is formed by the failure of the laminae of the fifth sacral vertebra to meet.
   d. The **sacral cornu** or horn is formed by the pedicles of the fifth sacral vertebra and is an important landmark for locating the sacral hiatus for the administration of caudal anesthesia.

6. **Coccyx**
   —is a wedge-shaped bone formed by the union of the four coccygeal vertebrae and provides attachment for the coccygeus and levator ani muscles.

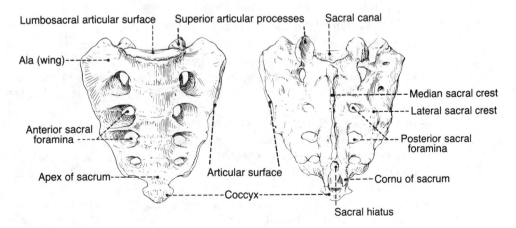

Figure 7.5. Sacrum.

D. Ligaments (see Figure 7.3)

  1. Anterior Longitudinal Ligament
  —runs from the skull to the sacrum on the anterior surface of the vertebral bodies and intervertebral disks.
  —is narrowest at the upper end but widens as it descends.
  —limits extension of the vertebral column, supports the anulus fibrosus
  anteriorly, and resists gravitational pull, preventing hyperextension of
  the vertebral column.

  2. Posterior Longitudinal Ligament
  —interconnects the vertebral bodies and intervertebral disks posteriorly
  and narrows as it descends.
  —supports the posterior aspect of the vertebral column and also supports
  the anulus fibrosus posteriorly.
  —limits flexion of the vertebral column and resists gravitational pull,
  preventing hyperflexion of the vertebral column.

  3. Ligamentum Flavum
  —connects the laminae of two adjacent vertebrae and functions to main-
  tain the upright posture.
  —may be pierced during lumbar (spinal) puncture.

  4. Ligamentum Nuchae
  —is a triangular-shaped median fibrous septum between the muscles on
  the two sides of the posterior aspect of the neck.
  —is an upward extension of the supraspinal ligament that extends from
  the seventh cervical vertebra to the external occipital protuberance and
  crest.
  —is attached to the posterior tubercle of the atlas and to the spinous
  processes of the other cervical vertebrae.

E. Vertebral Venous System
—is a valveless plexiform consisting of interconnecting channels.

  1. Internal Vertebral Venous Plexus
  —lies in the epidural space between the wall of the vertebral canal and
  the dura mater, forms anterior and posterior ladder-like configurations,
  and drains into segmental veins by the intervertebral veins that pass
  through the intervertebral and sacral foramina.

  2. External Vertebral Venous Plexus
  —its **anterior** part consists of anastomosing longitudinal and transverse
  veins, lying in front of the vertebral column. These veins communicate
  with the intervertebral veins, which leave through the intervertebral
  and sacral foramina, and the basivertebral veins, which lie within the
  vertebral bodies.
  —its **posterior** part lies on the vertebral arch and communicates with the
  intervertebral veins and the internal plexus.
  —also communicates above with the cranial dural sinus, below with the
  pelvic vein, and in the thoracic and abdominal regions with both the
  azygos and caval systems.
  —is thought to be the route of early metastasis of carcinoma from the
  lung, breast, and prostate gland to bones and the central nervous
  system.

V. Spinal Cord and Associated Structures

A. Spinal Cord

—is cylindrical, occupies about the upper two-thirds of the vertebral canal, and is enveloped by the three meninges.

—has cervical and lumbar enlargements for nerve supply of the upper and lower limbs, respectively.

—has gray matter located in the interior (in contrast to the cerebral hemispheres) and is surrounded by white matter.

—has a conical end known as the **conus medullaris**.

—grows much more slowly than the bony vertebral column during fetal development, and thus its end gradually shifts to a higher level.

—ends at the level of L2 in the adult and at the level of L3 in the newborn.

—receives blood from the anterior spinal artery and two posterior spinal arteries as well as from branches of the vertebral, cervical, and posterior intercostal and lumbar arteries.

B. Spinal Nerves

—have 31 pairs of nerves (8 cervical, 12 thoracic, 5 lumbar, 5 sacral, and 1 coccygeal).

—are divided into the **dorsal primary rami,** which innervate the skin and deep muscles of the back; the **ventral primary rami,** which form the plexuses (C1 to C4, cervical; C5 to T1, brachial; L1 to L4, lumbar; and L4 to S4, sacral); and the **intercostal** (T1 to T11) and **subcostal** (T12) **nerves**.

—are connected with the sympathetic chain ganglia by rami communicantes.

—are mixed nerves, containing all of the general functional components (i.e., general somatic afferent [GSA]; general somatic efferent [GSE]; general visceral afferent [GVA]; and general visceral efferent [GVE]).

—contain sensory (GSA and GVA) fibers with cell bodies in the dorsal root ganglion.

—contain motor (GSE) fibers with cell bodies in the anterior horn of the spinal cord.

—contain preganglionic sympathetic (GVE) fibers with cell bodies in the intermediolateral cell column in the lateral horn of the spinal cord (segments between T1 and L2).

—contain preganglionic parasympathetic (GVE) fibers with cell bodies in the intermediolateral cell column of the spinal cord segments between S2 to S4. These GVE fibers leave the sacral nerves via the pelvic splanchnic nerves.

C. Meninges and Spinal Cord (Figure 7.6)

1. Pia Mater

—is closely applied to the spinal cord, and thus it cannot be dissected from it. It also enmeshes blood vessels on the surfaces of the spinal cord.

—has lateral extensions (**denticulate ligaments**) between nerve roots of spinal nerves.

2. Arachnoid Mater

—is a filmy, transparent, spidery layer connected to the pia mater by web-like trabeculations.

—forms the **subarachnoid space**, the space between the arachnoid layer and the pia mater, that is filled with cerebrospinal fluid.

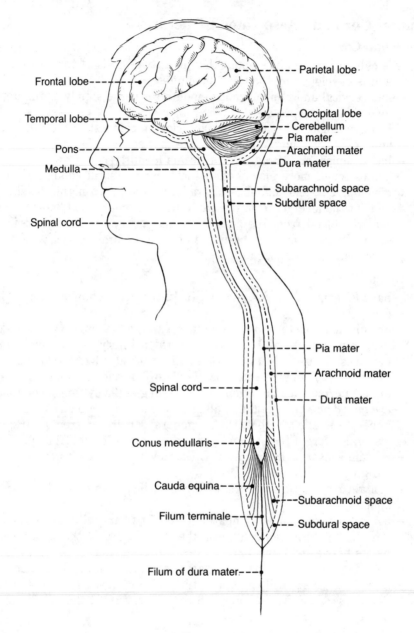

Frontal lobe

Parietal lobe

Temporal lobe

Occipital lobe

Cerebellum

Pia mater

Pons

Arachnoid mater

Medulla

Dura mater

Subarachnoid space

Subdural space

Spinal cord

Pia mater

Arachnoid mater

Spinal cord

Dura mater

Conus medullaris

Cauda equina

Subarachnoid space

Filum terminale

Subdural space

Filum of dura mater

Figure 7.6. Meninges.

3. Dura Mater

—is the tough, fibrous, outermost layer of the meninges.

—the **subdural space** is internal to it (between the arachnoid and dura).

—the **epidural space** is external to it and contains the internal vertebral venous plexus and epidural fat.

D. Structures Associated with the Spinal Cord

1. Cauda Equina ("Horse's Tail")

—is formed by a great lash of dorsal and ventral roots (of the lumbar and sacral spinal nerves) that surround the filum terminale.

—is located within the subarachnoid space, below the level of the conus medullaris.

—is free to float in the cerebrospinal fluid and therefore is not damaged during a spinal tap.

2. **Denticulate Ligaments**

—are a lateral extension of the pia through the arachnoid, attaching to the inner surface of the dura mater.

—consist of 21 pairs of tooth-like processes that attach to the dura mater between dorsal and ventral roots of the spinal nerves.

—help to suspend the spinal cord within the subarachnoid space.

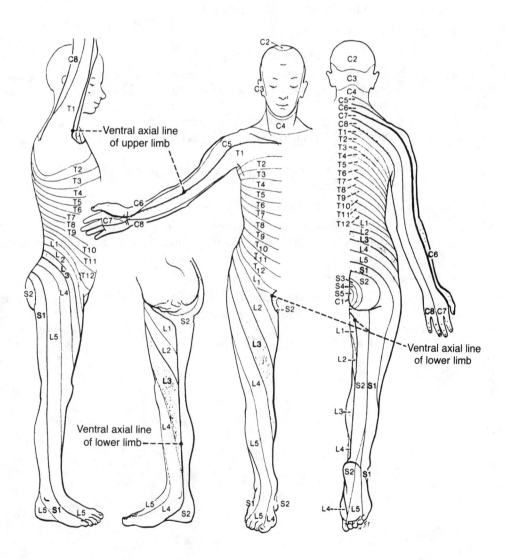

Figure 7.7. Dermatomes of the body.

3. Filum Terminale of the Spinal Cord

—is a prolongation of the pia mater from the tip (conus medullaris) of the spinal cord at the level of L2.

—lies in the midst of the cauda equina and ends at the level of S2 by attaching to the apex of the dural sac.

—blends with the dura at the apex of the dural sac, and then the dura continues downward as the **filum of the dura mater** (or **coccygeal ligament**), which is attached to the dorsum of the coccyx.

4. Cerebrospinal Fluid

—is contained in the subarachnoid space between the arachnoid and pia mater.

—is formed by vascular choroid plexuses in the ventricles of the brain.

—circulates through the ventricles, enters the subarachnoid space, and eventually filters into the venous system.

E. Clinical Considerations

1. Lumbar Puncture (Spinal Tap)

—is the tapping of the subarachnoid space in the lumbar region, usually between the laminae of the third and fourth or fourth and fifth lumbar vertebrae.

—allows the measurement of the pressure of the cerebrospinal fluid and withdrawal of some of the fluid for bacteriologic and chemical examinations.

—allows introduction of anesthesia, drugs, or radiopaque material into the subarachnoid space.

2. Caudal (Epidural) Anesthesia

—is used to block the spinal nerves in the epidural space by injection of local anesthetic agents via the sacral hiatus located between the sacral cornua.

F. Dermatome (Figure 7.7)

—is an area of skin innervated by sensory fibers derived from a particular spinal nerve or segment of the spinal cord. Knowledge of the segmental innervation is useful clinically to produce a region of anesthesia or to determine which nerve has been damaged.

Review Test

DIRECTIONS: Each of the numbered items or incomplete statements in this section is followed by answers or by completions of the statement. Select the **one** lettered answer or completion that is **best** in each case.

1. Each statement below concerning the suboccipital triangle is true EXCEPT

(A) the rectus capitis posterior major bounds the suboccipital triangle.
(B) the rectus capitis posterior minor does not bound the suboccipital triangle.
(C) the obliquus capitis superior and obliquus capitis inferior bound the suboccipital triangle.
(D) the greater occipital nerve innervates the muscles of the suboccipital triangle.
(E) within the triangle, the vertebral artery occupies the grooves on the superior surface of the posterior arch of the atlas.

2. Each muscle below is innervated by dorsal primary rami EXCEPT the

(A) semispinalis capitis.
(B) splenius.
(C) serratus posterior–superior.
(D) iliocostalis.
(E) spinalis.

3. In the spinal region, cerebrospinal fluid is found

(A) in the epidural space.
(B) in the subdural space.
(C) between the pia mater and the spinal cord.
(D) in the subarachnoid space.
(E) between the arachnoid and dura mater.

4. Each of the following statements concerning the multifidus muscle is true EXCEPT

(A) it is a transversospinalis muscle.
(B) it is heaviest in the lumbar region.
(C) it is attached to the spinous processes.
(D) it is innervated by the ventral primary rami of the spinal nerve.
(E) it lies deep to the semispinalis muscle.

5. Which of the following statements concerning the dura mater is true?

(A) It adheres closely to the surface of the brain and spinal cord.
(B) It is the innermost layer of the meninges.
(C) It lies internal to the epidural space.
(D) It forms the filum terminale.
(E) None of the above.

DIRECTIONS: Each group of items in this section consists of lettered options followed by a set of numbered items. For each item, select the **one** lettered option that is most closely associated with it. Each lettered option may be selected once, more than once, or not at all.

Questions 6–10

Match each description on the right with the most appropriate structure.

(A) Conus medullaris
(B) Dorsal root ganglion
(C) Cauda equina
(D) Internal vertebral venous plexus
(E) Subarachnoid space

6. Is found within the vertebral canal external to the dura mater of the spinal cord

7. Contains the cerebrospinal fluid

8. Its inferior limit is the lower border of the first lumbar vertebra

9. Contains the cell bodies of sensory nerve fibers

10. Is formed by the roots of the lumbar and sacral nerves

Questions 11–15

Match each description below with the most appropriate component or condition of the spinal column.

(A) Nucleus pulposus
(B) Scoliosis
(C) Anulus fibrosus
(D) Intervertebral disk
(E) Lordosis

11. Binds the vertebral column together; permits a small amount of movement; retains the nucleus pulposus

12. Is situated in the central portion of the intervertebral disk

13. Is situated between the bodies of two vertebrae

14. An abnormal accentuation of the lumbar curvature

15. A lateral deviation due to unequal growth of the spinal column

Questions 16–20

Match each description below with the most appropriate ligament.

(A) Longitudinal ligament of the cruciform ligament
(B) Apical ligament
(C) Alar ligament
(D) Ligamentum flavum
(E) Denticulate ligament

16. Extends from the apex of the dens to the anterior aspect of the foramen magnum

17. Extends from the dens of the axis to the anterior aspect of the foramen magnum and the body of the axis

18. Extends from the apex of the dens to the medial side of the occipital bone

19. A lateral extension of the pia mater

20. Connects the laminae of adjacent vertebrae

DIRECTIONS: Each question below contains four suggested answers of which **one or more** is correct. Choose answer

A if **1, 2, and 3** are correct
B if **1 and 3** are correct
C if **2 and 4** are correct
D if **4** is correct
E if **1, 2, 3, and 4** are correct

21. Which of the following ligaments are anterior to the spinal cord?

(1) Anterior longitudinal ligament
(2) Supraspinal and interspinal ligaments
(3) Posterior longitudinal ligament
(4) Ligamentum flavum

22. When withdrawing cerebrospinal fluid by lumbar puncture, which of the following structures are penetrated by the needle?

(1) Dura mater
(2) Anulus fibrosus
(3) Arachnoid layer
(4) Pia mater

23. Which of the following statements concerning the spinal epidural space are true?

(1) It may be entered via the sacral hiatus.
(2) It is continuous from the sacrum to the base of the skull.
(3) It contains a venous plexus.
(4) It contains cerebrospinal fluid.

24. Which of the following statements concerning the vertebral venous plexus are true?

(1) It communicates with the cranial venous sinuses.
(2) It is composed of thin-walled veins containing many valves.
(3) It communicates with veins of the thorax, abdomen, and pelvis.
(4) It is located in the subdural space.

25. Which of the following statements concerning the anterior longitudinal ligament are true?

(1) It lies between the intervertebral disk and the dura.
(2) It extends from the sacrum to the atlas.
(3) It ends superiorly as the tectorial membrane.
(4) It limits extension of the vertebral column.

26. Which of the following muscles are inner-vated by the suboccipital nerve?

(1) Rectus capitis posterior major
(2) Rectus capitis lateralis
(3) Semispinalis capitis
(4) Rectus capitis anterior

27. The atlanto-occipital joint is primarily in-volved in

(1) flexion of the head.
(2) rotation of the head.
(3) extension of the head.
(4) bending of the head.

Answers and Explanations

1–D. The suboccipital triangle is bounded by the rectus capitis posterior major muscle and the obliquus capitis superior and inferior muscles, which are innervated by the suboccipital nerve. The greater occipital nerve is a sensory nerve.

2–C. The serratus posterior–superior muscle is innervated by the anterior primary rami of the upper four thoracic nerves. The dorsal (posterior) primary rami of the spinal nerves innervate the deep muscles of the back.

3–D. Cerebrospinal fluid is found in the subarachnoid space, which is a wide interval between the arachnoid and the pia mater.

4–D. The multifidus muscle is innervated by the dorsal primary rami of the spinal nerves.

5–C. The dura mater is the tough, fibrous outermost layer of the meninges and is external to the subdural space, which contains a film of fluid to moisten its walls. The dura mater is internal to the epidural space, which contains the internal vertebral venous plexus and the middle meningeal arteries in the cranial cavity. The pia mater adheres closely to the surface of the brain and spinal cord and forms the filum terminale (internus) by its prolongation from the end of the conus medullaris.

6–D. The epidural space contains the internal vertebral venous plexus.

7–E. The cerebrospinal fluid is found in the subarachnoid space between the pia mater and the arachnoid mater.

8–A. The conus medullaris is a conical end of the spinal cord at the level of the first or second lumbar vertebra.

9–B. The dorsal root ganglion contains the cell bodies of the visceral and somatic sensory nerve fibers.

10–C. The dorsal and ventral roots of the lumbar and sacral nerves form the cauda equina.

11–C. The anulus fibrosus is the fibrocartilaginous ring that forms the circumference of the intervertebral disk; thus, retaining the nucleus pulposus. It binds the vertebral column together and permits a small amount of movement.

12–A. The nucleus pulposus is a fibrous semigelatinous material in the central portion of the intervertebral disk and is surrounded by the anulus fibrosus.

13–D. The intervertebral disk consists of the nucleus pulposus and the anulus fibrosus and lies between the bodies of two vertebrae.

14–E. Lordosis (sway back) is an abnormal accentuation of the lumbar curvature.

15–B. Scoliosis is a lateral deviation due to unequal growth of the spinal column, pathologic erosion of vertebral bodies, or asymmetric paralysis or weakness of vertebral muscles.

16–B. The apical ligament of the dens extends from the tip of the dens to the anterior margin of the foramen magnum of the occipital bone.

17–A. The longitudinal ligament of the cruciform ligament extends from the body of the axis, over the dens, to the occipital bone (anterior margin of the foramen magnum).

18–C. The alar ligament extends from the side of the apex of the dens to the medial side of the occipital condyle (lateral margin of the foramen magnum).

19–E. The denticulate ligament is a lateral extension of the pia mater. It helps anchor the spinal cord within the vertebral canal.

20–D. The ligamentum flavum extends between the laminae of adjacent vertebrae.

21–B. The anterior and posterior longitudinal ligaments lie anterior to the spinal cord; the supraspinal and interspinal ligaments lie posterior to the spinal cord. The ligamentum flavum connects the laminae of the two adjacent vertebrae and lies posterolateral to the spinal cord.

22–B. The cerebrospinal fluid is located in the subarachnoid space, between the arachnoid and pia mater. The anulus fibrosus is the fibrocartilaginous ring forming the circumference of the intervertebral disk.

23–A. The spinal epidural space is external to the dura mater and contains the internal vertebral venous plexus. It extends from the base of the skull to the sacrum and can be entered through the sacral hiatus for caudal (extradural) anesthesia.

24–B. The vertebral venous system consists of thin-walled veins; however, these veins do not have valves. The vertebral venous system lies in the epidural space and communicates with the cranial venous sinuses and paravertebral veins in the thorax, abdomen, and pelvis.

25–C. The anterior longitudinal ligament extends from the base of the skull to the sacrum and limits extension of the vertebral column.

26–B. The suboccipital nerve (dorsal primary ramus of C1) supplies the muscles of the suboccipital area (e.g., the rectus capitis posterior major) and the semispinalis capitis muscle. The rectus capitis anterior and rectus capitis lateralis muscles are innervated by the ventral primary rami of the first and second cervical nerves.

27–B. The atlanto-occipital joint is an ellipsoid (condyloid) synovial joint and is involved primarily in flexion and extension of the head.

8

Head and Neck

Structures of the Neck

I. Cervical Triangles (Figure 8.1)

A. Posterior Triangle

—is bounded by the posterior border of the sternocleidomastoid (sternomastoid) muscle, the anterior border of the trapezius muscle, and the superior border of the clavicle.

—its roof is formed by the platysma and the investing layer of the deep cervical fascia.

—its floor is formed by the splenius capitis and levator scapulae muscles, and the anterior, middle, and posterior scalene muscles.

—contains the accessory nerve, cutaneous branches of the cervical plexus, external jugular vein, posterior (inferior) belly of the omohyoid, roots and trunks of the brachial plexus, and transverse cervical and suprascapular vessels.

—is further divided into the occipital and subclavian (supraclavicular or omoclavicular) triangles by the omohyoid posterior belly.

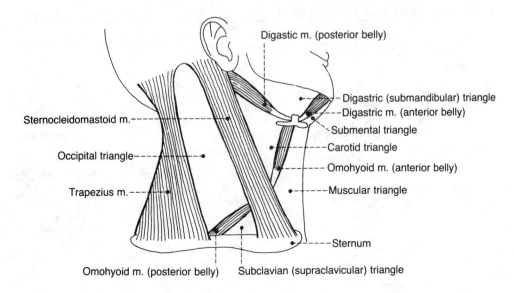

Figure 8.1. Subdivisions of the cervical triangle.

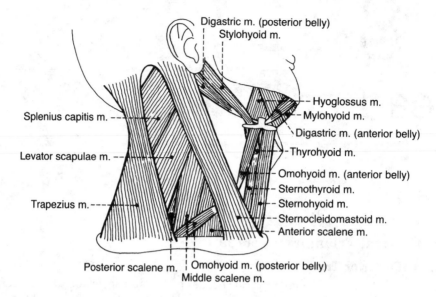

Figure 8.2. Muscles of the cervical triangle.

B. Anterior Triangle

—is bounded by the anterior border of the sternocleidomastoid, the antero-median line of the neck, and the inferior border of the mandible.

—its roof is formed by the platysma and the investing layer of the deep cervical fascia.

—is further divided by the omohyoid anterior belly and the digastric anterior and posterior bellies into the digastric (submandibular), submental (su-prahyoid), carotid, and muscular (inferior carotid) triangles.

II. Muscles of the Neck (Figure 8.2)

Muscle	Origin	Insertion	Nerve	Action
Cervical muscles:				
Platysma	Superficial fascia over upper part of deltoid and pectoralis major	Mandible; skin and muscles over mandible and angle of mouth	Facial n.	Depresses lower jaw and lip and angle of mouth; wrinkles skin of neck
Sternocleido-mastoid	Manubrium sterni and medial one-third of clavicle	Mastoid process and lateral one-half of superior nuchal line	Spinal accessory n.; C2–C3 (sensory)	Singly turns face toward opposite side; together flex head, raise thorax

(Continued on next page)

Muscle	Origin	Insertion	Nerve	Action
Suprahyoid muscles:				
Digastric	Anterior belly from digastric fossa of mandible; posterior belly from mastoid notch	Intermediate tendon attached to body of hyoid	Posterior belly by facial n.; anterior belly by mylohyoid n. of trigeminal n.	Elevates hyoid and tongue; depresses mandible
Mylohyoid	Mylohyoid line of mandible	Median raphe and body of hyoid bone	Mylohyoid n. of trigeminal n.	Elevates hyoid and tongue; depresses mandible
Stylohyoid	Styloid process	Body of hyoid	Facial n.	Elevates hyoid
Geniohyoid	Genial tubercle of mandible	Body of hyoid	C1 via hypoglossal n.	Elevates hyoid and tongue
Infrahyoid muscles:				
Sternohyoid	Manubrium sterni and medial end of clavicle	Body of hyoid	Ansa cervicalis	Depresses hyoid and larynx
Sternothyroid	Manubrium sterni; first costal cartilage	Oblique line of thyroid cartilage	Ansa cervicalis	Depresses thyroid cartilage and larynx
Thyrohyoid	Oblique line of thyroid cartilage	Body and greater horn of hyoid	C1 via hypoglossal n.	Depresses and retracts hyoid and larynx
Omohyoid	Inferior belly from medial lip of suprascapular notch and suprascapular ligament; superior belly from intermediate tendon	Inferior belly to intermediate tendon; superior belly to body of hyoid	Ansa cervicalis	Depresses and retracts hyoid and larynx

III. Nerves of the Neck (Figures 8.3 and 8.4)

A. Accessory Nerve

—is formed by the union of cranial and spinal roots.

—its cranial roots arise from the medulla oblongata below the roots of the vagus.

—its spinal roots arise from the cervical segment of the spinal cord between C1 and C3 (or C1 and C7) and unite to form a trunk that extends between the dorsal and ventral roots of the spinal nerves in the vertebral canal and passes through the foramen magnum.

—both spinal and cranial portions traverse the jugular foramen, where they interchange fibers.

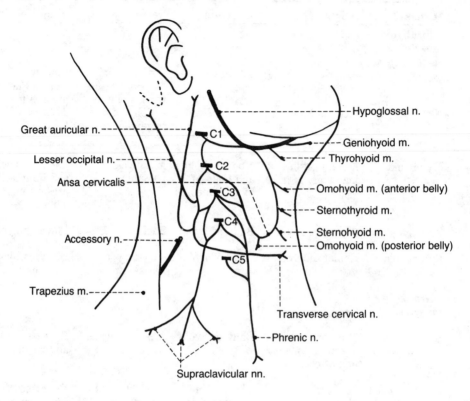

Great auricular n.

Lesser occipital n.

Ansa cervicalis

Accessory n.

Trapezius m.

Hypoglossal n.

Geniohyoid m.

Thyrohyoid m.

Omohyoid m. (anterior belly)

Sternothyroid m.

Sternohyoid m.
Omohyoid m. (posterior belly)

Transverse cervical n.

Phrenic n.

Supraclavicular nn.

C1
C2
C3
C4
C5

Figure 8.3. Cervical plexus.

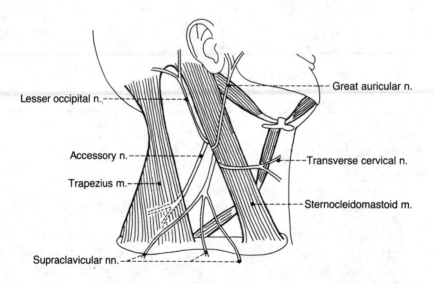

Lesser occipital n.

Accessory n.

Trapezius m.

Supraclavicular nn.

Great auricular n.

Transverse cervical n.

Sternocleidomastoid m.

Figure 8.4. Cutaneous branches of the cervical plexus.

—its cranial portion contains motor fibers that join the vagus nerve and supply the soft palate, pharyngeal constrictors, and larynx.

—its spinal portion supplies the sternocleidomastoid and trapezius muscles.

—lies on the levator scapulae muscle in the posterior cervical triangle and then passes deep to the trapezius muscle.

B. Cervical Plexus

—is formed by the ventral rami of the first four cervical nerves.

—gives off motor nerves, including the ansa cervicalis to the infrahyoid muscles, the phrenic nerve to the diaphragm, and twigs to the longus capitis and cervicis, sternomastoid, trapezius, levator scapulae, and scalene muscles.

—gives rise to cutaneous branches, including the lesser occipital, great auricular, transverse cervical, and supraclavicular nerves.

1. Lesser Occipital Nerve (C2)

—ascends along the posterior border of the sternocleidomastoid to the scalp behind the auricle.

2. Great Auricular Nerve (C2–C3)

—ascends on the sternocleidomastoid to supply the skin behind the auricle and on the parotid gland.

3. Transverse Cervical Nerve (C2–C3)

—turns around the posterior border of the sternocleidomastoid and supplies the skin on the anterior cervical triangle.

4. Supraclavicular Nerve (C3–C4)

—emerges as a common trunk from under the cover of the sternocleidomastoid muscle and then divides into anterior, middle, and posterior branches to the skin over the clavicle and the shoulder.

5. Ansa Cervicalis

—is a nerve loop formed by the union of the superior root (C1 or C1 and C2; descendens hypoglossi) and the inferior root (C2 and C3; descendens cervicalis).

—innervates the infrahyoid (or strap) muscles, such as the omohyoid, sternohyoid, and sternothyroid muscles, with the exception of the thyrohyoid muscle, which is innervated by C1 via the hypoglossal nerve.

6. Phrenic Nerve (C3–C5)

—lies on the anterior surface of the anterior scalene muscle and passes into the thorax deep to the subclavian vein.

—passes between the mediastinal pleura and fibrous pericardium and supplies the diaphragm, pericardium, and mediastinal pleura.

C. Brachial Plexus (see Figure 2.9)

—is formed by the union of the ventral primary rami of the lower four cervical nerves and the first thoracic nerve and passes between the anterior scalene and middle scalene muscles.

—its **roots** give rise to:

1. Dorsal Scapular Nerve (C5)

—emerges from behind the anterior scalene muscle and runs downward and backward through the middle scalene muscle and then deep to the trapezius muscle.

—passes deep to or through the levator scapulae muscle and descends along with the dorsal scapular artery on the deep surface of the rhomboid muscles along the medial border of the scapula, supplying the levator scapulae and rhomboid muscles.

2. Long Thoracic Nerve (C5–C7)

—pierces the middle scalene muscle, descends behind the brachial plexus, and enters the axilla to supply the serratus anterior muscle.

—its **upper trunk** gives rise to:

1. Suprascapular Nerve (C5–C6)

—passes deep to the trapezius muscle and joins the suprascapular artery in a course toward the shoulder.

—passes through the scapular notch under the superior transverse scapular ligament and supplies the supraspinatus and infraspinatus muscles.

2. Nerve to the Subclavius Muscle (C5)

—descends in front of the plexus and behind the clavicle to supply the subclavius muscle.

—often communicates with the phrenic nerve as the accessory phrenic nerve.

IV. Blood Vessels of the Neck (Figure 8.5)

A. Subclavian Artery

—is a branch of the brachiocephalic trunk on the right but arises directly from the arch of the aorta on the left.

—is divided into three parts by the anterior scalene muscle: The first part passes from the origin of the vessel to the medial margin of the anterior scalene; the second part lies behind this muscle; and the third part passes from the lateral margin of the muscle to the outer border of the first rib.

—its branches are the vertebral and internal thoracic arteries and the thyrocervical trunk from its first portion; the costocervical trunk from its second portion; and the dorsal scapular artery from its third portion.

1. Vertebral Artery

—arises from the first part of the subclavian artery and ascends between the anterior scalene muscle and the longus coli muscle.

—ascends through the transverse foramina of the upper six cervical vertebrae, winds around the superior articular process of the atlas, and passes through the foramen magnum into the cranial cavity.

2. Thyrocervical Trunk

—is a short trunk from the first part of the subclavian artery that divides into the **inferior thyroid, transverse cervical,** and **suprascapular arteries.**

a. Inferior Thyroid Artery

—ascends in front of the anterior scalene muscle, turns medially behind the carotid sheath but in front of the vertebral vessels, and then arches downward to the lower pole of the thyroid gland.

—gives off an **ascending cervical artery,** which ascends on the anterior scalene muscle medial to the phrenic nerve.

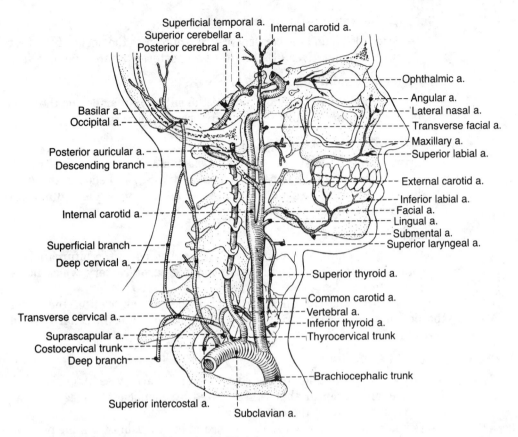

Figure 8.5. Subclavian and carotid arteries and their branches.

b. Transverse Cervical Artery

—runs laterally across the anterior scalene muscle, phrenic nerve, and trunks of the brachial plexus, passing deep to the trapezius muscle.
—divides into a superficial (ascending) branch and a deep (descending) branch, which is known as the dorsal (descending) scapular artery.

c. Suprascapular Artery

—passes in front of the anterior scalene muscle and the brachial plexus parallel to but below the transverse cervical artery.
—passes superior to the superior transverse scapular ligament, whereas the suprascapular nerve passes inferior to it.

3. Costocervical Trunk

—arises from the posterior aspects of the subclavian artery behind the anterior scalene muscle and divides into the **deep cervical** and **superior intercostal arteries.**

a. Deep Cervical Artery

—passes between the transverse process of the seventh cervical vertebra and the neck of the first rib, ascends between the semispinalis capitis and semispinalis cervicis muscles, and anastomoses with the deep branch of the descending branch of the occipital artery.

  b. Superior Intercostal Artery

   —descends behind the cervical pleura, anterior to the necks of the first two ribs, and gives off the first two posterior intercostal arteries.

 4. Dorsal (Descending) Scapular Artery

   —arises from either the third part of the subclavian artery or the deep (descending) branch of the transverse cervical artery.

 5. Internal Thoracic Artery

   —arises from the first part of the subclavian artery, descends through the thorax behind the upper six costal cartilages, and ends at the sixth intercostal space by dividing into the **superior epigastric** and **musculophrenic arteries**.

B. Common Carotid Artery

 —has the right common carotid artery, which begins at the bifurcation of the brachiocephalic artery, and the left common carotid artery, which arises from the aortic arch.

 —ascends within the carotid sheath and divides at the level of the upper border of the thyroid cartilage into the **external** and **internal carotid arteries**.

 1. Carotid Body

   —lies at the bifurcation of the common carotid artery as an ovoid body.
   —is a **chemoreceptor** that is stimulated by chemical changes (such as oxygen tension) in the circulating blood.
   —is innervated by the nerve to the carotid body arising from the pharyngeal branch of the vagus nerve and by the carotid sinus branch of the glossopharyngeal nerve.

 2. Carotid Sinus

   —is a spindle-shaped dilatation located at the origin of the internal carotid artery and functions as a **pressoreceptor (baroreceptor)**, stimulated by changes in blood pressure.
   —is innervated primarily by the carotid sinus branch of the glossopharyngeal nerve but also by the nerve to the carotid body arising from the pharyngeal branch of the vagus nerve.

C. Internal Carotid Artery

 —has no branches in the neck.
 —ascends within the carotid sheath in company with the vagus nerve and the internal jugular vein.
 —enters the cranium through the carotid canal in the petrous part of the temporal bone.
 —in the middle cranial fossa, gives off the ophthalmic artery and the anterior and middle cerebral arteries.

D. External Carotid Artery

 —extends from the level of the upper border of the thyroid cartilage to the neck of the mandible, where it ends in the substance of the parotid gland by dividing into the maxillary and superficial temporal arteries.
 —has eight named branches:

1. **Superior Thyroid Artery**
   —arises below the level of the greater horn of the hyoid bone.
   —descends obliquely forward in the carotid triangle and passes deep to the infrahyoid muscles to reach the superior pole of the thyroid gland.
   —gives rise to an **infrahyoid, sternocleidomastoid, superior laryngeal, cricothyroid,** and several **glandular branches**.

2. **Lingual Artery**
   —arises at the level of the tip of the greater horn of the hyoid bone and passes deep to the hyoglossus muscle to reach the tongue.
   —gives rise to the **suprahyoid, dorsal lingual, sublingual,** and **deep lingual branches**.

3. **Facial Artery**
   —arises just above the lingual artery and ascends forward deep to the posterior belly of the digastric and stylohyoid muscles.
   —hooks around the lower border of the mandible at the anterior margin of the masseter to enter the face (see p 271).

4. **Ascending Pharyngeal Artery**
   —arises from the deep surface of the external carotid artery in the carotid triangle and ascends between the internal carotid artery and the wall of the pharynx.
   —gives rise to the pharyngeal, palatine, inferior tympanic, and meningeal branches.

5. **Occipital Artery**
   —arises from the posterior surface of the external carotid artery, just above the level of the hyoid bone.
   —passes deep to the digastric posterior belly, occupies the groove on the mastoid process, and appears on the skin above the occipital triangle.
   —gives off:

   a. **Sternocleidomastoid Branch**
      —descends inferiorly and posteriorly over the hypoglossal nerve and enters the substance of the muscle.
      —anastomoses with the sternocleidomastoid branch of the superior thyroid artery.

   b. **Descending Branch**
      —its superficial branch anastomoses with the superficial (or ascending) branch of the transverse cervical artery.
      —its deep branch anastomoses with the deep cervical artery of the costocervical trunk.

6. **Posterior Auricular Artery**
   —arises from the posterior surface of the external carotid artery just above the digastric posterior belly.
   —ascends superficial to the styloid process and deep to the parotid gland and ends between the mastoid process and the external acoustic meatus.
   —gives off the stylomastoid, auricular, and occipital branches.

7. **Maxillary Artery**
   —arises behind the neck of the mandible as the larger terminal branch of the external carotid artery.

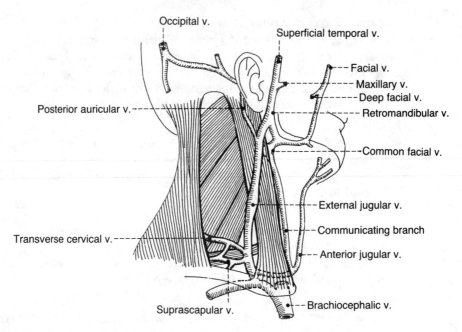

Figure 8.6. Veins of the cervical triangle.

—runs deep to the neck of the mandible and enters the infratemporal fossa (see p 277).

8. **Superficial Temporal Artery**
   —arises behind the neck of the mandible as the smaller terminal branch of the external carotid artery.
   —gives off the **transverse facial artery,** which runs between the zygomatic arch above and the parotid duct below.
   —ascends in front of the external acoustic meatus into the scalp, accompanying the auriculotemporal nerve and the superficial temporal vein.

E. **Retromandibular Vein** (Figure 8.6)
   —is formed by the superficial temporal and maxillary veins.
   —divides into an anterior branch, which joins the facial vein to form the common facial vein, and a posterior branch, which joins the posterior auricular vein to form the external jugular vein.

F. **External Jugular Vein** (see Figure 8.6)
   —is formed by the union of the posterior auricular vein and the posterior branch of the retromandibular vein.
   —crosses the sternomastoid obliquely, under the platysma, and ends in the subclavian (or sometimes the internal jugular) vein.
   —receives the suprascapular, transverse cervical, and anterior jugular veins.

G. **Internal Jugular Vein** (Figure 8.7)
   —begins in the jugular foramen as a continuation of the sigmoid sinus, descends in the carotid sheath, and ends in the brachiocephalic vein.
   —has the **superior bulb** at its beginning and the **inferior bulb** just above its termination.
   —receives blood from the brain, face, and neck.

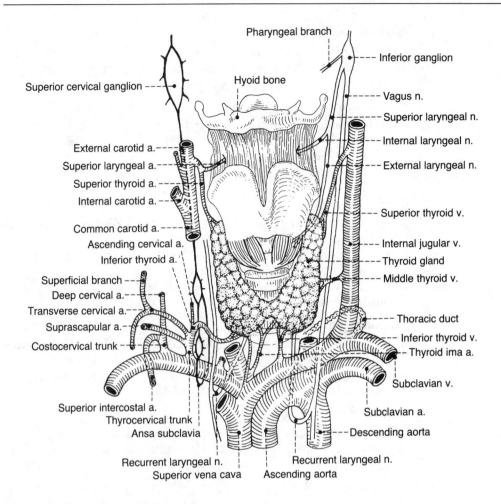

Figure 8.7. Deep structures of the neck.

V. Lymphatics of the Head and Neck

A. Superficial Lymph Nodes of the Head

1. Occipital Nodes

—lie on the occipital arteries between the attachments of the sternocleidomastoid and trapezius muscles.

—receive afferent lymphatics from the occipital part of the scalp.

—pass efferent vessels to the superior deep cervical nodes and the nodes along the accessory nerve.

2. Retroauricular Nodes

—lie on the mastoid process.

—their afferent lymphatics drain the posterior temporal and parietal regions and the back of the auricle.

—their efferent vessels pass to the superior deep cervical nodes and descend to the accessory nodes.

3. Anterior Auricular (Preauricular) Nodes

—lie in front of the tragus, receive lymph from the temporal and frontal parts of the scalp, and drain into the superior deep cervical nodes.

4. **Superficial Parotid Nodes**

   —lie in the parotid gland, receive lymph from the auricle, scalp, forehead, and eyelids, and drain into the deep parotid nodes and to the deep cervical nodes.

5. **Facial Nodes**

   —consist of the infraorbital, buccal, and mandibular nodes.

   —receive lymph from the conjunctiva, eyelids, nose, and cheeks and drain into the submandibular nodes.

B. **Superficial Lymph Nodes of the Neck**

1. **Submental Nodes**

   —lie on the mylohyoid muscles between the anterior bellies of the digastric muscles.

   —receive afferent lymphatics from the floor of the mouth, the mandible, the tip of the tongue, and the lower lip and chin.

   —pass efferent vessels to the submandibular nodes and the jugulo-omohyoid nodes.

2. **Submandibular Nodes**

   —lie in the submandibular region and receive their afferent lymphatics from the chin, cheeks, lips, nose, facial, and submental nodes.

   —pass efferent vessels to the superior deep cervical and jugulo-omohyoid nodes.

3. **External Jugular Nodes**

   —lie along the external jugular vein, receive afferent lymphatics from the lower part of the ear and the parotid region, and drain into the superior and inferior deep cervical nodes.

4. **Anterior Jugular Nodes**

   —lie along the anterior jugular vein, receive lymph from the anterior infrahyoid region, and drain into the inferior deep cervical nodes.

C. **Deep Lymph Nodes of the Neck**

1. **Superior Deep Cervical Nodes**

   —lie in the carotid triangle of the neck.

   —receive afferent lymphatics from the back of the head and neck, tongue, palate, nasal cavity, larynx, pharynx, trachea, thyroid gland, and esophagus.

   —their efferent vessels join those of the inferior deep cervical nodes to form the jugular trunk, which empties into the thoracic duct on the left and into the junction of the internal jugular and subclavian veins on the right.

2. **Retropharyngeal Nodes**

   —lie behind the nasopharynx and drain the nasal cavity, paranasal sinuses, palate, middle ear, and the nasopharynx and oropharynx.

3. **Deep Parotid Nodes**

   —lie in the deep surface of the parotid gland and receive lymph from the external acoustic meatus, auditory tube, tympanic membrane, soft palate, and the posterior part of the nasal cavity.

4. Accessory Nodes

—lie along the accessory nerve in the posterior cervical triangle, receive efferent vessels of the occipital and retroauricular nodes, join the transverse cervical nodes, and pass to the inferior cervical nodes.

5. Infrahyoid Nodes

—lie along the superior thyroid vessels on the thyrohyoid membrane, drain the superior half of the larynx, and pass to the deep cervical nodes.

6. Prelaryngeal Nodes

—lie on the cricothyroid ligament, drain the inferior half of the larynx and thyroid gland, and pass to the deep cervical nodes.

7. Pretracheal Nodes

—lie at the lower end of the thyroid gland, receive lymph from the trachea and the inferior half of the thyroid and larynx, and drain into the deep cervical nodes.

8. Paratracheal Nodes

—lie in the groove between the trachea and esophagus, drain the trachea and esophagus, and pass to the deep cervical nodes.

9. Jugulodigastric Nodes

—lie below the angle of the mandible and receive afferent lymphatics from the tongue and the palatine tonsil.

10. Jugulomyohyoid Nodes

—lie on the internal jugular vein just above the intermediate tendon of the omohyoid and receive lymph from the tip of the tongue and the submental region.

11. Transverse Cervical Nodes

—lie along the transverse cervical vessels and receive lymph from the apical axillary nodes, the lateral part of the neck, the anterior thoracic wall, and the mammary gland.

12. Inferior Deep Cervical Nodes

—lie on the internal jugular vein near the subclavian vein.
—receive afferent lymphatics from the anterior jugular, transverse cervical, and apical axillary nodes.
—their efferent vessels join those of the superior deep cervical nodes to form the jugular trunk, which empties into the thoracic duct on the left and into the junction of the internal jugular and subclavian veins on the right.

VI. Clinical Considerations

A. Injury of the Upper Trunk of the Brachial Plexus

—may be caused by a violent separation of the head from the shoulder, such as occurs in a fall from a motorcycle. In this condition, the arm is in medial rotation owing to the paralysis of the lateral rotators, resulting in a **"waiter's tip hand."**
—may be caused by stretching an infant's neck during a difficult delivery. This is referred to as **"birth palsy"** or **"obstetric paralysis."**

B. Neurovascular Compression Syndrome

—produces symptoms of nerve compression of the brachial plexus and the subclavian vessels.

—is caused by an abnormal insertion of the anterior and middle scalene muscles and by the cervical rib, which is the cartilaginous accessory rib attached to the seventh cervical vertebra. Such cases can be corrected by cutting the cervical rib or the anterior scalene muscle.

Deep Neck and Prevertebral Region

I. Deep Structures of the Neck (see Figure 8.7)

A. Thyroid Gland

—is an endocrine gland that produces thyroxin and thyrocalcitonin.

—consists of right and left lobes connected by the isthmus. (The muscular band descending from the hyoid bone to the isthmus is called the levator glandulae thyroideae.)

—is supplied by the superior and inferior thyroid arteries (and the arteria thyroidea ima, which is an inconsistent branch from the brachiocephalic trunk).

—its venous drainage is via the superior and middle thyroid veins to the internal jugular vein and via the inferior thyroid vein to the brachiocephalic vein.

B. Parathyroid Glands

—are endocrine glands that are essential to life.

—consist usually of four (can vary from two to six) small ovoid bodies that lie against the dorsum of the thyroid gland under its sheath but with their own capsule.

—are supplied chiefly by the inferior thyroid artery.

C. Thyroid Cartilage

—is a hyaline cartilage that forms a laryngeal prominence (the "Adam's apple").

—its superior horn is joined to the tip of the greater horn of the hyoid bone by the lateral thyroid ligament.

D. Sympathetic Trunk

—runs behind the carotid sheath and in front of the longus colli and longus capitis muscles.

—contains preganglionic and postganglionic sympathetic fibers, cell bodies of the postganglionic sympathetic fibers, and visceral afferent fibers with cell bodies in the upper thoracic dorsal root ganglia.

—receives gray rami communicantes but no white rami communicantes in the cervical region.

—bears the following cervical ganglia:

1. Superior Cervical Ganglion

—lies in front of the transverse processes of the first and second cervical vertebrae, posterior to the internal carotid artery and anterior to the longus capitis muscle.

—contains cell bodies of postganglionic sympathetic fibers that pass to the visceral structures of the head and neck.

—gives rise to the internal carotid nerve to form the internal carotid plexus; the external carotid nerve to form the external carotid plexus; the pharyngeal branches that join the pharyngeal branches of the glossopharyngeal and vagus nerves to form the pharyngeal plexus; and the superior cervical cardiac nerve to the heart.

2. Middle Cervical Ganglion

—lies at the level of the cricoid cartilage (sixth cervical vertebra).

—gives off a middle cervical cardiac nerve, which is the largest of the three cervical sympathetic cardiac nerves.

3. Inferior Cervical Ganglion

—fuses with the first thoracic ganglion to become the **cervicothoracic ganglion**.

—lies in front of the neck of the first rib and the transverse process of the seventh cervical vertebra and behind the dome of the pleura.

—gives rise to the inferior cervical cardiac nerve.

E. Ansa Subclavia

—is the cord connecting the middle and inferior cervical sympathetic ganglia, forming a loop around the first part of the subclavian artery.

II. Cervical Fasciae (Figure 8.8)

A. Superficial (Investing) Layer of Deep Cervical Fascia

—surrounds all of the deeper parts of the neck.

—splits to enclose the sternocleidomastoid and trapezius muscles.

—is attached superiorly along the mandible, mastoid process, external occipital protuberance, and superior nuchal line of the occipital bone.

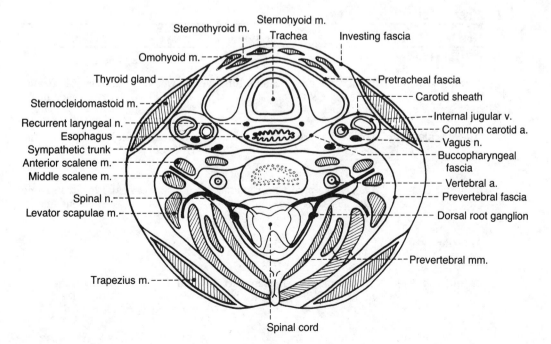

Figure 8.8. Cross section of the neck.

—is attached inferiorly along the acromion and spine of the scapula, clavicle, and manubrium sterni.

B. Prevertebral Fascia

—is cylindrical and encloses the vertebral column and its associated muscles.
—covers the scalene muscles and the deep muscles of the back.
—attaches to the external occipital protuberance and the basilar part of the occipital bone and becomes continuous with the endothoracic fascia and the anterior longitudinal ligament of the bodies of the vertebrae in the thorax.

C. Carotid Sheath

—contains the common and internal carotid arteries, internal jugular vein, and vagus nerve.
—does not contain the sympathetic trunk, which lies posterior to the carotid sheath and anterior to the prevertebral fascia.
—blends with the prevertebral, pretracheal, and investing fasciae and also attaches to the base of the skull.

D. Pretracheal Fascia

—is the layer of the cervical fascia investing the larynx and trachea, enclosing the thyroid gland, and contributing to the formation of the carotid sheath.

E. Buccopharyngeal Fascia

—covers the buccinator muscles and the pharynx.
—is attached to the pharyngeal tubercle and the pterygomandibular raphe.

F. Pharyngobasilar Fascia

—is the fibrous coat in the wall of the pharynx, situated between the mucous membrane and the pharyngeal constrictor muscles.

III. Prevertebral Muscles

Muscle	Origin	Insertion	Nerve	Action
Lateral vertebral:				
Anterior scalene	Transverse processes of CV3–CV6	Scalene tubercle on first rib	Lower cervical (C5–C8)	Elevates first rib; bends neck
Middle scalene	Transverse processes of CV2–CV7	Upper surface of first rib	Lower cervical (C5–C8)	Elevates first rib; bends neck
Posterior scalene	Transverse processes of CV4–CV6	Outer surface of second rib	Lower cervical (C6–C8)	Elevates second rib; bends neck
Anterior vertebral:				
Longus capitus	Transverse processes of CV3–CV6	Basilar part of occipital bone	C1–C4	Flexes and rotates head

(Continued on next page)

Muscle	Origin	Insertion	Nerve	Action
Longus colli (L. cervicis)	Transverse processes and bodies of CV3–TV3	Anterior tubercle of atlas; bodies of CV2–CV4; transverse process of CV5–CV6	C2–C6	Flexes and rotates head
Rectus capitis anterior	Lateral mass of atlas	Basilar part of occipital bone	C1–C2	Flexes and rotates head
Rectus capitis lateralis	Transverse process of atlas	Jugular process of occipital bone	C1–C2	Flexes head laterally

Face and Scalp

I. Skull (Figures 8.9 and 8.10; see 8.19)

—is the skeleton of the head, composed of 8 **cranial bones** (the frontal, occipital, 2 parietal, 2 temporal, the ethmoid, and the sphenoid), and 14 **facial bones** (2 lacrimal, 2 nasal, 2 palatine, 2 inferior turbinate, 2 maxillary, 2 zygomatic bones, and the vomer and the mandible).

A. Cranium

—is the portion of the skull excluding the mandible.

B. Calvaria

—is the skullcap or vault of the skull, excluding the facial bones; it consists of the superior portions of the frontal, parietal, and occipital bones.

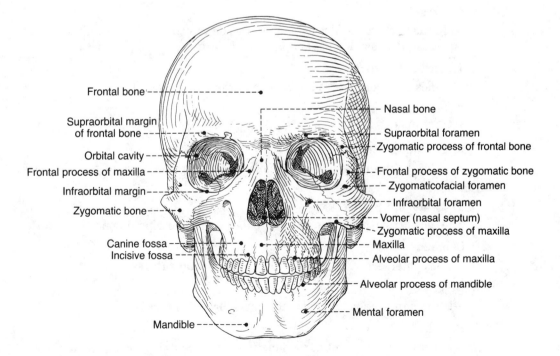

Figure 8.9. Anterior view of the skull.

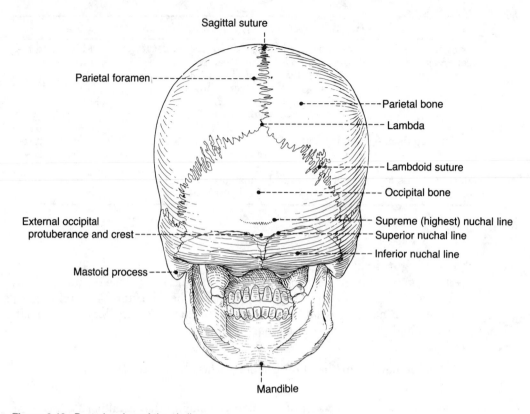

Sagittal suture

Parietal foramen

Parietal bone

Lambda

Lambdoid suture

Occipital bone

External occipital protuberance and crest

Supreme (highest) nuchal line

Superior nuchal line

Inferior nuchal line

Mastoid process

Mandible

Figure 8.10. Posterior view of the skull.

C. Sutures

—are the immovable fibrous joints between the bones of the skull: the coronal, sagittal, squamous, and lambdoid sutures.

1. Coronal Suture

—lies between the frontal bone and the two parietal bones.

2. Sagittal Suture

—lies between the two parietal bones.

3. Squamous Suture

—lies between the parietal bone and the squamous part of the temporal bone.

4. Lambdoid Suture

—lies between the two parietal bones and the occipital bone.

D. Lambda

—is the intersection of the lambdoid and sagittal sutures.

E. Bregma

—is the intersection of the sagittal and coronal sutures.

F. Pterion

—is a craniometric point at the junction of the frontal, parietal, and temporal bones and the great wing of the sphenoid bone.

II. Muscles of Facial Expression (Figure 8.11)

Muscle	Origin	Insertion	Nerve	Action
Occipitofrontalis	Superior nuchal line; upper orbital margin	Epicranial aponeurosis	Facial n.	Elevates eyebrows; wrinkles forehead (surprise)
Corrugator supercilii	Medial supraorbital margin	Skin of medial eyebrow	Facial n.	Draws eyebrows downward medially (anger)
Orbicularis oculi	Medial orbital margin; medial palpebral ligament; lacrimal bone	Skin and rim of orbit; tarsal plate; lateral palpebral raphe	Facial n.	Closes eyelids (squinting)
Procerus	Nasal bone and cartilage	Skin between eyebrows	Facial n.	Wrinkles skin over bones (sadness)
Nasalis	Maxilla lateral to incisive fossa	Ala of nose	Facial n.	Draws ala of nose toward septum
Depressor septi*	Incisive fossa of maxilla	Ala and nasal septum	Facial n.	Constricts nares
Orbicularis oris	Maxilla above incisor teeth	Skin of lip	Facial n.	Closes lips
Levator anguli oris	Canine fossa of maxilla	Angle of mouth	Facial n.	Elevates angle of mouth medially (disgust)
Levator labii superioris	Maxilla above infraorbital foramen	Skin of upper lip	Facial n.	Elevates upper lip; dilates nares (disgust)
Levator labii superioris alaeque nasi*	Frontal process of maxilla	Skin of upper lip	Facial n.	Elevates ala of nose and upper lip
Zygomaticus major	Zygomatic arch	Angle of mouth	Facial n.	Draws angle of mouth backward and upward (smile)
Zygomaticus minor	Zygomatic arch	Angle of mouth	Facial n.	Elevates upper lip
Depressor labii inferioris	Mandible below mental foramen	Orbicularis oris and skin of lower lip	Facial n.	Depresses lower lip
Depressor anguli oris	Oblique line of mandible	Angle of mouth	Facial n.	Depresses angle of mouth
Risorius	Fascia over masseter	Angle of mouth	Facial n.	Retracts angle of mouth (false smile)

(Continued on next page)

Muscle	Origin	Insertion	Nerve	Action
Buccinator	Mandible; pterygomandibular raphe; alveolar processes	Angle of mouth	Facial n.	Presses cheek to keep it taut
Mentalis	Incisive fossa of mandible	Skin of chin	Facial n.	Elevates and protrudes lower lip
Auricularis anterior, superior, and posterior*	Temporal fascia; epicranial aponeurosis; mastoid process	Anterior, superior, and posterior sides of auricle	Facial n.	Retract and elevate ear

* Indicates less important muscles.

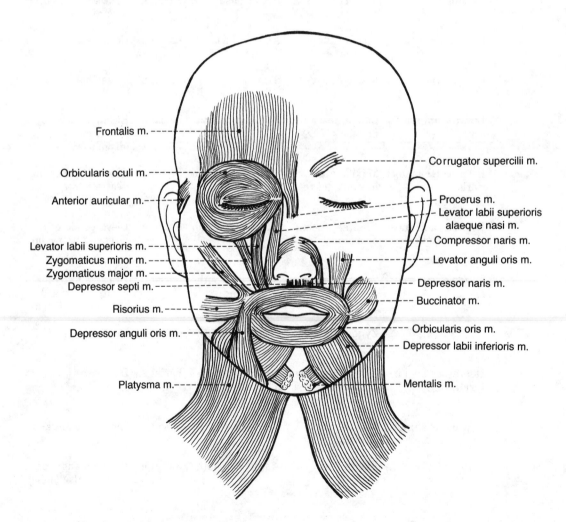

Figure 8.11. Muscles of facial expression.

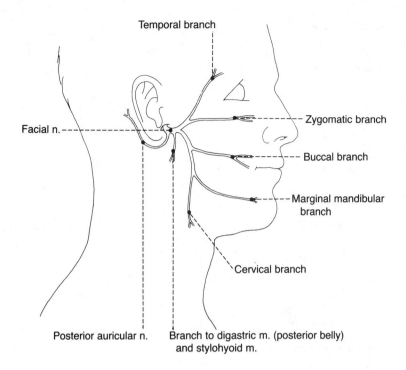

Figure 8.12. Distribution of the facial nerve.

III. Nerve Supply to the Face and Scalp (Figures 8.12–8.14)

A. Facial Nerve (see Figure 8.12 and p 290)

—comes through the **stylomastoid foramen** and appears posterior to the parotid gland.

—enters the parotid gland to give off five terminal branches, which radiate forward in the face as **temporal, zygomatic, buccal, mandibular,** and **cervical branches**.

—supplies the muscles of facial expression and sends the **posterior auricular branch** to muscles of the auricle and the occipitalis muscle.

—also supplies the digastric posterior belly and stylohyoid muscles.

B. Trigeminal Nerve

—is sensory to the skin of the face, whereas the facial nerve supplies the muscles of facial expression.

1. Ophthalmic Nerve (see p 298)

—supplies the area above the upper eyelid and dorsum of the nose.

—its branches are the supraorbital, supratrochlear, infratrochlear, external nasal, and lacrimal nerves.

2. Maxillary Nerve (see p 320)

—supplies the face below the level of the eyes and above the upper lip.

—its branches are the zygomaticofacial, zygomaticotemporal, and infraorbital nerves.

3. Mandibular Nerve (see p 276)

—supplies the face below the level of the lower lip.

—its branches are the auriculotemporal, buccal, and mental nerves.

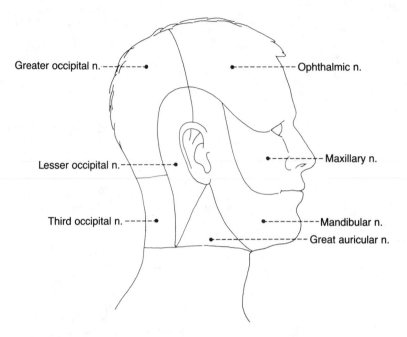

Figure 8.13. Sensory innervation of the face.

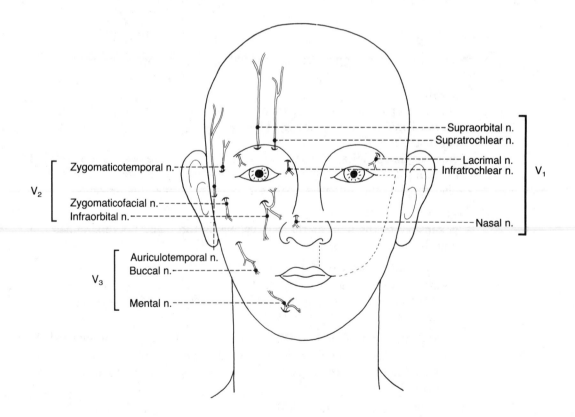

Figure 8.14. Cutaneous innervation of the face and scalp.

C. Clinical Considerations

1. Bell's Palsy (Facial Paralysis)

—is a unilateral paralysis of the facial muscles owing to a lesion of the facial nerve.

—is characterized by characteristic distortions of the face such as a sagging corner of the mouth and the inability to smile, whistle, or blow; and drooping of the upper eyelid, everting the lower eyelid, and the inability to close or blink the eye.

—causes decreased lacrimation (as a result of a lesion of the greater petrosal nerve), loss of taste in the anterior two-thirds of the tongue (chorda tympani), painful sensitivity to sounds (nerve to the stapedius), and deviation of the lower jaw and tongue (nerve to the digastric muscle).

2. Corneal Blink Reflex

—is closing of the eyes caused by blowing on the cornea or touching it with a wisp of cotton wool; it is caused by bilateral contraction of the orbicularis oculi muscles.

—its efferent limb (of the reflex arc) is the facial nerve, whereas its afferent limb is the nasociliary nerve.

3. Trigeminal Neuralgia (Tic Douloureux)

—is marked by a paroxysmal pain along the course of the trigeminal nerve.

—may be alleviated by sectioning the sensory root of the trigeminal nerve in the trigeminal (Meckel's) cave in the middle cranial fossa.

IV. Blood Vessels of the Face and Scalp (Figures 8.15 and 8.16)

A. Facial Artery

—arises from the external carotid artery just above the upper border of the hyoid bone.

—passes deep to the mandible, winds around the lower border of the mandible, and runs upward and forward on the face.

—gives off the **inferior labial, superior labial,** and **lateral nasal branches,** and terminates as an **angular artery** that anastomoses with the palpebral and dorsal nasal branches of the ophthalmic artery and thus establishes a communication between the external and internal carotid arteries.

B. Superficial Temporal Artery

—arises behind the neck of the mandible as the smaller terminal branch of the external carotid artery and ascends anterior to the external acoustic meatus into the scalp.

—accompanies the auriculotemporal nerve along its anterior surface.

—gives off the **transverse facial artery,** which passes forward across the masseter between the zygomatic arch above and the parotid duct below.

—also gives off the zygomatico-orbital, middle temporal, anterior auricular, frontal, and parietal branches.

C. Facial Vein

—begins as an angular vein by the confluence of the supraorbital and supratrochlear veins. The angular vein is continued at the lower margin of the orbital margin into the facial vein.

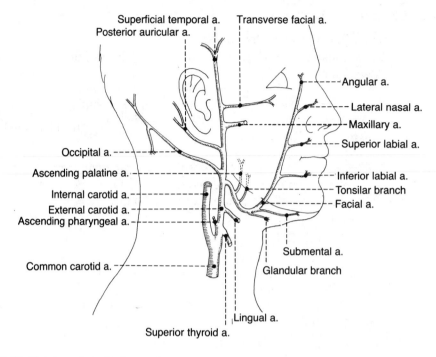

Figure 8.15. Blood supply to the face and scalp.

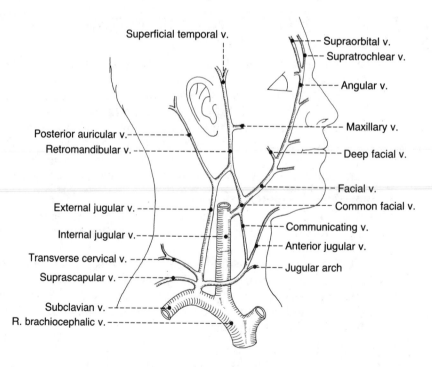

Figure 8.16. Veins of the head and neck.

—communicates with the infraorbital vein.

—receives tributaries corresponding to the branches of the artery and also receives the infraorbital and deep facial veins.

—drains either directly into the internal jugular vein or by joining the anterior branch of the retromandibular vein to form the common facial vein, which then enters the internal jugular vein.

—communicates with the superior ophthalmic vein and thus with the cavernous sinus, allowing a route of infection from the face to the cranial dural sinus.

D. Retromandibular Vein

—is formed by the union of the superficial temporal and maxillary veins behind the mandible.

—divides into an **anterior branch,** which joins the facial vein to form the common facial vein, and a **posterior branch,** which joins the posterior auricular vein to form the external jugular vein.

V. Scalp (Figures 8.17 and 8.18)

—is innervated by the supratrochlear, supraorbital, zygomaticotemporal, auriculotemporal, lesser occipital, greater occipital, and third occipital nerves.

—is supplied by the supratrochlear and supraorbital branches of the internal carotid and by the superficial temporal, posterior auricular, and occipital branches of the external carotid arteries.

—consists of five layers:

A. Skin

B. Connective Tissue (Close Subcutaneous Tissue)

—contains the larger blood vessels and nerves. Because of its toughness, the scalp gapes when cut and the blood vessels do not contract, which leads to severe bleeding.

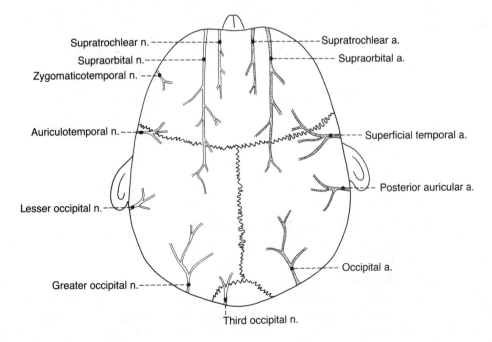

Figure 8.17. Nerves and arteries of the scalp.

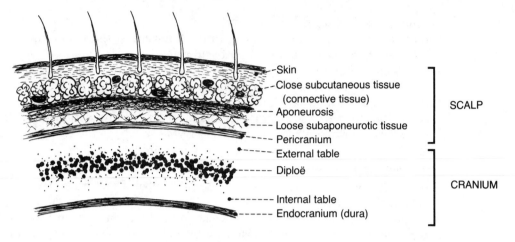

Figure 8.18. Layers of the scalp and cranium.

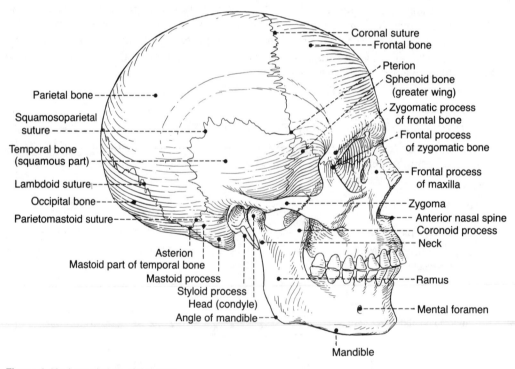

Figure 8.19. Lateral view of the skull.

C. Aponeurosis Epicranialis (Galea Aponeurotica)

—is a fibrous sheet that covers the vault of the skull and unites the occipitalis and frontalis muscles.

D. Loose Connective Tissue

—forms the subaponeurotic space and contains the emissary veins.

—is termed a dangerous area because infection (blood and pus) can spread easily in it or from the scalp to the intracranial sinuses by way of the emissary veins.

E. Pericranium

—is the periosteum over the surface of the skull.

Temporal and Infratemporal Fossae

I. Boundaries

A. Infratemporal Fossa (Figures 8.19 and 8.20)

—contains the lower portion of the temporalis muscle, the lateral and medial pterygoid muscles, the pterygoid plexus of veins, the mandibular nerve and its branches, the maxillary artery and its branches, the chorda tympani, and the otic ganglion.

—has the following boundaries:

1. **Anterior:** posterior surface of the maxilla.
2. **Posterior:** styloid process.
3. **Medial:** lateral pterygoid plate of the sphenoid bone.
4. **Lateral:** ramus and coronoid process of the mandible.
5. **Roof:** infratemporal surface of the greater wing of the sphenoid bone.

B. Temporal Fossa (see Figures 8.19 and 8.20)

—contains the temporalis muscle, the deep temporal nerves and vessels, the auriculotemporal nerve, and the superficial temporal vessels.

—has the following boundaries:

1. **Anterior:** zygomatic process of the frontal bone and the frontal process of the zygomatic bone.
2. **Posterior:** temporal line.
3. **Superior:** temporal line.
4. **Inferior:** zygomatic arch.
5. **Floor:** parts of the frontal, parietal, temporal, and greater wing of the sphenoid bone.

II. Muscles of Mastication (Figure 8.21)

Muscle	Origin	Insertion	Nerve	Action
Temporalis	Temporal fossa	Coronoid process and ramus of mandible	Trigeminal n.	Elevates and retracts mandible
Masseter	Lower border and medial surface of zygomatic arch	Lateral surface of coronoid process, ramus and angle of mandible	Trigeminal n.	Elevates mandible
Lateral pterygoid	Superior head from infratemporal surface of sphenoid; inferior head from lateral surface of lateral pterygoid plate	Neck of mandible; articular disk and capsule of temporomandibular joint	Trigeminal n.	Protracts (protrudes) and depresses mandible

(Continued on next page)

Muscle	Origin	Insertion	Nerve	Action
Medial pterygoid	Tuber of maxilla; medial surface of lateral pterygoid plate; pyramidal process of palatine bone	Medial surface of angle and ramus of mandible	Trigeminal n.	Protracts (protrudes) and elevates mandible

The jaws are opened by the lateral pterygoid muscle and are closed by the temporalis, masseter, and medial pterygoid muscles.

III. Nerves of the Infratemporal Region (see Figure 8.21)

A. Mandibular Nerve

—passes through the foramen ovale and supplies the tensor veli palatini and tensor tympani muscles, muscles of mastication (temporalis, masseter, and lateral and medial pterygoid), anterior belly of the digastric muscle, and the mylohyoid muscle.

—is sensory to the lower part of the face below the lower lip and the mouth as well as to the lower teeth.

1. Meningeal Branch

—accompanies the middle meningeal artery and enters the cranium through the foramen spinosum.

2. Masseteric, Deep Temporal, Medial Pterygoid, and Lateral Pterygoid Nerves

—supply the corresponding muscles of mastication.

3. Buccal Nerve

—descends between the two heads of the lateral pterygoid muscle.

—supplies skin and fascia on the buccinator muscle and penetrates this muscle to supply the mucous membrane of the cheek and gums.

4. Auriculotemporal Nerve

—arises by two roots that encircle the middle meningeal artery.

—supplies sensory branches to the temporomandibular joint.

—carries postganglionic parasympathetic and sympathetic fibers to the parotid gland.

—its terminal branches supply the skin of the auricle and the scalp.

5. Lingual Nerve

—descends deep to the lateral pterygoid muscle where it joins the chorda tympani, which conveys the preganglionic parasympathetic (secretomotor) fibers to the submandibular ganglion and taste fibers from the anterior two-thirds of the tongue.

—lies anterior to the inferior alveolar nerve on the medial pterygoid muscle, deep to the ramus of the mandible.

—crosses lateral to the styloglossus and hyoglossus muscles, passes deep to the mylohyoid nerve, and descends across the submandibular duct.

—supplies general sensation for the anterior two-thirds of the tongue.

6. Inferior Alveolar Nerve

—passes deep to the lateral pterygoid muscle and then between the sphenomandibular ligament and the ramus of the mandible.

—enters the mandibular canal through the mandibular foramen.

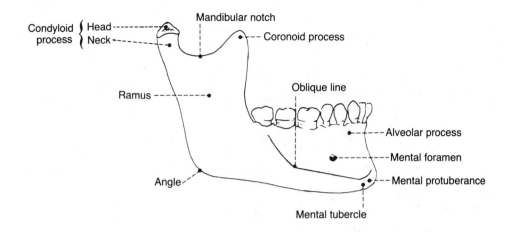

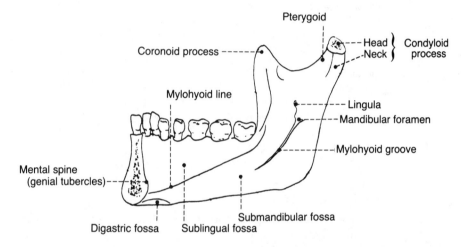

Figure 8.20. External (buccal) and internal (lingual) surfaces of the mandible.

—gives off the following branches:
a. The **mylohyoid nerve** to the mylohyoid and the anterior belly of the digastric muscle.
b. The **inferior dental branch** to the lower teeth.
c. The **mental nerve** to the skin over the chin.
d. The **incisive branch** to the canine and incisor teeth.

B. Otic Ganglion
—lies in the infratemporal fossa, just below the foramen ovale between the mandibular nerve and the tensor veli palatini muscle.
—receives preganglionic parasympathetic fibers that run in the glossopharyngeal nerve, tympanic plexus, and lesser petrosal nerve and synapse in this ganglion.
—sends postganglionic fibers that run in the auriculotemporal nerve and supply the parotid gland.

IV. Blood Vessels of the Infratemporal Region (see Figure 8.21)

A. Maxillary Artery
—arises from the external carotid artery at the posterior border of the ramus of the mandible.
—divides into three parts:

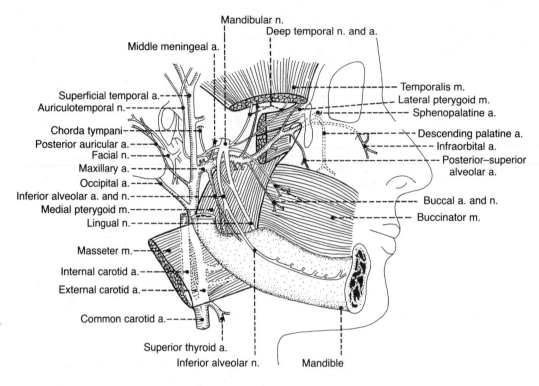

Figure 8.21. Infratemporal region.

1. **Mandibular Part**

—runs anteriorly between the neck of the mandible and the sphenomandibular ligament.

—gives rise to the following branches:

a. **Deep Auricular Artery**

—supplies the external acoustic meatus.

b. **Anterior Tympanic Artery**

—supplies the tympanic cavity and tympanic membrane.

c. **Middle Meningeal Artery**

—arises from the maxillary artery, is embraced by two roots of the auriculotemporal nerve, and enters the middle cranial fossa through the foramen spinosum.

—runs between the dura mater and the periosteum.

—damage to the artery results in epidural hematoma.

d. **Accessory Meningeal Artery**

—passes through the foramen ovale.

e. **Inferior Alveolar Artery**

—follows the inferior alveolar nerve between the sphenomandibular ligament and the ramus of the mandible.

—enters the mandibular canal through the mandibular foramen and supplies the tissues of the chin and lower teeth.

2. **Pterygoid Part**

   —runs anteriorly deep to the temporalis and lies superficial (or deep) to the lateral pterygoid muscle.

   —its branches include the **anterior and posterior deep temporal, pterygoid, masseteric,** and **buccal arteries,** which supply chiefly the muscles of mastication.

3. **Pterygopalatine Part**

   —runs between the two heads of the lateral pterygoid muscle and then through the pterygomaxillary fissure into the pterygopalatine fossa.

   —its branches include the following arteries: **posterior–superior alveolar, infraorbital, descending palatine, artery of the pterygoid canal, pharyngeal,** and **sphenopalatine.**

   a. **Posterior–Superior Alveolar Artery**

      —runs downward on the posterior surface of the maxilla and supplies the molar and premolar teeth and the maxillary sinus.

   b. **Infraorbital Artery**

      —runs upward and forward to enter the orbit through the inferior orbital fissure.

      —traverses the infraorbital groove and canal and emerges on the face through the infraorbital foramen.

      —divides into branches to supply the lower eyelid, lacrimal sac, upper lip, and cheek.

      —gives off the **anterior** and **middle superior alveolar branches** to the upper canine and incisor teeth and the maxillary sinus.

   c. **Descending Palatine Artery**

      —descends in the pterygopalatine fossa and the palatine canal.

      —gives off the **greater** and **lesser palatine arteries,** which pass through the greater and lesser palatine foramina, respectively, and supplies the soft and hard palates.

      —its greater palatine artery sends a branch to anastomose with the terminal (nasopalatine) branch of the sphenopalatine artery in the incisive canal or on the nasal septum.

   d. **Artery of the Pterygoid Canal**

      —passes through the pterygoid canal and supplies the upper part of the pharynx, auditory tube, and tympanic cavity.

   e. **Pharyngeal Artery**

      —supplies the roof of the nose and pharynx, sphenoid sinus, and auditory tube.

   f. **Sphenopalatine Artery**

      —is the terminal branch of the maxillary artery.

      —enters the nasal cavity through the sphenopalatine foramen in company with the nasopalatine branch of the maxillary nerve.

      —is the principal artery to the nasal cavity, supplying the conchae, meatus, and paranasal sinuses.

      —results in bleeding from the nose (epistaxis) when damaged.

B. Pterygoid Venous Plexus (Figure 8.22)

—lies in the infratemporal fossa, on the lateral surface of the medial pterygoid.

—communicates with the cavernous sinus via emissary veins and the inferior ophthalmic vein.

—communicates with the facial vein via the deep facial vein.

C. Retromandibular Vein

—is formed by the superficial temporal vein and the maxillary vein.

—divides into an anterior branch, which joins the facial vein to form the **common facial vein,** and a posterior branch, which joins the posterior auricular vein to form the **external jugular vein**.

V. Parotid Gland

—is separated from the submandibular gland by the stylomandibular ligament, which extends from the styloid process to the angle of the mandible. (Therefore, pus does not readily exchange between these two glands.)

—is innervated by parasympathetic (secretomotor) fibers of the glossopharyngeal nerve by way of the lesser petrosal nerve, otic ganglion, and auriculotemporal nerve.

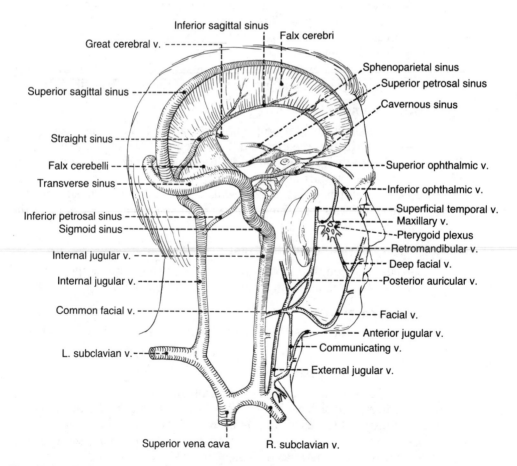

Figure 8.22. Cranial venous sinuses and veins of the head and neck.

—its duct crosses the masseter, pierces the buccinator muscle, and opens into the oral cavity opposite the second upper molar tooth.

—its complete surgical removal may damage the facial nerve.

—mumps (a viral infection) irritates the auriculotemporal nerve, causing severe pain because of the swelling of the gland and stretching of its capsule.

VI. Joints and Ligaments of the Infratemporal Region

A. Temporomandibular Joint

—is a synovial joint between the articular tubercle and the mandibular fossa of the temporal bone above and the head of the mandible below.

—combines a hinge and a gliding joint and has two (superior and inferior) synovial cavities, divided by an articular disk.

—has an articular capsule that extends from the articular tubercle and the margins of the mandibular fossa to the neck of the mandible.

—is reinforced by the **lateral (temporomandibular) ligament,** which extends from the tubercle on the zygoma to the neck of the mandible, and the **sphenomandibular ligament,** which extends from the spine of the sphenoid bone to the lingula of the mandible.

—is innervated by the auriculotemporal, masseteric, and deep temporal branches of the mandibular division of the trigeminal nerve.

—is supplied by the superficial temporal, maxillary (middle meningeal and anterior tympanic branches), and ascending pharyngeal arteries.

B. Pterygomandibular Raphe

—is a ligamentous band (or a tendinous inscription) between the buccinator muscle and the superior pharyngeal constrictor.

—extends between the pterygoid hamulus superiorly and the posterior end of the mylohyoid line of the mandible inferiorly.

C. Stylomandibular Ligament

—extends from the styloid process to the posterior border of the ramus of the mandible, near the angle of the mandible, separating the parotid from the submandibular gland.

Cranial Cavity

I. Structural Characteristics of the Cranium (Figure 8.23)

A. Foramen Cecum

—is a small pit in front of the crista galli between the ethmoid and frontal bones.

—may transmit an emissary vein from the nasal mucosa to the superior sagittal sinus.

B. Crista Galli

—is the triangular midline process of the ethmoid bone extending upward from the cribriform plate.

—gives attachment to the falx cerebri.

C. Cribriform Plate of the Ethmoid Bone

—supports the olfactory bulb and transmits olfactory nerve fibers from the olfactory mucosa to the olfactory bulb.

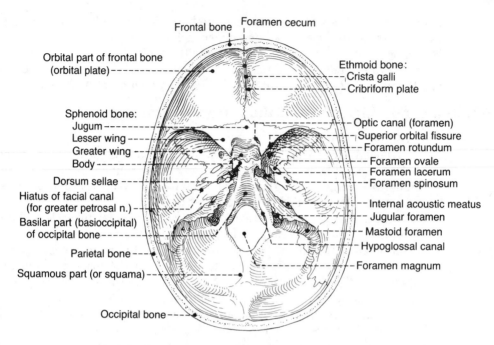

Figure 8.23. Interior of the base of the skull.

D. Anterior Clinoid Processes

—are two anterior processes of the lesser wing of the sphenoid bone.
—give attachment to the free border of the tentorium cerebelli.

E. Posterior Clinoid Processes

—are two tubercles from each side of the dorsum sellae.
—give attachment to the attached border of the tentorium cerebelli.

F. Lesser Wing of the Sphenoid Bone

—forms the anterior boundary of the middle cranial fossa.
—forms the sphenoidal ridge separating the anterior from the middle cranial fossa.
—forms the boundary of the superior orbital fissure (the space between the lesser and greater wings).

G. Greater Wing of the Sphenoid Bone

—forms the anterior wall and the floor of the middle cranial fossa.
—presents two openings, the foramen ovale and the foramen spinosum.

H. Sella Turcica (Turk's Saddle) of the Sphenoid Bone

—is bounded anteriorly by the **tuberculum sellae** and posteriorly by the **dorsum sellae.**
—its deep central depression is the **hypophyseal fossa,** which accommodates the pituitary gland or the hypophysis.
—lies directly above the sphenoid sinus located within the body of the sphenoid bone; its dural roof is formed by the **diaphragma sellae.**

I. Jugum Sphenoidale

—forms the roof for the sphenoidal air sinus.

II. Meninges of the Brain

A. Pia Mater

—is a delicate investment that is closely applied to the brain and dips into fissures and sulci.

—enmeshes blood vessels on the surfaces of the brain.

B. Arachnoid

—is a filmy, transparent, spidery layer and is connected to the pia mater by web-like trabeculations.

—is separated from the pia mater by the subarachnoid space, which is filled with **cerebrospinal fluid** (may contain blood upon hemorrhage of a cerebral artery).

—projects into the venous sinuses to form arachnoid villi, which serve as sites where the cerebrospinal fluid diffuses into the venous blood.

—may have **arachnoid granulations,** which are tuft-like collections of highly folded arachnoid that project through the dura mater.

1. Cerebrospinal Fluid (CSF)

—is formed by vascular choroid plexuses in the ventricle of the brain and is contained in the subarachnoid space between the arachnoid and pia mater.

—circulates through the ventricles, enters the subarachnoid space, and eventually filters into the venous system.

2. Arachnoid Granulations

—are tuft-like collections of highly folded arachnoid that project into the superior sagittal sinus and other dural sinuses.

—absorb the CSF into the dural sinuses and often produce erosion or pitting of the inner surface of the calvaria.

C. Dura Mater

—is the tough, fibrous, outermost layer of the meninges external to the **subdural space,** which is the space between the arachnoid and the dura.

—lies internal to the **epidural space,** which is a potential space and contains the middle meningeal arteries in the cranial cavity.

—forms the dural venous sinuses, which are spaces between the periosteal and meningeal layers or between duplications of the meningeal layers.

—is innervated by the following branches:

1. **Anterior and posterior ethmoidal branches** of the ophthalmic division of the trigeminal nerve in the anterior cranial fossa.

2. **Meningeal branches** of the maxillary and mandibular divisions of the trigeminal nerve in the middle cranial fossa.

3. **Meningeal branches** of the vagus and hypoglossal nerves in the posterior cranial fossa.

III. Projections of the Dura Mater (see Figure 8.22)

A. Falx Cerebri

—is the sickle-shaped double layer of the dura mater, lying between the two cerebral hemispheres.

—is attached anteriorly to the crista galli and posteriorly to the tentorium cerebelli.

—its inferior concave border is free and contains the inferior sagittal sinus, and its upper convex margin encloses the superior sagittal sinus.

B. Falx Cerebelli

—is a small sickle-shaped projection between the two cerebellar hemispheres.
—is attached to the posterior and inferior parts of the tentorium.
—contains the occipital sinus in its posterior border.

C. Tentorium Cerebelli

—is a crescentic fold of dura mater that supports the occipital lobes of the cerebral hemispheres and covers the cerebellum.
—its internal concave border is free and bounds the tentorial notch, whereas its external convex border encloses the transverse sinus posteriorly and the superior petrosal sinus anteriorly.
—the free border is anchored to the anterior clinoid process, whereas the attached border is attached to the posterior clinoid process.

D. Diaphragma Sellae

—is a circular, horizontal fold of dura that forms the roof of the sella turcica, covering the pituitary gland or the hypophysis.
—has a central aperture for the hypophyseal stalk or infundibulum.

IV. Cranial Venous Channels (Figure 8.24; see Figure 8.22)

A. Superior Sagittal Sinus

—lies in the midline along the convex border of the falx cerebri.
—begins at the crista galli and receives the cerebral, diploic, meningeal, and parietal emissary veins.

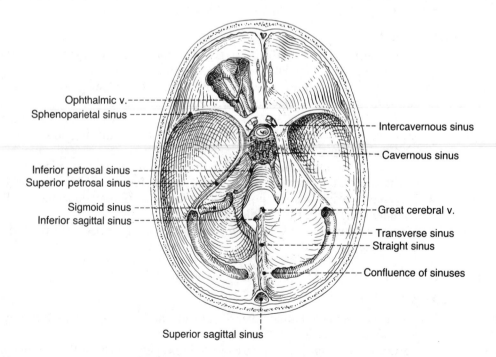

Figure 8.24. Cranial venous sinuses.

B. Inferior Sagittal Sinus

—lies in the free edge of the falx cerebri and is joined by the great cerebral vein (the great vein of Galen) to form the straight sinus.

C. Straight Sinus

—runs along the line of attachment of the falx cerebri to the tentorium cerebelli.

D. Transverse Sinus

—runs laterally from the confluence along the edge of the tentorium cerebelli.

E. Sigmoid Sinus

—is a continuation of the transverse sinus and arches downward and medially in an S-shaped groove on the mastoid part of the temporal bone.
—enters the superior bulb of the internal jugular vein.

F. Cavernous Sinuses

—are located on each side of the sella turcica and the body of the sphenoid bone and lie between the meningeal and periosteal layers of the dura mater.
—the internal carotid artery and the abducens nerve pass through these sinuses.
—the oculomotor, trochlear, ophthalmic, and maxillary nerves pass forward in the lateral wall of these sinuses.
—communicate with the pterygoid venus plexus by emissary veins and receive the superior ophthalmic vein.

G. Superior Petrosal Sinus

—lies in the margin of the tentorium cerebelli, running from the posterior end of the cavernous sinus to the transverse sinus.

H. Inferior Petrosal Sinus

—drains the cavernous sinus into the bulb of the internal jugular vein.
—runs in a groove between the petrous part of the temporal bone and the basilar part of the occipital bone.

I. Sphenoparietal Sinus

—lies along the posterior edge of the lesser wing of the sphenoid bone and drains into the cavernous sinus.

J. Occipital Sinus

—lies in the falx cerebelli and drains into the confluence of sinuses.

K. Basilar Plexus

—consists of interconnecting venous channels on the basilar part of the occipital bone and connects the two inferior petrosal sinuses.
—communicates with the internal vertebral venous plexus.

L. Diploic Veins

—lie in the diploë of the skull and are connected with the cranial dura sinuses by the emissary veins.

M. Emissary Veins

—are small veins connecting the venous sinuses of the dura with the diploic veins and the veins of the scalp.

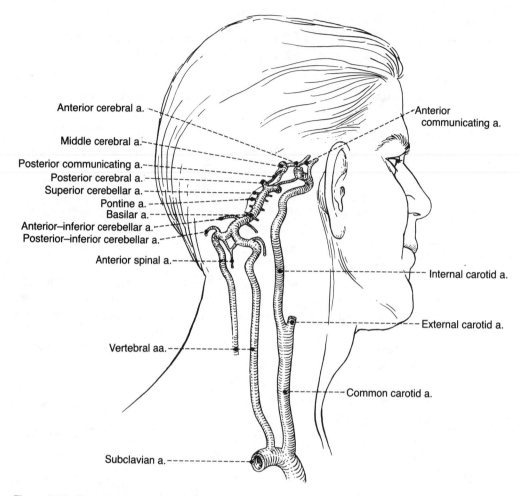

Anterior cerebral a.

Middle cerebral a.

Posterior communicating a.
Posterior cerebral a.
Superior cerebellar a.
Pontine a.
Basilar a.
Anterior–inferior cerebellar a.
Posterior–inferior cerebellar a.

Anterior spinal a.

Vertebral aa.

Subclavian a.

Anterior communicating a.

Internal carotid a.

External carotid a.

Common carotid a.

Figure 8.25. Formation of the circle of Willis.

V. Blood Supply of the Brain (Figures 8.25 and 8.26)

A. Internal Carotid Artery

—gives no branches to the neck.

—enters the carotid canal in the petrous portion of the temporal bone.

—is separated from the tympanic cavity by a thin bony structure.

—lies within the cavernous sinus and gives rise to small twigs to the wall of the cavernous sinus, to the hypophysis, and to the semilunar ganglion of the trigeminal nerve.

—pierces the dural roof of the cavernous sinus between the anterior and middle clinoid processes.

—gives off the **superior hypophyseal, ophthalmic, posterior communicating, anterior choroid, anterior cerebral,** and **middle cerebral arteries.**

B. Vertebral Artery

—is the first branch of the first part of the subclavian artery and ascends through the transverse foramina of the upper six cervical vertebrae.

—curves posteriorly behind the lateral mass of the atlas, pierces the dura mater into the vertebral canal, and then enters the cranial cavity through the foramen magnum.

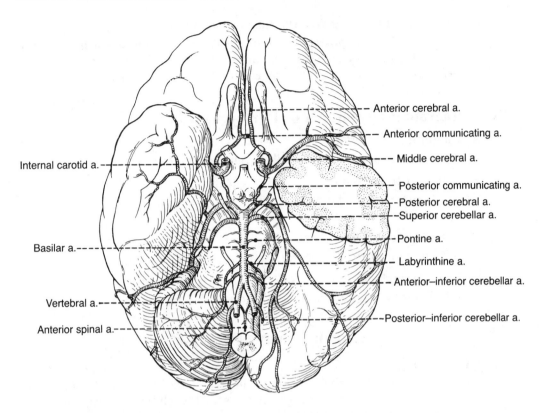

Internal carotid a.

Basilar a.

Vertebral a.

Anterior spinal a.

Anterior cerebral a.

Anterior communicating a.

Middle cerebral a.

Posterior communicating a.

Posterior cerebral a.

Superior cerebellar a.

Pontine a.

Labyrinthine a.

Anterior–inferior cerebellar a.

Posterior–inferior cerebellar a.

Figure 8.26. Arterial circle on the inferior surface of the brain.

—gives rise to the posterior–inferior cerebellar artery and the anterior and posterior spinal arteries and joins to form the basilar artery.

1. **Anterior Spinal Artery**
   —arises as two roots from the vertebral arteries shortly before two vertebral arteries join in the basilar artery.
   —descends in front of the medulla, and the two roots unite to form a single median trunk at the level of the foramen magnum.

2. **Posterior Spinal Artery**
   —arises from the vertebral artery or the posterior–inferior cerebellar artery and descends on the side of the medulla; the right and left roots unite at the lower cervical region.

3. **Posterior–Inferior Cerebellar Artery**
   —is the largest branch of the vertebral artery, distributes to the posterior–inferior surface of the cerebellum, and gives rise to the posterior spinal artery.

C. **Basilar Artery**
   —is formed by the union of the two vertebral arteries at the lower border of the pons.
   —gives rise to the **pontine, anterior–inferior cerebellar, labyrinthine,** and **superior cerebellar arteries;** ends near the upper border of the pons by dividing into the **right** and **left posterior cerebral arteries.**

D. **Circle of Willis (Circulus Arteriosus)** [see Figure 8.26]

—is formed by the **posterior cerebral, posterior communicating, internal carotid, anterior cerebral,** and **anterior communicating arteries**.

—forms an important means of collateral circulation in the event of obstruction.

VI. Intracranial Hematoma or Hemorrhage

A. Epidural Hematoma

—is due to rupture of the middle meningeal artery.

B. Subdural Hematoma

—is due to rupture of cerebral veins as they pass from the brain surface into one of the venous sinuses.

C. Subarachnoid Hemorrhage

—is due to rupture of cerebral arteries.

D. Pial Hemorrhage

—is due to damage to the small vessels of the pia and brain tissue.

Cranial Nerves (Figure 8.27)

I. Olfactory Nerves

—consist of about 20 bundles of unmyelinated special visceral afferent (SVA) fibers that arise from olfactory neurons in the olfactory area, the upper one-third of the nasal mucosa.

—pass through the foramina in the cribriform plate of the ethmoid bone and enter the olfactory bulb, where they synapse.

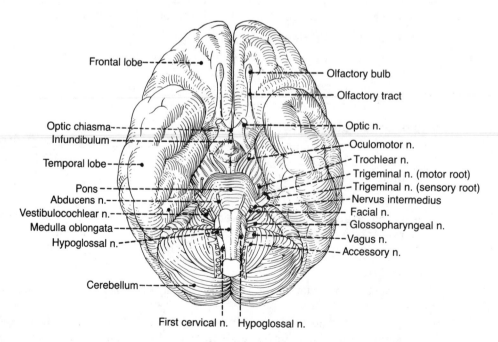

Figure 8.27. Cranial nerves on the base of the brain.

II. Optic Nerve

—is formed by the axons of ganglion cells of the retina, which converge at the optic disk. This nerve carries special somatic afferent (SSA) fibers from the retina to the brain.

—leaves the orbit through the optic canal and forms the optic chiasma (chiasma means "cross"), where fibers from the nasal side of either retina cross over to the opposite side of the brain.

III. Oculomotor Nerve

—enters the orbit through the superior orbital fissure within the tendinous ring.

—supplies general somatic efferent (GSE) fibers to the extraocular muscles (i.e., medial, superior, and inferior recti; inferior oblique; and levator palpebrae superioris).

—contains preganglionic parasympathetic general visceral efferent (GVE) fibers with cell bodies located in the Edinger-Westphal nucleus, and postganglionic fibers derived from the ciliary ganglion that run in the short ciliary nerves to supply the sphincter pupillae and the ciliary muscle.

IV. Trochlear Nerve

—passes through the lateral wall of the cavernous sinus during its course.

—enters the orbit by passing through the superior orbital fissure and supplies GSE fibers to the superior oblique muscle.

—is the smallest cranial nerve and the only cranial nerve that emerges from the dorsal aspect of the brainstem.

V. Trigeminal Nerve

—is the first branchiomeric nerve and supplies the first branchial arch.

—is sensory (general somatic afferent [GSA]) to the face, scalp, auricle, external auditory meatus, nose, paranasal sinuses, mouth (except the posterior one-third of the tongue), parts of the nasopharynx, auditory tube, and cranial dura mater.

—has a ganglion (semilunar or trigeminal) that consists of cell bodies of GSA fibers and occupies the trigeminal impression on the petrous portion of the temporal bone.

—divides into the ophthalmic, maxillary, and mandibular divisions.

A. Ophthalmic Nerve (see p 298)

—runs in the dura of the lateral wall of the cavernous sinus and enters the orbit through the superior orbital fissure.

—is sensory to the eyeball, tip of the nose, and skin of the face above the eye.

—mediates the afferent limb of the corneal reflex by way of the nasociliary branch.

B. Maxillary Nerve (see p 320)

—passes through the lateral wall of the cavernous sinus and through the foramen rotundum.

—is sensory to the midface (below the eye but above the upper lip), palate, paranasal sinuses, and maxillary teeth, with cell bodies in the trigeminal ganglion.

—mediates the afferent limb of the sneeze reflex.

C. Mandibular Nerve (see p 276)

—passes through the foramen ovale and supplies SVE fibers to the tensor veli palatini, tensor tympani, muscles of mastication (temporalis, masseter, and lateral and medial pterygoid), and the anterior belly of the digastric and mylohyoid muscles.

—is sensory to the lower part of the face (below the lower lip and mouth), scalp, jaw, mandibular teeth, and anterior two-thirds of the tongue.

—mediates the afferent limb of the jaw jerk reflex.

VI. Abducens Nerve

—pierces the dura on the dorsum sellae of the sphenoid bone.

—passes through the cavernous sinus, enters the orbit through the superior orbital fissure, and supplies GSE fibers to the lateral rectus muscle.

VII. Facial Nerve (Figure 8.28)

—consists of a larger root, which contains special visceral efferent (SVE) fibers to supply the muscles of facial expression, and a smaller root, termed the **nervus intermedius,** which contains SVA fibers from the anterior two-thirds of the tongue; preganglionic parasympathetic GVE fibers for the lacrimal, submandibular, sublingual, nasal, and palatine glands; general visceral afferent (GVA) fibers from the palate; and general somatic afferent (GSA) fibers from the external acoustic meatus and the auricle.

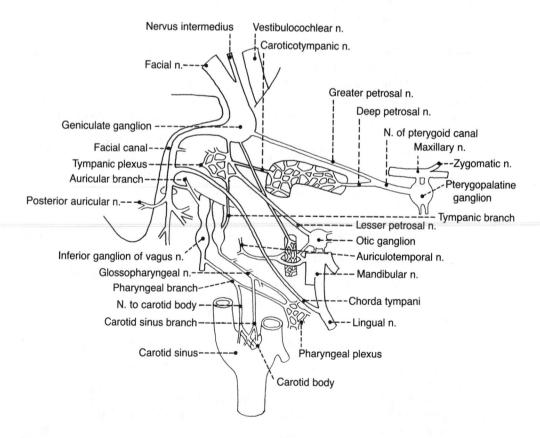

Figure 8.28. Facial nerve and its connections with other nerves.

—enters the internal acoustic meatus, the facial canal in the temporal bone, and emerges from the stylomastoid foramen.

—has a sensory ganglion, the **geniculate ganglion,** which lies at the knee-shaped bend or genu (L., "knee"). It gives off the greater petrosal nerve (see p 332), which carries preganglionic parasympathetic fibers to the pterygopalatine ganglion (for the lacrimal, nasal, and palatine glands); a branch to the tympanic plexus; and a branch to the sympathetic plexus on the middle meningeal artery.

—gives off the following branches:

A. Greater Petrosal Nerve

—contains preganglionic parasympathetic GVE fibers and joins the deep petrosal nerve (containing postganglionic sympathetic fibers) to form the nerve of the pterygoid canal (vidian nerve).

—also contains SVA (taste) and GVA fibers, which pass from the palate through the pterygopalatine ganglion, the nerve of the pterygoid canal, and the greater petrosal nerve to the geniculate ganglion (where cell bodies are located).

B. Communicating Branch

—joins the lesser petrosal nerve.

C. Stapedial Nerve

—supplies motor (SVE) fibers to the stapedius muscle.

D. Chorda Tympani

—arises in the descending part of the facial canal, crosses the medial aspect of the tympanic membrane, passing between the handle of the malleus and the long process of the incus.

—exits the skull through the petrotympanic fissure and joins the lingual nerve in the infratemporal fossa.

—contains preganglionic parasympathetic GVE fibers that synapse on the postganglionic cell bodies in the submandibular ganglion. Their postganglionic fibers innervate the submandibular, sublingual, and lingual glands.

—also contains taste (SVA) fibers from the anterior two-thirds of the tongue and the soft palate, with cell bodies located in the geniculate ganglion.

—may communicate with the otic ganglion below the base of the skull.

E. Muscular Branches

—supply motor (SVE) fibers to the stylohyoid and the posterior belly of the digastric muscle.

F. Fine Branch

—joins the auricular branch of the vagus nerve and the glossopharyngeal nerve to supply GSA fibers to the external ear.

G. Posterior Auricular Nerve

—runs behind the auricle with the posterior auricular artery.

—supplies SVE fibers to the muscles of the auricle and the occipital belly of the occipitofrontalis muscle.

H. Terminal Branches

—arise in the parotid gland and radiate onto the face.

—include the temporal, zygomatic, buccal, marginal mandibular, and cervical branches, which supply motor (SVE) fibers to the muscles of facial expression.

VIII. Vestibulocochlear (Acoustic) Nerve

—enters the internal acoustic meatus and remains within the temporal bone to supply SSA fibers to the cochlea, the ampullae of the semicircular ducts, and the utricle and saccule.

—is split into a cochlear portion for hearing and a vestibular portion for equilibrium.

IX. Glossopharyngeal Nerve (see Figure 8.28)

—passes through the jugular foramen and gives rise to the following branches:

A. Tympanic Nerve

—forms the tympanic plexus (see p 290) on the medial wall of the middle ear with sympathetic fibers from the internal carotid plexus (caroticotympanic nerves) and a branch from the geniculate ganglion of the facial nerve. The plexus supplies GVA fibers to the tympanic cavity, the mastoid antrum and air cells, and the auditory tube.

—continues beyond the plexus as the **lesser petrosal nerve** (see p 333), which transmits preganglionic parasympathetic GVE fibers to the otic ganglion.

B. Communicating Branch

—joins the auricular branch of the vagus nerve and provides GSA fibers.

C. Pharyngeal Branch

—supplies GVA fibers to the pharynx and forms the **pharyngeal plexus** (see p 315) on the middle constrictor muscle along with the pharyngeal branch of the vagus nerve and branches from the sympathetic trunk.

D. Carotid Sinus Branch

—supplies GVA fibers to the carotid sinus, a pressoreceptor, and also to the carotid body, a chemoreceptor.

E. Tonsilar Branches

—supply GVA fibers to the palatine tonsil and the soft palate.

F. Motor Branch

—supplies SVE fibers to the stylopharyngeus muscle.

G. Lingual Branch

—supplies GVA and SVA fibers to the posterior one-third of the tongue and SVA fibers to the vallate papillae.

X. Vagus Nerve (see p 136)

—supplies the fourth to sixth branchial arches during development.

—passes through the jugular foramen.

—is motor (GVE) to smooth muscle and cardiac muscle, secretory to all glands, and afferent (GVA) from all mucous membranes in the thoracic and abdominal visceral organs (except for the descending colon, sigmoid colon, rectum, and other pelvic organs).

—is branchiomotor (SVE) to all muscles of the larynx, pharynx (except the stylopharyngeus), and palate (except the tensor veli palatini).

—gives off the following branches:

A. Meningeal Branch

—arises from the superior ganglion and supplies the dura mater of the posterior cranial fossa.

B. Auricular Branch

—is joined by a branch from the glossopharyngeal nerve and the facial nerve and supplies GSA fibers to the external acoustic meatus.

C. Pharyngeal branch

—supplies motor (SVE) fibers to all muscles of the pharynx, except the stylopharyngeus muscle, by way of the pharyngeal plexus and all muscles of the palate except the tensor veli palatini muscle.

—gives off the nerve to the carotid body, which supplies the carotid body and the carotid sinus.

D. Superior, Middle, and Inferior Cardiac Branches

—pass to the cardiac plexuses.

E. Superior Laryngeal Nerve

—divides into internal and external laryngeal nerves.

1. Internal Laryngeal Nerve

—is sensory (GVA) to the larynx above the vocal cord.

—supplies SVA fibers to the taste buds on the root of the tongue near the epiglottis.

2. External Laryngeal Nerve

—supplies motor (SVE) fibers to the cricothyroid and inferior pharyngeal constrictor muscles.

F. Recurrent Laryngeal Nerve

—hooks around the subclavian artery on the right and around the arch of the aorta lateral to the ligamentum arteriosum on the left.

—ascends in the groove between the trachea and the esophagus.

—is sensory (GVA) to the larynx below the vocal cord and is motor to all muscles of the larynx except the cricothyroid muscle.

—becomes the inferior laryngeal nerve at the lower border of the cricoid cartilage.

XI. Accessory Nerve

—passes through the jugular foramen.

—its spinal roots unite to form the trunk that ascends between dorsal and ventral roots of the spinal nerves and passes through the foramen magnum.

—is branchiomotor (SVE) to the sternocleidomastoid and trapezius muscles.

—its cranial portion contains motor fibers that join the vagus nerve.

XII. Hypoglossal Nerve

—passes through the hypoglossal canal.

—loops around the occipital artery and passes between the external carotid and internal jugular vessels.

—passes above the hyoid bone on the lateral surface of the hyoglossus muscle and deep to the mylohyoid muscle.

—supplies GSE fibers to all of the intrinsic and extrinsic muscles of the tongue except the palatoglossus muscle, which is supplied by the vagus nerve.

XIII. Cranial Nerves

Nerve	Cranial Exit	Cell Bodies	Components	Chief Functions
I: Olfactory	Cribriform plate	Nasal mucosa	SVA	Smell
II: Optic	Optic canal	Ganglion cells of retina	SSA	Vision
III: Oculomotor	Superior orbital fissure	Nucleus CN III (midbrain)	GSE	Eye movements (superior, inferior, and medial recti, inferior oblique, and levator palpebrae superioris mm.)
		Edinger-Westphal nucleus (midbrain)	GVE	Constriction of pupil (sphincter pupillae m.) and accommodation (ciliary m.)
IV: Trochlear	Superior orbital fissure	Nucleus CN IV (midbrain)	GSE	Eye movements (superior oblique m.)
V: Trigeminal	Superior orbital fissure; foramen rotundum and foramen ovale	Motor nucleus CN V (pons)	SVE	Muscles of mastication, (mylohyoid, anterior belly of digastric, tensor veli palatini, and tensor tympani mm.)
		Trigeminal ganglion	GSA	Sensation in head (skin and mucous membranes of face and head)
VI: Abducens	Superior orbital fissure	Nucleus CN VI (pons)	GSE	Eye movement (lateral rectus m.)
VII: Facial	Stylomastoid foramen	Motor nucleus CN VII (pons)	SVE	Muscle of facial expression (posterior belly of digastric, stylohyoid, and stapedius mm.)
		Salivatory nucleus (pons)	GVE	Lacrimal and salivary secretion
		Geniculate ganglion	SVA	Taste from anterior two-thirds of tongue and palate
		Geniculate ganglion	GVA	Sensation from palate
		Geniculate ganglion	GSA	Sensation from external acoustic meatus

(Continued on next page)

Nerve	Cranial Exit	Cell Bodies	Components	Chief Functions
VIII: Vestibulocochlear	Does not leave skull	Vestibular ganglion Spiral ganglion	SSA SSA	Equilibrium Hearing
IX: Glossopharyngeal	Jugular foramen	Nucleus ambiguus (medulla)	SVE	Elevation of pharynx (stylopharyngeus m.)
		Dorsal nucleus (medulla)	GVE	Secretion of saliva (parotid gland)
		Inferior ganglion	GVA	Sensation in carotid sinus and body, tongue, and pharynx
		Inferior ganglion	SVA	Taste from posterior one-third of tongue
		Inferior ganglion	GSA	Sensation in external and middle ear
X: Vagus	Jugular foramen	Nucleus ambiguus (medulla)	SVE	Muscles of movements of pharynx, larynx, and palate
		Dorsal nucleus (medulla)	GVE	Involuntary muscle and gland control in thoracic and abdominal viscerae
		Inferior ganglion	GVA	Sensation in pharynx, larynx, and other viscerae
		Inferior ganglion	SVA	Taste from root of tongue and epiglottis
		Superior ganglion	GSA	Sensation in external ear and external acoustic meatus
XI: Accessory	Jugular foramen	Spinal cord (cervical)	SVE	Movement of head and shoulder (sternocleidomastoid and trapezius mm.)
XII: Hypoglossal	Hypoglossal canal	Nucleus CN XII (medulla)	GSE	Muscles of movements of tongue

Cranial Foramina

I. Interior of the Skull (see Figure 8.23)

A. Anterior Cranial Fossa

1. **Cribriform plate:** olfactory nerves.

2. **Foramen cecum:** occasional small emissary vein from nasal mucosa to superior sagittal sinus.

B. **Middle Cranial Fossa**

1. **Optic canal:** optic nerve and ophthalmic artery.

2. **Superior orbital fissure:** oculomotor, trochlear, and abducens nerves; ophthalmic division of trigeminal nerve; and ophthalmic vein.

3. **Foramen rotundum:** maxillary division of trigeminal nerve.

4. **Foramen ovale:** mandibular division of trigeminal nerve, accessory meningeal artery, and occasionally lesser petrosal nerve.

5. **Foramen spinosum:** middle meningeal artery.

6. **Foramen lacerum:** upper part traversed by internal carotid artery and greater and deep petrosal nerves enroute to the pterygoid canal.

7. **Carotid canal:** internal carotid artery and sympathetic nerves (carotid plexus).

C. **Posterior Cranial Fossa**

1. **Internal auditory meatus:** facial and vestibulocochlear nerves and labyrinthine artery.

2. **Jugular foramen:** glossopharyngeal nerve, vagus nerve, spinal accessory nerve, and beginning of internal jugular vein.

3. **Hypoglossal canal:** hypoglossal nerve and meningeal artery.

4. **Foramen magnum:** spinal cord, spinal accessory nerve, vertebral arteries, venous plexus of vertebral canal, and anterior and posterior spinal arteries.

5. **Condyloid foramen:** condyloid emissary vein.

6. **Mastoid foramen:** branch of occipital artery to dura mater and mastoid emissary vein.

II. Front of the Skull (see Figure 8.9)

A. **Zygomaticofacial foramen:** zygomaticofacial nerve.

B. **Supraorbital notch:** supraorbital nerve and vessels.

C. **Infraorbital foramen:** infraorbital nerve and vessels.

D. **Mental foramen:** mental nerve and vessels.

III. Base of the Skull (Figure 8.29)

A. **Foramen ovale:** mandibular nerve.

B. **Foramen spinosum:** middle meningeal artery.

C. **Foramen lacerum:** upper part traversed by internal carotid artery and greater and deep petrosal nerves.

D. **Carotid canal:** internal carotid artery and sympathetic nerve (carotid plexus).

E. **Jugular foramen:** cranial nerves IX, X, and XI and jugular vein.

F. **Hypoglossal canal:** hypoglossal nerve and meningeal artery.

G. **Foramen magnum:** spinal cord, spinal accessory nerve, vertebral artery, and venous plexus.

H. **Petrotympanic fissure:** chorda tympani.

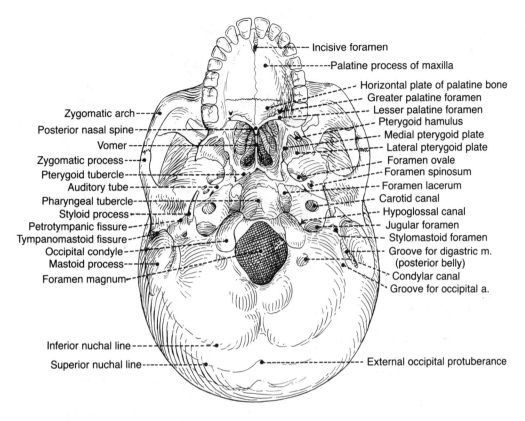

Figure 8.29. Base of the skull.

    I. Stylomastoid foramen: facial nerve.

    J. Incisive canal: nasopalatine nerve.

    K. Greater palatine canal: greater palatine nerve and vessels.

    L. Lesser palatine canal: lesser palatine nerve and vessels.

Orbit

I. Bony Orbit (Figure 8.30)

A. Orbital Margin

—is formed by the frontal, maxillary, and zygomatic bones.

B. Walls of the Orbit

1. **Superior wall or roof:** orbital part of frontal bone and lesser wing of sphenoid bone.

2. **Lateral wall:** zygomatic bone (frontal process) and greater wing of sphenoid bone.

3. **Inferior wall or floor:** maxilla (orbital surface), zygomatic, and palatine bones.

4. **Medial wall:** ethmoid bone (orbital plate), frontal lacrimal bone, and body of sphenoid bone.

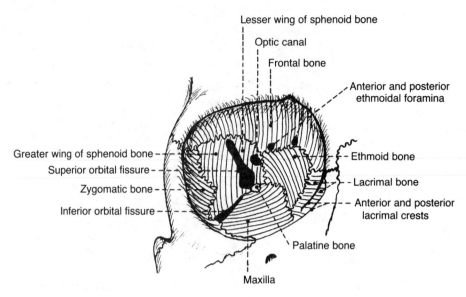

Figure 8.30. Bony orbit.

C. Fissures, Canals, and Foramina

1. Superior Orbital Fissure

—communicates with the middle cranial fossa and is bounded by the greater and lesser wings of the sphenoid.

—transmits the oculomotor, trochlear, abducens, and ophthalmic nerves (three branches), as well as the ophthalmic veins.

2. Interior Orbital Fissure

—communicates with the infratemporal and pterygopalatine fossae.

—is bounded by the greater wing of the sphenoid (above) and the maxillary and palatine bones (below).

—transmits the maxillary nerve and its zygomatic branch and the infraorbital vessels.

3. Optic Canal

—connects the orbit with the middle cranial fossa.

—is formed by the two roots of the lesser wing of the sphenoid and is situated in the posterior part of the roof of the orbit.

—transmits the optic nerve and ophthalmic artery.

4. Infraorbital Groove and Infraorbital Foramen

—transmit the infraorbital nerve and vessels.

5. Supraorbital Notch or Foramen

—transmits the supraorbital nerve and vessels.

6. Anterior and Posterior Ethmoidal Foramina

—transmit the anterior and posterior ethmoidal nerves and vessels, respectively.

7. Nasolacrimal Canal

—is formed by the maxilla, lacrimal bone, and inferior nasal concha.

—transmits the nasolacrimal duct from the lacrimal sac to the inferior nasal meatus.

II. Nerves of the Orbit (Figures 8.31 and 8.32)

A. Ophthalmic Nerve

—enters the orbit through the superior orbital fissure and divides into three branches: lacrimal nerve, frontal nerve, and nasociliary nerve.

1. Lacrimal Nerve

—enters the orbit through the superior orbital fissure.

—enters the lacrimal gland, giving off branches to the lacrimal gland, the conjunctiva, and the skin of the upper eyelid.

2. Frontal Nerve

—enters the orbit through the superior orbital fissure.

—runs superior to the levator palpebrae superioris muscle.

—divides into the **supraorbital nerve,** which passes through the supraorbital notch or foramen and supplies the scalp, forehead, frontal sinus, and upper eyelid; and the **supratrochlear nerve,** which passes through the trochlea and supplies the scalp, forehead, and upper eyelid.

3. Nasociliary Nerve

—is the sensory nerve to the eye.

—enters the orbit through the superior orbital fissure, within the common tendinous ring.

—gives off:

a. A **communicating branch** to the ciliary ganglion.

b. **Short ciliary nerves,** which carry postganglionic parasympathetic and sympathetic fibers to the ciliary body and iris.

c. **Long ciliary nerves,** which transmit postganglionic sympathetic fibers to the dilator pupillae and afferent fibers from the iris and cornea.

d. The **posterior ethmoidal nerve,** which passes through the posterior ethmoidal foramen to the sphenoidal and posterior ethmoidal sinuses.

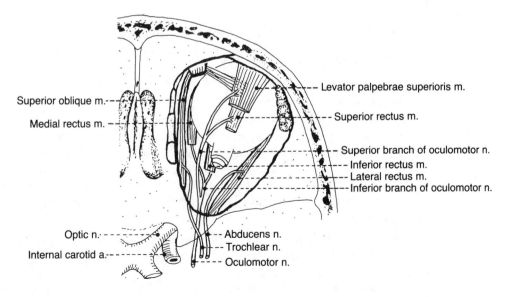

Figure 8.31. Motor nerves of the orbit.

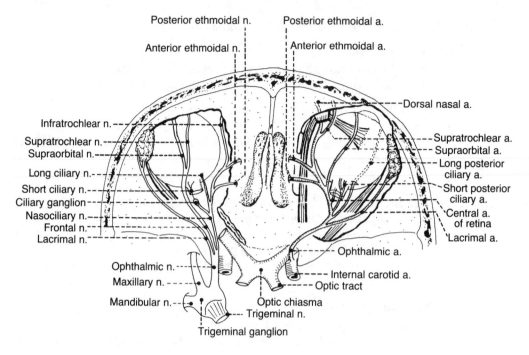

Posterior ethmoidal n.

Posterior ethmoidal a.

Anterior ethmoidal n.

Anterior ethmoidal a.

Dorsal nasal a.

Infratrochlear n.

Supratrochlear n.

Supraorbital n.

Long ciliary n.

Short ciliary n.

Ciliary ganglion

Nasociliary n.

Frontal n.

Lacrimal n.

Supratrochlear a.

Supraorbital a.

Long posterior ciliary a.

Short posterior ciliary a.

Central a. of retina

Lacrimal a.

Ophthalmic a.

Ophthalmic n.

Maxillary n.

Mandibular n.

Internal carotid a.

Optic tract

Optic chiasma

Trigeminal n.

Trigeminal ganglion

Figure 8.32. Branches of the ophthalmic nerve and ophthalmic artery.

    e. The **anterior ethmoidal nerve,** which passes through the anterior ethmoidal foramen to supply the anterior ethmoidal air cells. It divides into internal nasal branches, which supply the septum and lateral walls of the nasal cavity, and external nasal branches, which supply the skin of the tip of the nose.

    f. The **infratrochlear nerve** to the eyelids, conjunctiva, skin of the nose, and lacrimal sac.

B. Optic Nerve

—consists of the axons of the ganglion cells of the retina.

—leaves the orbit by passing through the optic canal and carries afferent optic fibers from the retina to the brain.

—joins the optic nerve from the corresponding eye to form the optic chiasma.

C. Oculomotor Nerve

—leaves the cranium through the superior orbital fissure.

—divides into a superior division, which innervates the superior rectus and levator palpebrae superioris muscles, and an inferior division, which innervates the medial rectus, inferior rectus, and inferior oblique muscles.

—its inferior division also carries preganglionic parasympathetic fibers (with cell bodies located in the Edinger-Westphal nucleus) to the ciliary ganglion.

D. Trochlear Nerve

—passes through the lateral wall of the cavernous sinus during its course.

—enters the orbit by passing through the superior orbital fissure and supplies the superior oblique muscle.

—is the smallest cranial nerve and the only cranial nerve that emerges from the dorsal aspect of the brainstem.

E. **Abducens Nerve** (see p 290)

—enters the orbit through the superior orbital fissure and supplies the lateral rectus muscle.

F. **Ciliary Ganglion** (see p 331)

—is a parasympathetic ganglion situated behind the eyeball, between the optic nerve and the lateral rectus muscle.

III. Blood Vessels of the Orbit (see Figure 8.32)

A. Ophthalmic Artery

—is a branch of the internal carotid artery and enters the orbit through the optic canal beneath the optic nerve.

—gives off the **ocular vessels**, such as the artery of the retina, the short and long posterior ciliary and anterior ciliary arteries, and the **orbital vessels**, such as the lacrimal, muscular, supraorbital, anterior and posterior ethmoidal, medial palpebral, supratrochlear, and dorsal nasal arteries.

—ends by dividing into the dorsal nasal and supratrochlear arteries.

1. Central Artery of the Retina

—is the most important branch of the ophthalmic artery.

—travels in the optic nerve; it divides into superior and inferior branches at the optic disk, and each of those further divides into temporal and nasal branches.

—is an end artery that does not anastomose with other arteries, and thus its occlusion results in blindness.

2. Long Posterior Ciliary Arteries

—pierce the sclera and supply the ciliary body and the iris.

3. Short Posterior Ciliary Arteries

—pierce the sclera and supply the choroid.

4. Lacrimal Artery

—passes along the superior border of the lateral rectus and supplies the lacrimal gland, conjuntiva, and eyelids.

—gives off two lateral palpebral arteries, which contribute to arcades in the upper and lower eyelids.

5. Medial Palpebral Arteries

—contribute to arcades in the upper and lower eyelids.

6. Muscular Branches

—supply orbital muscles and give off the anterior ciliary arteries, which supply the iris.

7. Supraorbital Artery

—passes through the supraorbital notch (or foramen) and supplies the forehead and the scalp.

8. Posterior Ethmoidal Artery

—passes through the posterior ethmoidal foramen to the posterior ethmoidal air cells.

9. Anterior Ethmoidal Artery

—passes through the anterior ethmoidal foramen to the anterior and middle ethmoidal air cells, frontal sinus, nasal cavity, and external nose.

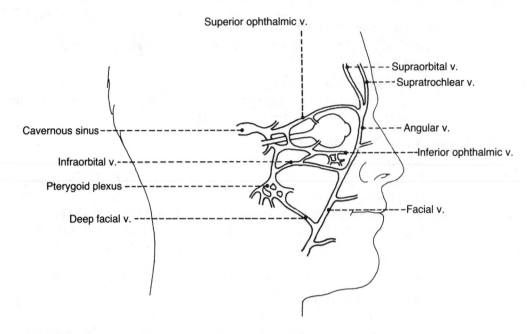

Figure 8.33. Ophthalmic veins.

10. Supratrochlear Artery
—passes to the supraorbital margin and supplies the forehead and the scalp.

11. Dorsal nasal artery
—supplies the side of the nose and the lacrimal sac.

B. Ophthalmic Veins (Figure 8.33)

1. Superior Ophthalmic Vein
—is formed by the union of the supraorbital, supratrochlear, and angular veins.

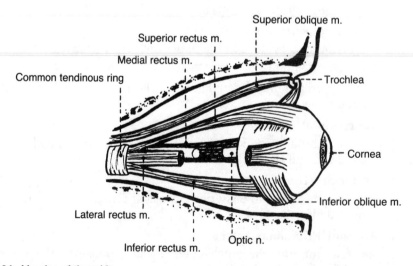

Figure 8.34. Muscles of the orbit.

—receives branches corresponding to most of those of the ophthalmic artery and, in addition, receives the inferior ophthalmic vein before draining into the cavernous sinus.

2. Inferior Ophthalmic Vein
—begins by the union of small veins in the floor of the orbit.
—communicates with the pterygoid venous plexus and often with the infraorbital vein and terminates directly or indirectly in the cavernous sinus.

IV. Muscles of Eye Movement (Figures 8.34–8.36)

Muscle	Origin	Insertion	Nerve	Actions
Superior rectus	Common tendinous ring	Sclera just behind cornea	Oculomotor n.	Elevates eyeball
Inferior rectus	Common tendinous ring	Sclera just behind cornea	Oculomotor n.	Depresses eyeball
Medial rectus	Common tendinous ring	Sclera just behind cornea	Oculomotor n.	Adducts eyeball
Lateral rectus	Common tendinous ring	Sclera just behind cornea	Abducens n.	Abducts eyeball
Levator palpebrae superioris	Lesser wing of sphenoid above and anterior to optic canal	Tarsal plate and skin of upper eyelid	Oculomotor n.	Elevates upper eyelid
Superior oblique	Body of sphenoid bone above optic canal	Sclera beneath superior rectus	Trochlear n.	Rotates downward and medially; depresses adducted eye
Inferior oblique	Floor of orbit lateral to lacrimal groove	Sclera beneath lateral rectus	Oculomotor n.	Rotates upward and laterally; elevates adducted eye

A. Innervation of Muscles of the Eyeball
—can be summarized as **SO$_4$**, **LR$_6$**, and **Remainder$_3$**, which means that the superior oblique muscle is innervated by the trochlear nerve, the lateral rectus by the abducens nerve, and the remainder by the oculomotor nerve.

B. Intorsion
—is an inward rotation of the upper pole of the vertical corneal meridians, caused by the inferior oblique and inferior rectus muscles.

C. Extorsion
—is an outward rotation of the upper pole of the vertical corneal meridians, caused by the inferior oblique and inferior rectus muscles.

V. Lacrimal Apparatus (Figure 8.37)

A. Lacrimal Gland
—lies in the upper lateral region of the orbit on the lateral rectus and the levator palpebrae superioris muscles.

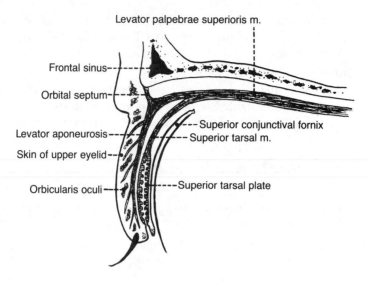

Figure 8.35. Structure of the upper eyelid.

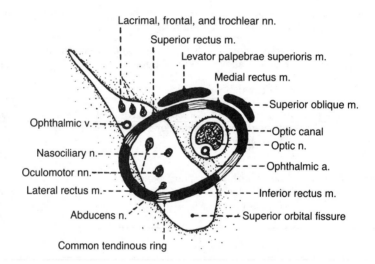

Figure 8.36. Common tendinous ring.

—is drained by 12 lacrimal ducts, which open into the superior conjunctival fornix.

B. Lacrimal Canaliculi

—are two curved canals, which begin as a **lacrimal punctum** (or pore) in the margin of the eyelid and open into the lacrimal sac.

C. Lacrimal Sac

—is the upper dilated end of the **nasolacrimal duct,** which opens into the inferior meatus of the nasal cavity.

D. Tears

—are produced by the lacrimal gland.
—pass through excretory ductules into the superior conjunctival fornix.

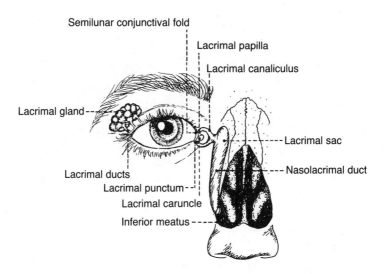

Semilunar conjunctival fold
Lacrimal papilla
Lacrimal canaliculus
Lacrimal gland
Lacrimal sac
Lacrimal ducts
Nasolacrimal duct
Lacrimal punctum
Lacrimal caruncle
Inferior meatus

Figure 8.37. Lacrimal apparatus.

—are spread evenly over the eyeball by blinking movements and accumulate in the area of the **lacrimal lake.**

—enter the **lacrimal canaliculi** through their **lacrimal puncta** (which is on the summit of the lacrimal papilla) before draining into the lacrimal sac, nasolacrimal duct, and finally the inferior nasal meatus.

VI. Characteristics of the Eyeball (Figure 8.38)

A. Optic Disk (Blind Spot)

—consists merely of optic nerve fibers and marks their exit.

—is located nasal (or medial) to the fovea centralis and the posterior pole of the eye, has no receptors, and is insensitive to light.

—has a depression in its center termed the **"physiological cup."**

B. Macula (Yellow Spot or Macula Lutea)

—is a yellowish area of the retina on the temporal side of the optic disk.

—contains the **fovea centralis.**

C. Fovea Centralis

—is a central depression (foveola) in the macula.

—is avascular and is nourished by the choriocapillary lamina of the choroid.

—has cones only (no rods), each of which is connected with only one ganglion cell, and functions in detailed vision.

D. Rods

—are 100 to 120 million in number and are most numerous about 0.5 cm from the fovea centralis.

—contain rhodopsin, a visual purple pigment, and are specialized for vision in dim light.

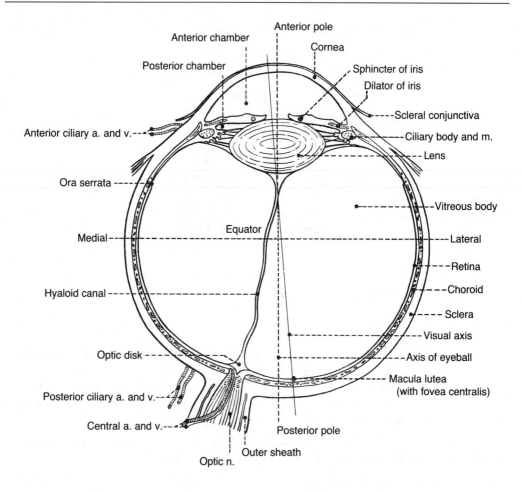

Figure 8.38. Horizontal section of the eyeball.

E. Cones

—are 7 million in number and are most numerous in the foveal region.
—are associated with visual acuity (sharpest vision) and color vision.

VII. Additional Characteristics

A. Structures Passing Through the Common Tendinous Ring (see Figure 8.36)

1. The **oculomotor, nasociliary,** and **abducens nerves** enter the orbit through the superior orbital fissure and the common tendinous ring.
2. The **optic nerve, ophthalmic artery,** and **central artery** of the retina enter the orbit through the optic canal and the tendinous ring.
3. The **superior ophthalmic vein** and the **trochlear, frontal,** and **lacrimal nerves** enter the orbit through the superior orbital fissure but outside the tendinous ring.

B. Horner's Syndrome

—is caused by injury to cervical sympathetic fibers.
—is characterized by:

1. **Miosis:** constriction of a pupil as a result of paralysis of the associated dilator muscle of the pupil.

2. **Ptosis:** drooping of an upper eyelid as a result of paralysis of the smooth muscle component of the levator palpebrae superioris muscle.
3. **Enophthalmos:** retraction of an eyeball as a result of paralysis of the tarsal muscle.
4. **Anhidrosis:** absence of sweating.
5. **Vasodilation:** dilation of arteries leading to increased blood flow in the facial and cervical regions.

Oral Cavity and Salivary Glands

I. Oral Cavity (Figure 8.39)
—its roof is formed by the palate, and its floor is formed by the tongue and the mucosa, supported by the geniohyoid and mylohyoid muscles.
—its lateral and anterior walls are formed by an outer fleshy wall (cheeks and lips) and an inner bony wall (teeth and gums). (The vestibule is between the walls, and the oral cavity proper is the area inside the teeth and gums.)

II. Palate (see Figure 8.39)
—forms the roof of the mouth and the floor of the nasal cavity.
—consists of the hard palate (anterior two-thirds) and the soft palate (posterior one-third).

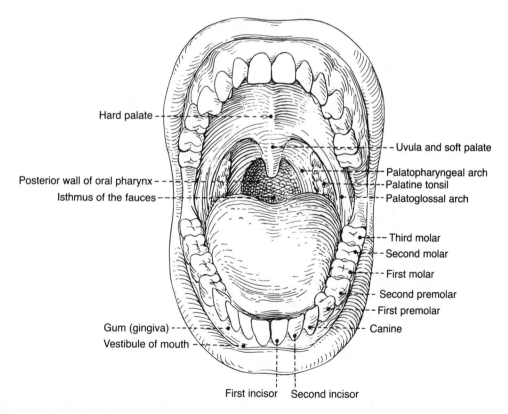

Figure 8.39. Oral cavity.

A. Hard Palate

—forms a bony partition between the nasal and oral cavities.

—consists of the palatine processes of the **maxillae** and horizontal parts of the **palatine bones.**

—contains the incisive foramen in its median plane anteriorly and the greater and lesser palatine foramina posteriorly.

B. Soft Palate

—is a fibromuscular fold extending from the posterior border of the hard palate.

—moves posteriorly against the pharyngeal wall to close the oropharyngeal (faucial) isthmus when swallowing or speaking.

—is continuous with the palatoglossal and palatopharyngeal folds.

—receives blood from the greater and lesser palatine arteries of the descending palatine artery of the maxillary artery, the ascending palatine artery of the facial artery, and the palatine branch of the ascending pharyngeal artery.

—receives sensory innervation through the greater and lesser palatine nerves.

C. Muscles of the Palate (see Figure 8.43)

Muscle	Origin	Insertion	Nerve	Action
Tensor veli palatini	Scaphoid fossa; spine of sphenoid; cartilage of auditory tube	Tendon hooks around hamulus of medial pterygoid plate to insert into aponeurosis of soft palate	Mandibular branch of trigeminal n.	Tenses soft palate
Levator veli palatini	Petrous part of temporal bone; cartilage of auditory tube	Aponeurosis of soft palate	Vagus n. via pharyngeal plexus	Elevates soft palate
Palatoglossus	Aponeurosis of soft palate	Dorsolateral side of tongue	Vagus n. via pharyngeal plexus	Elevates tongue
Palatopharyngeus	Aponeurosis of soft palate	Thyroid cartilage and side of pharynx	Vagus n. via pharyngeal plexus	Elevates pharynx; closes nasopharynx
Musculus uvulae	Posterior nasal spine of palatine bone; palatine aponeurosis	Mucous membrane of uvula	Vagus n. via pharyngeal plexus	Elevates uvula

III. Tongue (Figure 8.40)

—is attached by muscles to the hyoid bone, mandible, styloid process, palate, and pharynx.

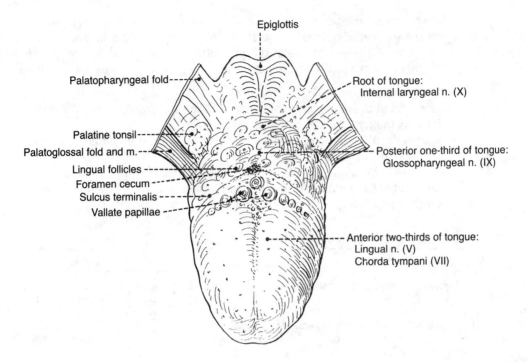

Epiglottis

Palatopharyngeal fold

Root of tongue:
Internal laryngeal n. (X)

Palatine tonsil

Palatoglossal fold and m.

Lingual follicles

Foramen cecum

Sulcus terminalis

Vallate papillae

Posterior one-third of tongue:
Glossopharyngeal n. (IX)

Anterior two-thirds of tongue:
Lingual n. (V)
Chorda tympani (VII)

Figure 8.40. Tongue.

—is divided by a V-shaped **sulcus terminalis** into two parts, an anterior two-thirds and a posterior one-third, which differ developmentally, structurally, and in innervation.

—has the **foramen cecum** at the apex of the V, which indicates the site of origin of the embryonic thyroglossal duct.

A. **Lingual Papillae**

—are small, nipple-shaped projections on the anterior two-thirds of the dorsum of the tongue.

—include the **vallate, fungiform, filiform,** and **foliate papillae**.

1. **Vallate Papillae**

—are arranged in the form of a V in front of the sulcus terminalis.

—are studded with numerous taste buds and are innervated by the glossopharyngeal nerve.

2. **Foliate Papillae**

—are found in certain animals but are rudimentary in humans.

B. **Lingual Tonsil**

—is the collection of nodular masses of lymphoid follicles on the posterior one-third of the dorsum of the tongue.

C. **Lingual Innervation**

—the extrinsic and intrinsic muscles of the tongue are innervated by the hypoglossal nerve except for the palatoglossus muscle, which is supplied by the vagus nerve.

—the anterior two-thirds of the tongue is innervated by the lingual nerve for general sensation and by the chorda tympani for special (taste) sensation.

—the posterior one-third of the tongue and the vallate papillae are innervated by the glossopharyngeal nerve for both general and special sensation.

—the root of the tongue near the epiglottis is innervated by the internal laryngeal nerve of the vagus nerve for both general and special sensation.

D. Lingual Artery

—arises from the external carotid artery at the level of the tip of the greater horn of the hyoid bone in the carotid triangle.

—passes deep to the hyoglossus muscle and lies on the middle pharyngeal constrictor muscle.

—gives off the **suprahyoid, dorsal lingual,** and **sublingual arteries** and terminates as the **deep lingual artery,** which ascends between the genioglossus and inferior pharyngeal constrictor muscles.

E. Muscles of the Tongue

Muscle	Origin	Insertion	Nerve	Action
Styloglossus	Styloid process	Side and inferior aspect of tongue	Hypoglossal n.	Retracts and elevates tongue
Hyoglossus	Body and greater horn of hyoid bone	Side and inferior aspect of tongue	Hypoglossal n.	Depresses and retracts tongue
Genioglossus	Genial tubercle of mandible	Inferior aspect of tongue; body of hyoid bone	Hypoglossal n.	Protrudes and depresses tongue
Palatoglossus	Aponeurosis of soft palate	Dorsolateral side of tongue	Vagus n. via pharyngeal plexus	Elevates tongue

IV. Teeth and Gums (Gingivae)

A. Structure of the Teeth

1. **Enamel:** the hardest substance that covers the crown.
2. **Dentine:** a hard substance that is nurtured through the fine dental tubules of odontoblasts lining the central pulp space.
3. **Pulp:** fills the central cavity, which is continuous with the root canal. It contains numerous blood vessels, nerves, and lymphatics, which enter the pulp through an apical foramen at the apex of the root.

B. Parts of the Teeth

1. **Crown:** projects above the gingival surface and is covered by enamel.
2. **Neck:** the constricted area of junction of the crown and root.
3. **Root:** embedded in the alveolar part of the maxilla or mandible. It is covered with cement, which is connected to the bone of the alveolus by a layer of modified periosteum, the periodontal ligament.

C. Basic Types of Teeth

1. **Incisors:** are chisel-shaped and are used for cutting or biting.
2. **Canines:** have a single prominent cone and are used for tearing.
3. **Premolars:** usually have two cusps and are used for grinding.
4. **Molars:** usually have three cusps and are used for grinding.

D. Two Sets of Teeth

1. **Deciduous teeth:** two incisors, one canine, and two molars in each quadrant, for a total of 20.
2. **Permanent teeth:** two incisors, one canine, two premolars, and three molars in each quadrant, for a total of 32.

E. Innervation of the Teeth (Figure 8.41)

1. **Maxillary teeth:** are innervated by the anterior, middle, and posterior–superior alveolar branches of the maxillary nerve.
2. **Mandibular teeth:** are innervated by the inferior alveolar branch of the mandibular nerve.

F. Innervation of the Gingivae (Gums) [see Figure 8.41]

1. **Maxillary Gingiva**

   a. **Outer (buccal) surface:** posterior, middle and anterior–superior alveolar and infraorbital nerves.

   b. **Inner (lingual) surface:** greater palatine and nasopalatine nerves.

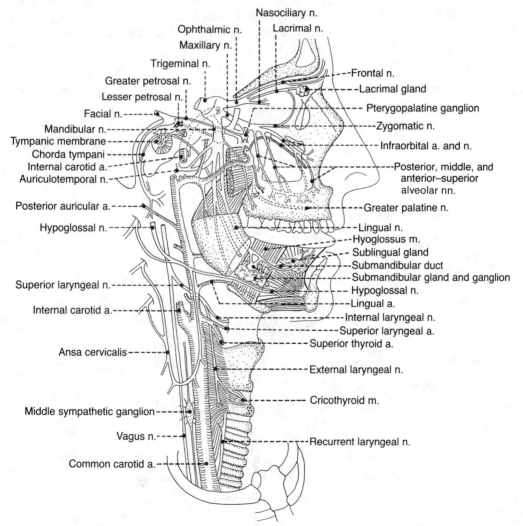

Figure 8.41. Branches of the trigeminal nerve and their relationship with other structures.

2. **Mandibular Gingiva**

a. **Outer (buccal) surface:** buccal and mental nerves.

b. **Inner (lingual) surface:** lingual nerve.

V. Salivary Glands (see Figure 8.41)

A. Submandibular Gland

—is ensheathed by the investing layer of the deep cervical fascia and lies in the submandibular triangle.

—its superficial portion is situated superficial to the mylohyoid muscle.

—its deep portion is located between the hyoglossus and styloglossus muscles medially and the mylohyoid muscle laterally and between the lingual nerve above and the hypoglossal nerve below.

—its duct (Wharton's) arises from the deep portion and runs forward between the mylohyoid laterally and the hypoglossus medially, where it is crossed laterally by the lingual nerve. It then runs between the sublingual gland and the genioglossus muscle and empties into the mouth on the summit of the sublingual papilla (caruncle) at the side of the frenulum of the tongue.

—is innervated by parasympathetic secretomotor fibers from the facial nerve, which run in the chorda tympani and in the lingual nerve and synapse in the submandibular ganglion (see p 333).

B. Sublingual Gland

—is located in the floor of the mouth between the mucous membrane above and the mylohyoid muscle below.

—surrounds the terminal portion of the submandibular duct.

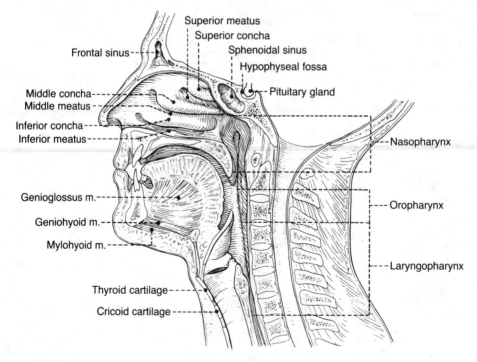

Figure 8.42. Pharynx.

—empties mostly into the floor of the mouth along the sublingual fold by 12 short ducts, some of which enter the submandibular duct.

—is supplied by postganglionic parasympathetic (secretomotor) fibers from the submandibular ganglion either directly or through the lingual nerve.

Pharynx

I. Pharynx (Figure 8.42)

—is a funnel-shaped fibromuscular tube that extends from the base of the skull to the inferior border of the cricoid cartilage.

—conducts food to the esophagus and air to the larynx and lungs.

II. Subdivisions of the Pharynx (see Figure 8.42)

A. Nasopharynx

—is situated behind the nasal cavity above the soft palate and communicates with the nasal cavities through the choanae.

—has the **pharyngeal tonsils** in its posterior wall.

—is connected with the tympanic cavity through the **auditory (eustachian) tube**.

B. Oropharynx

—extends between the soft palate above and the superior border of the epiglottis below and communicates with the mouth through the oropharyngeal isthmus.

—contains the **palatine tonsils,** which are lodged in the tonsilar fossae and are bounded by the palatoglossal and palatopharyngeal folds.

C. Laryngopharynx (Hypopharynx)

—extends from the upper border of the epiglottis to the lower border of the cricoid cartilage.

—contains the piriform recesses, one on each side of the opening of the larynx, in which swallowed foreign bodies may be lodged.

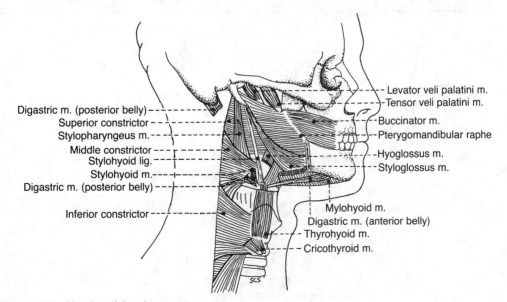

Figure 8.43. Muscles of the pharynx.

III. Muscles of the Pharynx (Figures 8.43 and 8.44)

Muscle	Origin	Insertion	Nerve	Action
Circular muscles:				
Superior constrictor	Medial pterygoid plate; pterygoid hamulus; pterygomandibular raphe; mylohyoid line of mandible; side of tongue	Median raphe and pharyngeal tubercle of skull	Vagus n. via pharyngeal plexus	Constricts upper pharynx
Middle constrictor	Greater and lesser horns of hyoid; stylohyoid ligament	Median raphe	Vagus n. via pharyngeal plexus	Constricts lower pharynx
Inferior constrictor	Arch of cricoid and oblique line of thyroid cartilages	Median raphe of pharynx	Vagus n. via pharyngeal plexus, recurrent and external laryngeal n.	Constricts lower pharynx
Longitudinal muscles:				
Stylopharyngeus	Styloid process	Thyroid cartilage and muscles of pharynx	Glossopharyngeal n.	Elevates pharynx and larynx
Palatopharyngeus	Hard palate; aponeurosis of soft palate	Thyroid cartilage and muscles of pharynx	Vagus n. via pharyngeal plexus	Elevates pharynx and closes nasopharynx
Salpingopharyngeus	Cartilage of auditory tube	Muscles of pharynx	Vagus n. via pharyngeal plexus	Elevates nasopharynx; opens auditory tube

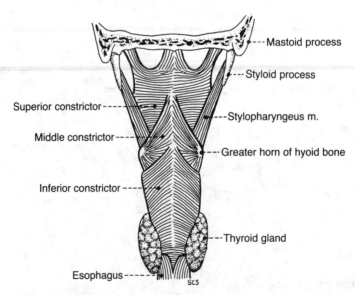

Figure 8.44. Pharyngeal constrictors.

IV. Innervation and Blood Supply of the Pharynx (Figure 8.45)

A. Pharyngeal Plexus
—lies on the middle pharyngeal constrictor.
—is formed by the pharyngeal branches of the glossopharyngeal and vagus nerves and the sympathetic branches from the superior cervical ganglion.
—its vagal branch innervates all of the muscles of the pharynx with the exception of the **stylopharyngeus,** which is supplied by the glossopharyngeal nerve.
—its glossopharyngeal component supplies sensory fibers to the pharyngeal mucosa.

B. Arteries
—are the ascending pharyngeal artery, ascending palatine branch of the facial artery, descending palatine arteries, pharyngeal branches of the maxillary artery, and branches of the superior and inferior thyroid arteries.

V. Swallowing (Deglutition)
—is described in several stages:
A. The tongue pushes the bolus of food back into the **fauces,** which is the passage from the mouth to the oropharynx.
B. The palatoglossus and palatopharyngeus muscles contract to squeeze the bolus backward into the oropharynx. The tensor veli palatini and levator veli palatini muscles elevate the soft palate to close the entrance into the nasopharynx.
C. The walls of the pharynx are raised by the palatopharyngeus, stylopharyngeus, and salpingopharyngeus muscles to receive the food. The suprahy-

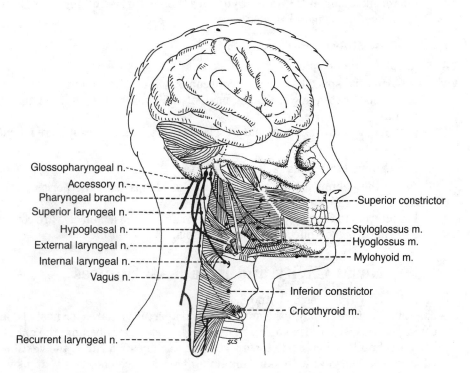

Glossopharyngeal n.
Accessory n.
Pharyngeal branch
Superior laryngeal n.
Hypoglossal n.
External laryngeal n.
Internal laryngeal n.
Vagus n.
Recurrent laryngeal n.

Superior constrictor
Styloglossus m.
Hyoglossus m.
Mylohyoid m.
Inferior constrictor
Cricothyroid m.

Figure 8.45. Nerve supply to the pharynx.

oid muscles elevate the hyoid bone and the larynx to close the opening into the larynx, thus preventing the food from entering the respiratory passageways.

    D. The serial contraction of the superior, middle, and inferior pharyngeal constrictor muscles moves the food through the oropharynx and the laryngopharynx into the esophagus where it is propelled by peristalsis.

VI. Tonsils

A. Pharyngeal Tonsil

—is found in the posterior wall and roof of the nasopharynx.

—is called the **adenoid** when enlarged. It causes difficulty in nasal breathing as a result of obstruction of the nasopharynx. The adenoid or chronic inflammation of the pharyngeal tonsil may cause deafness because of its close relationship with the auditory tube.

B. Palatine Tonsil

—lies on each side of the oropharynx in an interval between the palatoglossal and palatopharyngeal folds.

—receives blood from the ascending palatine and tonsilar branches of the **facial artery,** the descending palatine branch of the **maxillary artery,** a palatine branch of the **ascending pharyngeal artery,** and the dorsal lingual branches of the **lingual artery.**

—is supplied by branches of the **glossopharyngeal nerve** and the lesser palatine branch of the **maxillary nerve.**

C. Tubal Tonsil

—lies near the pharyngeal opening of the auditory tube.

D. Waldeyer's Ring

—is a tonsillar ring at the oropharyngeal isthmus, formed of lingual, palatine, tubal, and pharyngeal tonsils.

VII. Fascia and Space (see Figure 8.8)

A. Retropharyngeal Space

—is a potential space between the buccopharyngeal fascia and the prevertebral fascia, extending from the base of the skull to the superior mediastinum.

—permits movement of the pharynx, larynx, trachea, and esophagus during swallowing.

B. Pharyngobasilar Fascia

—forms the submucosa of the pharynx and blends with the periosteum of the base of the skull.

—lies internal to the muscular coat of the pharynx; these muscles are covered externally by the buccopharyngeal fascia.

Nasal Cavity and Paranasal Sinuses

I. Nasal Cavity (Figures 8.46 and 8.47)

—opens on the face through the anterior nasal apertures (nares, or nostrils) and communicates with the nasopharynx through a posterior opening, the **choana.**

—has a slight dilatation inside the aperture of each nostril, called the **vestibule,** which is lined largely with skin containing hair, sebaceous glands, and sweat glands.

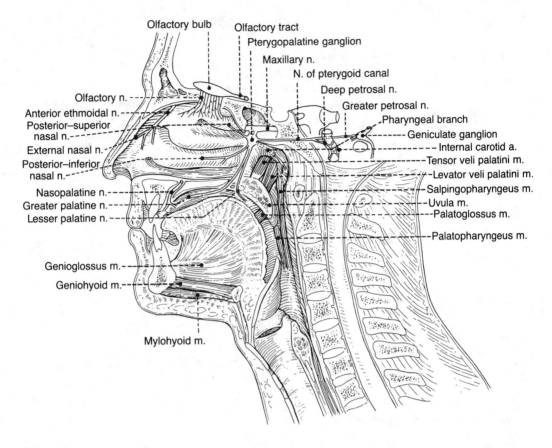

Figure 8.46. Nasal cavity.

A. Roof

—is formed by the nasal, frontal, ethmoid (cribriform plate), and sphenoid (body) bones. The cribriform plate transmits the olfactory nerves.

B. Floor

—is formed by the palatine process of the **maxilla** and the horizontal plate of the **palatine bone**.

—has an incisive foramen, which transmits the nasopalatine nerve and terminal branches of the sphenopalatine artery.

C. Medial Wall (or Nasal Septum)

—is formed primarily by the perpendicular plate of the ethmoid bone, vomer (bone), and septal cartilage.

—is also formed by the processes of the palatine, maxillary, frontal, sphenoid, and nasal bones.

D. Lateral Wall

—is formed by the superior and middle conchae of the ethmoid bone and the inferior concha (a bone itself).

—is also formed by the nasal bone, frontal process and nasal surface of the maxilla, lacrimal bone, perpendicular plate of the palatine bone, and medial pterygoid plate of the sphenoid bone.

—has the following structures, which receive openings:

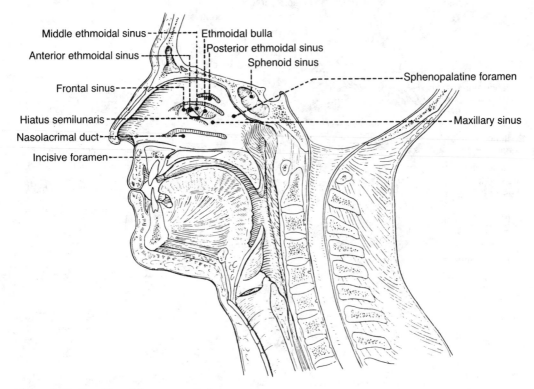

Figure 8.47. Openings of the paranasal sinuses.

1. **Sphenoethmoidal recess:** opening of the sphenoid sinus.
2. **Superior meatus:** opening of the posterior ethmoidal air cells.
3. **Middle meatus:** opening of the frontal sinus into the infundibulum; openings of the middle ethmoidal air cells on the ethmoidal bulla; openings of the anterior ethmoidal air cells and maxillary sinus in the hiatus semilunaris.
4. **Inferior meatus:** opening of the nasolacrimal duct.
5. **Sphenopalatine foramen:** opening into the pterygopalatine fossa; transmits the sphenopalatine artery.

II. Mucous Membrane

A. Respiratory Region

—consists of the lower two-thirds of the nasal cavity.

B. Olfactory Region

—consists of the superior nasal concha and the upper one-third of the nasal septum.

—is innervated by olfactory nerves, which enter the cranial cavity through the cribriform plate of the ethmoid bone to end in the olfactory bulb.

III. Blood Supply and Hemorrhage

A. Blood Supply to the Nasal Cavity

—is from the anterior and posterior ethmoidal branches of the **ophthalmic artery;** the sphenopalatine branch of the **maxillary artery;** the greater palatine branch (its terminal branch reaching the lower part of the nasal septum through the incisive canal) of the descending palatine artery of the **maxillary artery;** and the septal branch of the superior labial artery of the **facial artery**.

B. Epistaxis

—is a nosebleed (nasal hemorrhage) resulting from the rupture of the **sphenopalatine artery;** also occurs from nose picking, which tears the veins in the vestibule of the nose.

IV. Nerve Supply

A. Special Visceral Afferent (SVA) Sensation

—is supplied by the olfactory nerves for the olfactory area.

B. General Somatic Afferent (GSA) Sensation

—is supplied by the anterior and posterior ethmoidal branches of the **ophthalmic nerve;** the posterior–superior and inferior lateral nasal branches of the **maxillary nerve;** and the anterior–superior alveolar branch of the **infraorbital nerve**.

V. Paranasal Sinuses (see Figure 8.47)

A. Ethmoidal Sinus

—consists of ethmoidal air cells, which are numerous small cavities within the ethmoidal labyrinth between the orbit and the nasal cavity.

—can be subdivided into the following groups:

1. **Posterior ethmoidal air cells:** drain into the superior nasal meatus.
2. **Middle ethmoidal air cells:** drain into the summit of the ethmoidal bulla of the middle nasal meatus.
3. **Anterior ethmoidal air cells:** drain into the anterior aspect of the hiatus semilunaris in the middle nasal meatus.

B. Frontal Sinus

—lies in the frontal bone and opens into the anterior part of the **middle nasal meatus** by way of the frontonasal duct.

—is innervated by the supraorbital branch of the ophthalmic nerve.

C. Maxillary Sinus

—is the largest of the paranasal air sinuses and is the only paranasal sinus that may be present at birth.

—lies in the maxilla on each side, lateral to the lateral wall of the nasal cavity and inferior to the floor of the orbit, and drains into the posterior aspect of the hiatus semilunaris in the **middle nasal meatus**.

D. Sphenoidal Sinus

—is contained within the body of the sphenoid bone.

—opens into the **sphenoethmoidal recess** of the nasal cavity.

—is supplied by branches from the maxillary nerve and by the posterior ethmoidal branch of the nasociliary nerve.

Pterygopalatine Fossa

I. Boundaries and Openings

A. Anterior wall: posterior surface of the maxilla or the posterior wall of the maxillary sinus (no openings).

B. **Posterior wall:** pterygoid process and greater wing of the sphenoid.
—**Foramen rotundum** to middle cranial cavity: maxillary nerve.
—**Pterygoid canal** to foramen lacerum: nerve of the pterygoid canal.
—**Palatovaginal** (pharyngeal or pterygopalatine) **canal** to choana: pharyngeal branch of the maxillary artery and pharyngeal nerve from the pterygopalatine ganglion.

C. **Medial wall:** perpendicular plate of the palatine.
—**Sphenopalatine foramen** to nasal cavity: sphenopalatine artery and nasopalatine nerve.

D. **Lateral wall:** open (pterygomaxillary fissure to the infratemporal fossa).

E. **Roof:** greater wing and body of the sphenoid.
—**Inferior orbital fissure** to the orbit: maxillary nerve.

F. **Floor:** fusion of the maxilla and the pterygoid process of the sphenoid.
—**Greater palatine foramen** to the palate: greater palatine nerve and vessels.

II. Contents

A. **Maxillary Nerve** (see Figure 8.41)
—passes through the lateral wall of the cavernous sinus and enters the pterygopalatine fossa through the foramen rotundum.
—is sensory to the skin of the face below the eye but above the upper lip.
—gives off the following branches:

1. **Meningeal Branch**
—innervates the dura mater of the middle cranial fossa.

2. **Pterygopalatine Nerves (Communicating Branches)**
—are connected to the pterygopalatine ganglion.
—contain sensory fibers from the trigeminal ganglion.

3. **Posterior–Superior Alveolar Nerves**
—descend through the pterygopalatine fissure and enter the posterior-superior alveolar canals.
—innervate the cheeks, gums, molar teeth, and maxillary sinus.

4. **Zygomatic Nerve**
—enters the orbit through the inferior orbital fissure, divides into the zygomaticotemporal and zygomaticofacial branches, which supply the skin over the temporal region and over the zygomatic bone, respectively. It joins the lacrimal nerve in the orbit and transmits postganglionic parasympathetic fibers to the lacrimal gland.

5. **Infraorbital Nerve**
—enters the orbit through the inferior orbital fissure and runs through the infraorbital groove and canal.
—emerges through the infraorbital foramen and divides on the face into the inferior palpebral, nasal, and superior labial branches.
—gives off the **middle** and **anterior–superior alveolar nerves,** which supply the maxillary sinus, teeth, and gums.

6. **Branches (Sensory) Via the Pterygopalatine Ganglion**

a. **Orbital Branches**
—supply the periosteum of the orbit and the mucous membrane of the posterior ethmoidal and sphenoidal sinuses.

    b. Pharyngeal Branch

        —runs in the pharyngeal canal and supplies the roof of the pharynx and the sphenoidal sinuses.

    c. Posterior–Superior Lateral Nasal Branches

        —enter the nasal cavity through the sphenopalatine foramen and supply the posterior part of the septum, the posterior ethmoidal air cells, and the superior and middle conchae.

    d. Greater Palatine Nerve

        —descends through the palatine canal and emerges through the greater palatine foramen to supply the palate.
        —gives off the inferior lateral nasal branches.

    e. Lesser Palatine Nerve

        —descends through the palatine canal and emerges through the lesser palatine foramen to supply the soft palate and the palatine tonsil.

    f. Nasopalatine Nerve

        —runs obliquely downward and forward on the septum, supplying the septum, and passes through the incisive canal.

B. Pterygopalatine Ganglion (see Figures 8.28 and 8.55)

    —lies in the pterygopalatine fossa just below the maxillary nerve, lateral to the sphenopalatine foramen and anterior to the pterygoid canal.
    —receives preganglionic parasympathetic fibers from the facial nerve by way of the greater petrosal nerve and the nerve of the pterygoid canal.
    —sends postganglionic parasympathetic fibers to the nasal and palatine glands and to the lacrimal gland by way of the maxillary, zygomatic, and lacrimal nerves.
    —also receives postganglionic sympathetic fibers (by way of the deep petrosal nerve and the nerve of the pterygoid canal), which are distributed with the postganglionic parasympathetic fibers.

C. Pterygopalatine Part of the Maxillary Artery (see p 279)

    —supplies blood to the maxilla and maxillary teeth, nasal cavities, and palate.
    —gives off the posterior–superior alveolar artery, infraorbital artery (which gives off anterior–superior alveolar branches), descending palatine artery (which gives off the lesser palatine and greater palatine branches), artery of the pterygoid canal, pharyngeal artery, and sphenopalatine artery (see p 279).

Larynx

I. Larynx

    —extends from the lower part of the pharynx to the trachea.
    —acts as a valve to prevent the passage of foods into the airway and closes the glottis for building up air pressure for coughing.
    —controls the airway during swallowing and vocalization (or phonation).

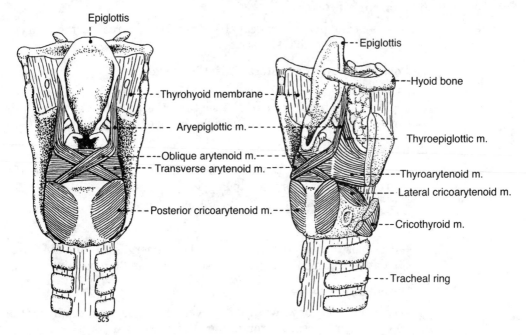

Figure 8.48. Cartilages of the larynx.

—regulates the flow of air to and from the lungs for sound production.

—forms a framework of cartilage for the attachment of ligaments and muscles.

II. Cartilages of the Larynx (Figure 8.48)

A. Thyroid Cartilage

—is a **single hyaline cartilage** that forms a median elevation called the laryngeal prominence ("Adam's apple").

—its superior horn is attached to the tip of the greater horn of the hyoid bone, and its inferior horn articulates with the cricoid cartilage.

—has an oblique line on the lateral surface of its lamina, which gives attachment for the inferior pharyngeal constrictor, sternothyroid, and thyrohyoid muscles.

B. Cricoid Cartilage

—is a **single hyaline cartilage,** which is shaped like a signet ring.

—marks the end of the pharynx and larynx at its lower border.

C. Epiglottis

—is a **single elastic cartilage** and is a spoon-shaped plate that lies behind the root of the tongue.

—its lower end is attached to the back of the thyroid cartilage.

D. Arytenoid Cartilages

—are **paired cartilages** composed of **elastic** and **hyaline cartilages**.

—are shaped liked pyramids, and their bases articulate with the cricoid cartilage.

—have **vocal processes,** which give attachment to the vocal ligament and vocalis muscle, and **muscular processes,** which give attachment to the thyroarytenoid, lateral, and posterior cricoarytenoid muscles.

E. Corniculate Cartilages

—are **paired elastic cartilages** that lie on the apices of the arytenoid cartilages.

—are enclosed within the aryepiglottic folds of mucous membrane.

F. Cuneiform Cartilages

—are **paired elastic cartilages** that lie in the aryepiglottic folds anterior to the corniculate cartilages.

III. Ligaments of the Larynx

A. Thyrohyoid Membrane

—extends from the thyroid cartilage to the medial surface of the hyoid bone.

—its middle (thicker) part is called the middle thyrohyoid ligament; its lateral portion is pierced by the internal laryngeal nerve and the superior laryngeal vessels.

B. Cricothyroid Ligament

—extends from the arch of the cricoid cartilage to the thyroid cartilage and the vocal processes of the arytenoid cartilages.

C. Vocal Ligament

—extends from the posterior surface of the thyroid cartilage to the vocal process of the arytenoid cartilage.

—is considered as the upper border of the conus elasticus.

D. Vestibular (Ventricular) Ligament

—extends from the thyroid cartilage to the anterior lateral surface of the arytenoid cartilage.

E. Conus Elasticus (Cricovocal Ligament)

—is the paired lateral portion of the fibroelastic membrane that extends upward from the entire arch of the cricoid cartilage to the vocal ligaments.

—is formed by the cricothyroid, median cricothyroid, and vocal ligaments.

IV. Cavities and Folds of the Larynx (Figure 8.49)

—The larynx is divided into three portions by the vestibular and vocal folds:

A. Vestibule

—extends from the laryngeal inlet to the vestibular (ventricular) folds.

B. Ventricles

—extend between the ventricular fold and the vocal fold.

C. Infraglottic Cavity

—extends from the rima glottidis to the lower border of the cricoid cartilage.

D. Rima Glottidis

—is the space between the vocal folds and arytenoid cartilages.

—is the narrowest part of the laryngeal cavity.

E. Vestibular (Ventricular) Folds or False Vocal Cords

—extend from the thyroid cartilage above the vocal ligament to the arytenoid cartilage.

F. Vocal Folds or True Vocal Cords

—extend from the angle of the thyroid cartilage to the vocal processes of the arytenoid cartilages.

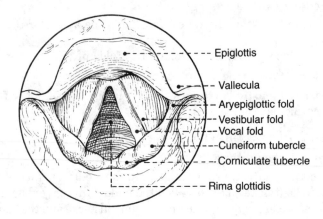

Figure 8.49. Interior view of the larynx.

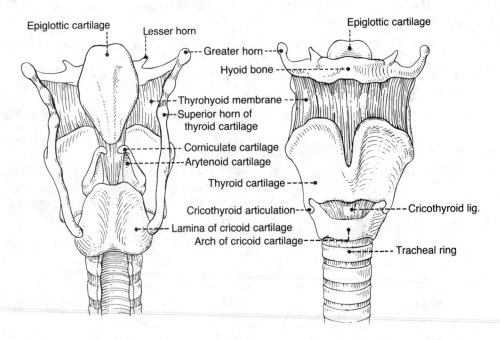

Figure 8.50. Muscles of the larynx.

—contain the vocal ligament near their free margin and the vocalis muscle, which forms the bulk of the vocal fold.

—are important in voice production because they control the stream of air passing through the rima glottidis.

—alter the shape and size of the rima glottidis by movement of the arytenoids to facilitate respiration and phonation. (The rima glottidis is wide during inspiration and narrow and wedge-shaped during expiration and sound production.)

V. Muscles of the Larynx (Figure 8.50)

Muscle	Origin	Insertion	Nerve	Action on Vocal Cords
Cricothyroid	Arch of cricoid cartilage	Inferior horn and lower lamina of thyroid cartilage	External laryngeal n.	Tenses
Posterior cricoarytenoid	Posterior surface of lamina of cricoid cartilage	Muscular process of arytenoid cartilage	Recurrent laryngeal n.	Abducts
Lateral cricoarytenoid	Arch of cricoid cartilage	Muscular process of arytenoid cartilage	Recurrent laryngeal n.	Adducts
Transverse arytenoid	Posterior surface of arytenoid cartilage	Opposite arytenoid cartilage	Recurrent laryngeal n.	Adducts
Oblique arytenoid	Muscular process of arytenoid cartilage	Apex of opposite arytenoid	Recurrent laryngeal n.	Adducts
Aryepiglottic	Apex of arytenoid cartilage	Side of epiglottic cartilage	Recurrent laryngeal n.	Adducts
Thyroarytenoid	Inner surface of thyroid lamina	Anterolateral surface of arytenoid cartilage	Recurrent laryngeal n.	Adducts
Thyroepiglottic	Anteromedial surface of lamina of thyroid cartilage	Lateral margin of epiglottic cartilage	Recurrent laryngeal n.	Adducts
Vocalis	Anteromedial surface of lamina of thyroid cartilage	Vocal process	Recurrent laryngeal n.	Adducts and tenses

VI. Innervation of the Larynx (Figure 8.51)

A. Recurrent (Inferior) Laryngeal Nerve

—supplies all of the intrinsic muscle of the larynx except the cricothyroid, which is innervated by the external laryngeal branch of the superior laryngeal branch of the vagus nerve.

—supplies the sensory innervation below the vocal cord.

—its terminal portion above the lower border of the cricoid cartilage is called the **inferior laryngeal nerve**.

B. Internal Laryngeal Nerve

—innervates the mucous membrane above the vocal cord.

—is accompanied by the superior laryngeal artery and pierces the thyrohyoid membrane.

C. External Laryngeal Nerve

—innervates the cricothyroid and inferior pharyngeal constrictor muscles.

—is accompanied by the superior thyroid artery.

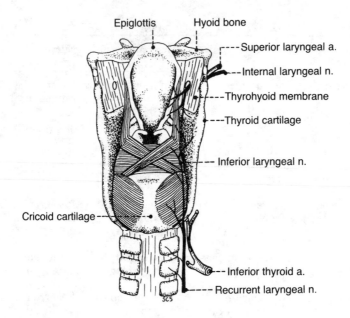

Figure 8.51. Nerve supply to the larynx.

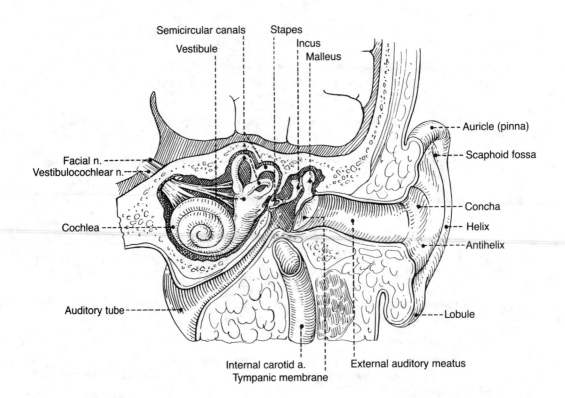

Figure 8.52. External, middle, and inner ear.

Ear

I. External Ear (Figure 8.52)

A. Auricle

—consists of cartilage connected to the skull by ligaments and muscles and is covered by skin.

—funnels sound waves into the external auditory meatus.

—receives sensory nerves from the auricular branch of the vagus nerve and the greater auricular, auriculotemporal, and lesser occipital nerves.

—receives blood from the superficial temporal and posterior auricular arteries.

—has the following features:

1. **Helix:** the slightly curved rim of the auricle.

2. **Antihelix:** a broader curved eminence internal to the helix, which divides the auricle into an outer scaphoid fossa and the deeper concha.

3. **Concha:** the deep cavity in front of the antihelix.

4. **Tragus:** a small projection from the anterior portion of the external ear anterior to the concha.

5. **Lobule:** a structure made up of areolar tissue and fat but no cartilage.

B. External Acoustic (Auditory) Meatus

—is about 2.5 cm long, extending from the concha to the tympanic membrane.

—its external one-third is formed by cartilage, and its internal two-thirds is formed by bone. The cartilaginous portion is wider than the bony portion and has numerous ceruminous glands that produce ear wax.

—is supplied by the auriculotemporal branch of the trigeminal nerve and the auricular branch of the vagus nerve, which is joined by a branch of the facial nerve and the glossopharyngeal nerve (see pp 290–292).

—receives blood from the superficial temporal, posterior auricular, and maxillary arteries (a deep auricular branch).

C. Tympanic Membrane (Eardrum)

—lies obliquely across the end of the meatus; thus, the anterior and inferior walls are longer than the posterior and superior walls.

—consists of three layers: an outer (cutaneous), an intermediate (fibrous), and an inner (mucous) layer.

—has a thickened fibrocartilaginous ring at the greater part of its circumference, which is fixed in the tympanic sulcus at the inner end of the meatus.

—its small triangular portion between the anterior and posterior malleolar folds is called the **pars flaccida,** and the remainder of the membrane is called the **pars tensa.**

—contains the **cone of light,** which is a triangular reflection of light seen in the anterior–inferior quadrant.

—its most depressed center point of the concavity is called the **umbo.**

—conducts sound waves to the middle ear.

—its lateral (outer) concave surface is covered by skin and is supplied by the auriculotemporal branch of the trigeminal nerve and the auricular branch of the vagus nerve. The auricular branch may contain fibers from the glossopharyngeal and facial nerves.

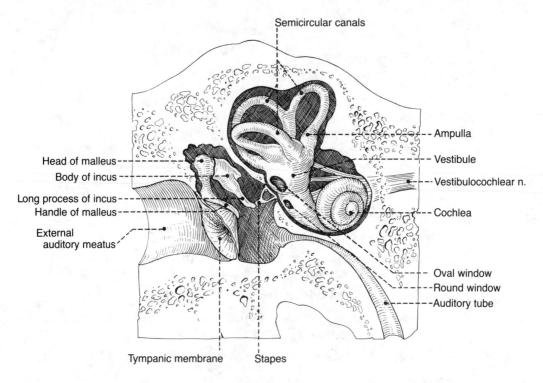

Semicircular canals

Head of malleus

Body of incus

Long process of incus

Handle of malleus

External
auditory meatus

Ampulla

Vestibule

Vestibulocochlear n.

Cochlea

Oval window

Round window

Auditory tube

Tympanic membrane   Stapes

Figure 8.53. Middle and inner ear.

—its medial (inner) surface is covered by a mucous membrane, is innervated by the tympanic branch of the glossopharyngeal nerve, and serves as an attachment for the handle of the malleus.

—its inner surface is supplied by the auricular branch of the occipital artery and the anterior tympanic artery; its outer surface is supplied by the deep auricular artery.

II. Middle Ear (Figures 8.53 and 8.54)

A. Tympanic (or Middle Ear) Cavity

—includes the **tympanic cavity proper,** the space internal to the tympanic membrane; and the **epitympanic recess,** the space superior to the tympanic membrane that contains the head of the malleus and the body of the incus.

—communicates anteriorly with the nasopharynx via the auditory (eustachian) tube and posteriorly with the mastoid air cells and the mastoid antrum through the **aditus ad antrum.**

—is traversed by the chorda tympani and lesser petrosal nerve.

1. Boundaries of the Tympanic Cavity

a. Roof: tegmen tympani.

b. Floor: jugular fossa.

c. Anterior: carotid canal.

d. Posterior: mastoid air cells and mastoid antrum through the **aditus ad antrum.**

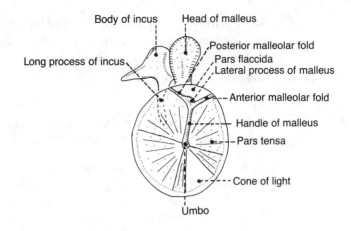

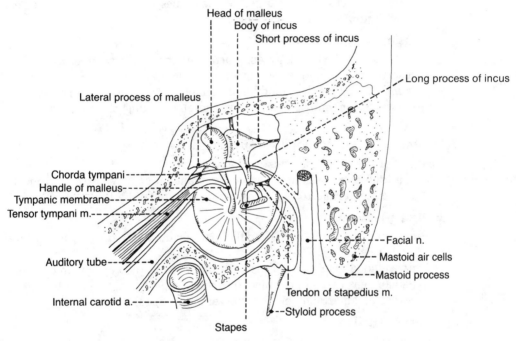

Figure 8.54. Ossicles of the middle ear and tympanic membrane.

    e. Lateral: tympanic membrane.

    f. Medial: lateral wall of the inner ear, presenting the promontory formed by the basal turn of the cochlea; the fenestra vestibuli (oval window); the fenestra cochleae (round window); and the prominence of the facial canal.

2. Oval Window (Fenestra Vestibuli)

    —is pushed back and forth by the footplate of the stapes and transmits the sonic vibrations of the ossicles to the perilymph of the scala vestibuli in the inner ear.

3. Round Window (Fenestra Cochlea or Tympani)

—is closed by the mucous membrane of the middle ear and accommodates the pressure waves transmitted to the perilymph of the scala tympani.

B. Muscles of the Middle Ear

1. Stapedius Muscle

—is the smallest of the skeletal muscles in the human body.
—arises within the pyramidal eminence, and its tendon emerges from the eminence.
—inserts on the neck of the stapes.
—is innervated by a branch of the facial nerve.
—pulls the head of the stapes posteriorly, thereby tilting the base of the stapes in the vestibular window.
—prevents excessive oscillation of the stapes and protects the inner ear from injury during a loud noise.

2. Tensor Tympani Muscle

—arises from the cartilaginous portion of the auditory tube.
—inserts on the handle (manubrium) of the malleus.
—is innervated by the mandibular branch of the trigeminal nerve.
—draws the manubrium medially, thereby making the tympanic membrane taut.

C. Auditory Ossicles (see Figures 8.53 and 8.54)

—consist of the malleus, incus, and stapes.
—form a bridge in the middle ear cavity, transmit sonic vibrations from the tympanic membrane to the inner ear, and amplify the force.

1. Malleus (Hammer)

—consists of a head, neck, handle (manubrium), and anterior and lateral processes.
—its rounded head articulates with the incus in the epitympanic recess.
—its handle is fused to the medial surface of the tympanic membrane and serves as an attachment for the tensor tympani muscle.

2. Incus (Anvil)

—consists of a body and two processes (crura).
—its long process descends vertically, parallel to the handle of the malleus, and articulates with the stapes.
—its short process extends horizontally backward to the fossa of the incus and provides the attachment for the posterior ligament of the incus.

3. Stapes (Stirrup)

—consists of a head and neck, two processes (crura), and a base (footplate).
—its neck provides insertion of the stapedius muscle.
—has a hole through which the stapedial artery is transmitted in the embryo; this hole is obturated by a thin membrane in the adult.
—its base (footplate) is attached by the annular ligament to the margin of the fenestra vestibuli. Abnormal ossification between the footplate and the fenestra vestibuli (**otosclerosis**) limits the movement of the stapes, causing deafness.

D. Auditory Tube

—connects the middle ear to the nasopharynx.

—allows air to enter or leave the middle ear cavity and thus balances the pressure in the middle ear with atmospheric pressure, allowing free movement of the tympanic membrane.

—its cartilaginous portion remains closed except during swallowing or yawning.

—is opened by the simultaneous contraction of the tensor veli palatini and salpingopharyngeus muscles.

E. Sensory Nerve and Blood Supply to the Middle Ear

—is innervated by the auriculotemporal branch of the trigeminal nerve, tympanic branch of the glossopharyngeal nerve, and auricular branch of the vagus nerve.

—is supplied from the stylomastoid branch of the posterior auricular artery and the anterior tympanic branch of the maxillary artery.

III. Inner Ear (see Figure 8.53)

—consists of the **cochlea,** for auditory sense, and the **utricle** and **saccule** and the vestibular apparatus of the **semicircular canals** for the sense of equilibrium.

—consists of the bony labyrinth and the membranous labyrinth.

A. Bony Labyrinth

—consists of three parts: the vestibule, the three semicircular canals, and the cochlea, all of which contain the **perilymph,** in which the membranous labyrinth is suspended.

B. Membranous Labyrinth

—is enclosed within the bony labyrinth and is filled with **endolymph** and contains the sensory organs.

—has comparable parts and arrangement as the bony labyrinth.

Autonomics of the Head and Neck

I. Parasympathetic Ganglia and Associated Nerves (Figure 8.55)

A. Ciliary Ganglion

—is situated behind the eyeball, between the optic nerve and the lateral rectus muscle.

—receives preganglionic parasympathetic fibers (with cell bodies in the Edinger-Westphal nucleus of CN III in the mesencephalon), which run in the inferior division of the oculomotor nerve.

—sends its postganglionic parasympathetic fibers to the sphincter pupillae (which constricts the pupil) and the ciliary muscle (which affects the shape of the lens in visual accommodation) via the short ciliary nerves.

—receives postganglionic sympathetic fibers (derived from the superior cervical ganglion) that reach the dilator pupillae muscle (which increases the diameter of the pupil) by way of the sympathetic plexus on the internal carotid artery, the long ciliary nerve and/or the ciliary ganglion (without synapsing), and the short ciliary nerves.

B. Pterygopalatine Ganglion

—lies in the pterygopalatine fossa just below the maxillary nerve, lateral to the sphenopalatine foramen and anterior to the pterygoid canal.

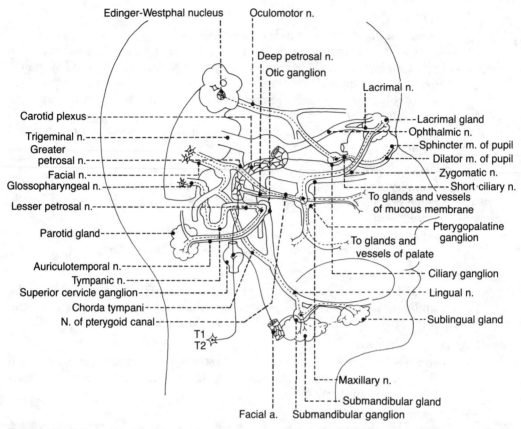

Figure 8.55. Autonomics of the head and neck.

—receives its preganglionic parasympathetic fibers from the facial nerve by way of the greater petrosal nerve and the nerve of the pterygoid canal.

—sends its postganglionic parasympathetic fibers to the nasal and palatine glands and to the lacrimal glands by way of the maxillary, zygomatic, and lacrimal nerves.

—also receives postganglionic sympathetic fibers (derived from the superior cervical ganglion) by way of the plexus on the internal carotid artery, the deep petrosal nerve, and the nerve of the pterygoid canal. The fibers merely pass through the ganglion and are distributed with the postganglionic parasympathetic fibers.

1. **Greater Petrosal Nerve**

—arises from the facial nerve adjacent to the geniculate ganglion.

—emerges at the hiatus of the canal for the greater petrosal nerve in the middle cranial fossa.

—contains preganglionic parasympathetic fibers and joins the deep petrosal nerve (containing postganglionic sympathetic fibers) to form the nerve of the pterygoid canal (or vidian nerve).

—also contains taste fibers, which pass from the palate nonstop through the pterygopalatine ganglion, the nerve of the pterygoid canal, and the greater petrosal nerve to the geniculate ganglion (where cell bodies are found).

2. Deep Petrosal Nerve

—arises from the plexus on the internal carotid artery.

—contains postganglionic sympathetic fibers with cell bodies located in the superior cervical ganglion. These fibers run in the nerve of the pterygoid canal, pass through the pterygopalatine ganglion without synapsing, and then join the postganglionic parasympathetic fibers in supplying the lacrimal gland and the nasal and oral mucosa.

3. Nerve of the Pterygoid Canal (Vidian Nerve)

—consists of preganglionic parasympathetic fibers from the greater petrosal nerve and postganglionic sympathetic fibers from the deep petrosal nerve.

—passes through the pterygoid canal and ends in the pterygopalatine ganglion, which is slung from the maxillary nerve. The postganglionic parasympathetic fibers have cell bodies located in the pterygopalatine ganglion, and the postganglionic sympathetic fibers are distributed to the lacrimal, nasal, and palatine glands.

—also contains SVA and GVA fibers from the palate.

C. Submandibular Ganglion

—lies on the lateral surface of the hyoglossus muscle but deep to the mylohyoid muscle and is suspended from the lingual nerve.

—receives preganglionic parasympathetic (secretomotor) fibers that run in the facial nerve, chorda tympani (see p 291), and lingual nerve.

—sends its postganglionic fibers to supply the submandibular gland mostly, although some of the fibers join the lingual nerve to reach the sublingual and lingual glands.

D. Otic Ganglion

—lies in the infratemporal fossa, just below the foramen ovale, between the mandibular nerve and the tensor veli palatini.

—receives preganglionic parasympathetic fibers that run in the glossopharyngeal nerve, tympanic plexus, and lesser petrosal nerve and synapse in the otic ganglion.

—sends postganglionic fibers that run in the auriculotemporal nerve and supply the parotid gland.

1. Tympanic Nerve

—contains preganglionic parasympathetic (secretomotor) fibers for the parotid gland.

—arises from the inferior ganglion of the glossopharyngeal nerve.

—passes through a small canal between the jugular foramen and the carotid canal into the tympanic cavity.

—enters the tympanic plexus on the promontory of the medial wall of the tympanic cavity.

2. Lesser Petrosal Nerve

—is a continuation of the tympanic nerve beyond the tympanic plexus.

—runs just lateral to the greater petrosal nerve and leaves the middle cranial fossa through either the foramen ovale or the fissure between the petrous bone and the great wing of the sphenoid to enter the otic ganglion.

—contains preganglionic parasympathetic (secretomotor) fibers that run in the glossopharyngeal and tympanic nerves before synapsing in the otic ganglion. (The postganglionic fibers arising from the ganglion are passed to the parotid gland by the auriculotemporal nerves.)

—also transmits postganglionic sympathetic fibers to the parotid gland.

II. Summary of Autonomic Ganglia of the Head and Neck

Ganglion	Location	Parasympathetic Fibers	Sympathetic Fibers	Chief Distribution
Ciliary	Lateral to optic n.	Oculomotor n. and its inferior division	Internal carotid plexus	Ciliary muscle, and sphincter pupillae (parasympathetic); dilator pupillae and tarsal mm. (sympathetic)
Pterygopalatine	In pterygopalatine fossa	Facial n., greater petrosal n., and n. of pterygoid canal	Internal carotid plexus	Lacrimal gland and glands in palate and nose
Submandibular	On hyoglossus	Facial n., chorda tympani, and lingual n.	Plexus on facial a.	Submandibular and sublingual glands
Otic	Below foramen ovale	Glossopharyngeal n., its tympanic branch, and lesser petrosal n.	Plexus on middle meningeal a.	Parotid gland

Review Test

DIRECTIONS: Each of the numbered items or incomplete statements in this section is followed by answers or by completions of the statement. Select the **one** lettered answer or completion that is **best** in each case.

1. Which of the following arteries is functionally replaced when a dorsal scapular artery arises from the third part of the subclavian artery?
(A) Superficial branch of the transverse cervical artery
(B) Deep branch of the transverse cervical artery
(C) Subscapular artery
(D) Suprascapular artery
(E) Internal thoracic artery

2. Each of the following vessels may be encountered in a low tracheotomy performed below the isthmus of the thyroid EXCEPT the
(A) inferior thyroid vein or its tributaries.
(B) jugular arch.
(C) costocervical trunk.
(D) thyroidea ima artery.
(E) left brachiocephalic vein.

3. Damage to the external laryngeal nerve during thyroid surgery will result in the patient's inability to
(A) relax the vocal cords.
(B) rotate the arytenoid cartilages.
(C) tense the vocal cords.
(D) widen the rima glottidis.
(E) abduct the vocal cords.

4. Each of the following branches of the cervical plexus contains cutaneous fibers EXCEPT the
(A) phrenic nerve.
(B) greater auricular nerve.
(C) transverse cervical nerve.
(D) supraclavicular nerve.
(E) lesser occipital nerve.

5. Each of the following structures empties or opens into the middle nasal meatus EXCEPT the
(A) middle ethmoidal air cells.
(B) maxillary sinus.
(C) sphenoid sinus.
(D) anterior ethmoidal air cells.
(E) frontal sinus.

6. Which of the following nerves accompanies the superior laryngeal artery?
(A) External laryngeal nerve
(B) Internal laryngeal nerve
(C) Superior laryngeal nerve
(D) Hypoglossal nerve
(E) Vagus nerve

7. Which of the following nerves provides sensory innervation to the mucosa of the pharynx?
(A) Transverse cervical nerves
(B) Phrenic nerves
(C) Vagus nerves
(D) Glossopharyngeal nerve
(E) Ansa cervicalis (hypoglossi)

8. Each statement below concerning nerves of the neck is true EXCEPT
(A) cutaneous nerves of the neck are branches of the cervical plexus.
(B) the transverse cervical nerve provides sensory innervation to the anterior and lateral parts of the neck.
(C) the supraclavicular nerves innervate the platysma muscle.
(D) the motor nerves for most infrahyoid muscles are branches of the ansa cervicalis.
(E) anterior primary rami of the second, third, and fourth cervical nerves contribute to the cervical plexus.

9. Which of the following arteries is a branch of the costocervical trunk?
(A) Inferior thyroid artery
(B) Transverse cervical artery
(C) Suprascapular artery
(D) Deep cervical artery
(E) Ascending cervical artery

10. On each side of the body, the phrenic nerve in the neck passes
(A) across the anterior surface of the subclavian vein.
(B) across the posterior surface of the subclavian artery.
(C) across the deep surface of the anterior scalene muscle.
(D) medial to the common carotid artery.
(E) across the superficial surface of the anterior scalene muscle.

11. Which of the following muscles is a landmark for locating the glossopharyngeal nerve in the neck?

(A) Inferior pharyngeal constrictor muscle
(B) Stylopharyngeus muscle
(C) Posterior belly of the digastric muscle
(D) Longus colli muscle
(E) Rectus capitis anterior muscle

12. The external laryngeal branch of the superior laryngeal nerve innervates the cricothyroid muscle and sends fibers to the

(A) inferior pharyngeal constrictor muscle.
(B) thyrohyoid muscle.
(C) sternohyoid muscle.
(D) stylohyoid muscle.
(E) sternothyroid muscle.

13. Which of the following statements concerning the larynx is true?

(A) The inlet to the larynx is formed by the aryepiglottic folds.
(B) The true vocal folds are inferior to the ventricle of the larynx.
(C) Afferent nerve fibers from the larynx are carried by the vagus nerve.
(D) The larynx extends inferiorly to the level of the sixth cervical vertebra.
(E) All of the above.

14. Each of the following nerves passes through the superior orbital fissure EXCEPT

(A) the abducens nerve.
(B) the ophthalmic nerve (or its branches).
(C) the oculomotor nerve.
(D) the trochlear nerve.
(E) the optic nerve.

15. Which statement below concerning the anterior and posterior ethmoidal nerves is true?

(A) Both have terminal branches called external nasal nerves.
(B) Both are branches of the infraorbital nerve.
(C) Both leave the orbit by way of the posterior ethmoidal foramina.
(D) Both send branches to the ethmoidal air sinuses.
(E) One leaves the orbit with the zygomaticotemporal nerve.

16. Postganglionic parasympathetic fibers that innervate glands in the palate originate in the

(A) nodose ganglion.
(B) otic ganglion.
(C) pterygopalatine ganglion.
(D) submandibular ganglion.
(E) ciliary ganglion.

17. All of the following statements concerning glands in the oral cavity are true EXCEPT

(A) the duct of the submandibular gland opens into the sublingual region.
(B) the duct of the submandibular gland pierces the mylohyoid muscle.
(C) the duct of the parotid gland pierces the buccinator muscle.
(D) the duct of the parotid gland opens into the vestibule of the mouth.
(E) part of the course of the parotid duct parallels the transverse facial artery.

18. All of the following statements concerning the nasal cavity are true EXCEPT

(A) the conchae are attached to its lateral wall.
(B) the ethmoid bone contributes to its roof and medial and lateral borders.
(C) its septum is partially cartilaginous.
(D) most of its floor is formed by the palatine bone.
(E) part of its roof is formed by the vomer bone.

19. Bell's palsy can involve corneal inflammation and subsequent corneal ulceration, which results from

(A) sensory loss of the cornea and conjunctiva.
(B) lack of secretion of the salivary glands.
(C) absence of the blinking reflex due to paralysis of the muscles that close the eyelid.
(D) absence of the blinking reflex due to paralysis of the muscles that open the eyelid.
(E) constriction of the pupil due to paralysis of the dilator pupillae.

20. Which of the following structures are direct tributaries to the straight dural sinus in the cranial cavity?

(A) Transverse and sigmoid sinuses
(B) Inferior sagittal sinus and greater cerebral vein
(C) Superior sagittal sinus and greater cerebral vein
(D) Superior and inferior petrosal sinuses
(E) Cavernous sinus and basilar plexus

21. Which of the following conditions results from severance of the abducens nerve proximal to its entrance into the orbit?

(A) Ptosis of the upper eyelid
(B) Loss of the pupil's ability to dilate
(C) External strabismus (lateral deviation)
(D) Loss of visual accommodation
(E) Loss of abduction of the eye

22. Death may result from bilateral severance of the

(A) trigeminal nerve.
(B) facial nerve.
(C) vagus nerve.
(D) accessory nerve.
(E) hypoglossal nerve.

23. When the middle meningeal artery is ruptured but the meninges remain intact, blood enters the

(A) subarachnoid space.
(B) subdural space.
(C) epidural space.
(D) subpial space.
(E) dural sinuses.

24. Which of the following locations contains preganglionic neurons of the parasympathetic nervous system?

(A) Cervical and sacral spinal cord
(B) Cervical and thoracic spinal cord
(C) Brainstem and cervical spinal cord
(D) Thoracic and lumbar spinal cord
(E) Brainstem and sacral spinal cord

25. Nerve fibers that innervate the anterior two-thirds of the tongue for general sensation originate from cranial nerve

(A) V.
(B) VII.
(C) IX.
(D) X.
(E) XII.

26. The pituitary gland lies in the sella turcica, immediately posterior and superior to the

(A) frontal sinus.
(B) maxillary sinus.
(C) ethmoid cells.
(D) mastoid cells.
(E) sphenoidal sinus.

27. Each statement below concerning the carotid sheath is true EXCEPT

(A) it is partially reinforced by cervical visceral fascia.
(B) it contains the vagus nerve.
(C) it continues into the mediastinum as adventitia of the great vessels.
(D) it encloses a portion of the ansa cervicalis.
(E) it contains the sympathetic trunk.

28. The sella turcica is part of the

(A) frontal bone.
(B) ethmoid bone.
(C) temporal bone.
(D) basioccipital bone.
(E) sphenoid bone.

29. Which of the following statements concerning accommodation of the eye is true?

(A) The ciliary muscle contracts to make the lens thinner.
(B) The ciliary muscle contracts when its parasympathetic fibers are stimulated.
(C) During accommodation for objects close to the eye, the lens becomes thinner.
(D) Accommodation for objects close to the eye is mediated by sympathetic nerve action.
(E) During accommodation, the lens does not change shape; it moves forward or backward.

30. A horizontal cut through the neck that severs the inferior thyroid arteries will also sever all of the structures below EXCEPT the

(A) recurrent laryngeal nerves.
(B) external carotid arteries.
(C) inferior thyroid veins.
(D) vagus nerves.
(E) trachea.

31. The dural venous sinus nearest the pituitary gland is the

(A) straight sinus.
(B) cavernous sinus.
(C) superior petrosal sinus.
(D) sigmoid sinus.
(E) confluence of sinuses.

32. Which of the following statements concerning the pterygopalatine fossa is true?

(A) It is located in the petrous portion of the temporal bone.
(B) It contains the otic ganglion.
(C) It communicates with the middle cranial cavity through the foramen rotundum.
(D) It is located directly inferior to the maxillary sinus.
(E) It communicates with the infratemporal fossa through the sphenopalatine foramen.

33. Each statement below concerning the tensor tympani muscle is true EXCEPT

(A) it inserts on the handle of the malleus and functions to tighten the tympanic membrane.
(B) it is innervated by a branch of the facial nerve.
(C) it runs parallel to the auditory tube.
(D) it arises chiefly from the cartilaginous portion of the auditory tube.
(E) it is innervated by the mandibular branch of the trigeminal nerve.

34. Which of the following muscles causes abduction of the vocal cords during quiet breathing?

(A) Vocalis muscles
(B) Cricothyroid muscles
(C) Oblique arytenoid muscles
(D) Posterior cricoarytenoid muscles
(E) Thyroarytenoid muscles

35. Each of the following conditions can result from severance of the oculomotor nerve EXCEPT

(A) partial ptosis.
(B) adduction of the eyeball.
(C) a dilated pupil.
(D) impaired lacrimal secretion.
(E) paralysis of the ciliary muscle.

36. Each nerve below supplies striated muscles and is of branchiomeric origin EXCEPT the

(A) vagus nerve.
(B) glossopharyngeal nerve.
(C) oculomotor nerve.
(D) facial nerve.
(E) trigeminal nerve.

37. Which of the following pairs of muscles is most instrumental in preventing food from entering the larynx and trachea during swallowing?

(A) Sternohyoid and sternothyroid muscles
(B) Oblique arytenoid and aryepiglottic muscles
(C) Inferior pharyngeal constrictor and thyrohyoid muscles
(D) Levator veli palatini and tensor veli palatini muscles
(E) Uvulae and geniohyoid muscles

38. Each of the following structures is located within the carotid sheath EXCEPT the

(A) vagus nerve.
(B) common carotid artery.
(C) internal jugular vein.
(D) sympathetic trunk.
(E) internal carotid artery.

39. All of the following muscles are innervated by the facial nerve EXCEPT the

(A) buccinator muscle.
(B) stylohyoid muscle.
(C) posterior belly of the digastric muscle.
(D) tensor tympani muscle.
(E) zygomaticus major muscle.

40. All of the following structures communicate with the middle meatus of the nasal cavity EXCEPT the

(A) maxillary sinus.
(B) frontal sinus.
(C) posterior ethmoidal air cells.
(D) middle ethmoidal air cells.
(E) anterior ethmoidal air cells.

41. Each of the following statements concerning the internal laryngeal nerve is true EXCEPT

(A) it is a branch of the superior laryngeal nerve.
(B) it is an indirect branch of the vagus nerve.
(C) it is sensory to the mucosa of the larynx.
(D) it is motor to the cricothyroid muscle.
(E) it is accompanied by the superior laryngeal artery.

42. Veins of the brain are direct tributaries of

(A) the emissary veins.
(B) the pterygoid venous plexus.
(C) the diploid veins.
(D) the dural sinuses.
(E) all of the above.

43. Which of the following conditions or actions results from stimulation of the parasympathetic fibers to the eyeball?

(A) Enhanced vision for distant objects
(B) Dilation of the pupil
(C) Contraction of capillaries in the iris
(D) Contraction of the ciliary muscle
(E) None of the above

44. Each structure below is embedded in the wall of the cavernous sinus along part of its course EXCEPT

(A) oculomotor nerves.
(B) abducens nerves.
(C) trochlear nerves.
(D) the mandibular division of the trigeminal nerve.
(E) ophthalmic nerves.

45. Which of the following statements concerning the tympanic nerve is true?

(A) It is a branch of the facial nerve.
(B) It is the deep petrosal nerve.
(C) It synapses with fibers of the lesser petrosal nerve.
(D) It is a branch of the glossopharyngeal nerve.
(E) It is a branch of the vagus nerve.

DIRECTIONS: Each group of items in this section consists of lettered options followed by a set of numbered items. For each item, select the **one** lettered option that is most closely associated with it. Each lettered option may be selected once, more than once, or not at all.

Questions 46 and 47

Match each structure below with the appropriate lettered structure in this radiograph of a lateral view of the head.

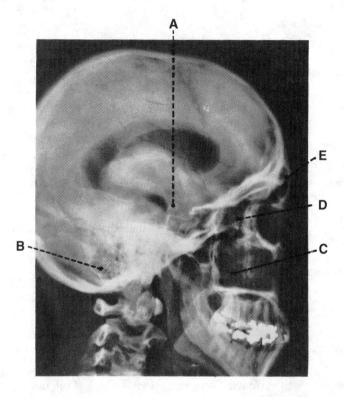

46. Maxillary sinus

47. Mastoid air cells

Questions 48–52

Match each description below with the most appropriate sinus.
(A) Cavernous sinus
(B) Sigmoid sinus
(C) Superior sagittal sinus
(D) Transverse sinus
(E) Straight sinus

48. Lies at the junction of the falx cerebri and the tentorium cerebelli

49. Curves laterally and forward in the convex outer border of the tentorium cerebelli

50. Lies in the superior convex border of the falx cerebri

51. Communicates directly with the ophthalmic veins

52. Becomes continuous with the internal jugular vein at the jugular foramen

Questions 53–57

Match each description below with the most appropriate nerve.
(A) Hypoglossal nerve
(B) Vagus nerve
(C) Chorda tympani nerve
(D) Lingual nerve
(E) Glossopharyngeal nerve

53. Provides motor innervation to the intrinsic muscles of the larynx

54. Carries general sensation from the anterior two-thirds of the tongue

55. Provides parasympathetic innervation for the parotid gland

56. Provides motor innervation to the intrinsic muscles of the tongue

57. Carries general somatic sensory information to the tympanic cavity (or the middle ear).

Questions 58–62

Match each statement below with the most appropriate fossa of the middle cranium.
(A) Foramen ovale
(B) Foramen spinosum
(C) Foramen rotundum
(D) Foramen lacerum
(E) Superior orbital fissure

58. Transmits the ophthalmic nerve

59. Is related to the internal carotid artery

60. Transmits the middle meningeal artery

61. Injury to the nerve passing through this structure causes a general loss of sensation of the maxillary teeth

62. Injury to the nerve passing through this structure causes a loss of sensation of the temporomandibular joints

Questions 63–67

Match each statement below with the most appropriate opening of the skull.
(A) Stylomastoid foramen
(B) Pterygoid canal
(C) Pterygomaxillary fissure
(D) Sphenopalatine foramen
(E) Jugular foramen

63. Transmits the nerve that innervates the muscles of facial expression

64. Traversed by sensory fibers from the posterior one-third of the tongue

65. Traversed by sensory fibers from the mucosa of the nasal septum, posterior lateral nasal wall, and anterior portion of the hard palate

66. Provides a passage between the infratemporal fossa and the pterygopalatine fossa for the terminal part of the maxillary artery

67. Traversed by preganglionic parasympathetic fibers to the lacrimal gland

DIRECTIONS: Each question below contains four suggested answers of which **one or more** is correct. Choose answer

> **A** if **1, 2, and 3** are correct
> **B** if **1 and 3** are correct
> **C** if **2 and 4** are correct
> **D** if **4** is correct
> **E** if **1, 2, 3, and 4** are correct

68. Which of the following cavities are separated from the middle cranial fossa by a thin layer of bone?
(1) Auditory tube
(2) Middle ear cavity
(3) Sigmoid sinus
(4) Sphenoid sinus

69. Which of the following statements concerning the palatine tonsil are true?
(1) It is bounded by the palatoglossal and palatopharyngeal folds.
(2) It receives sensory innervation from the glossopharyngeal nerve.
(3) It is supplied by branches of the facial artery.
(4) It is located on each side of the oropharynx.

70. Which of the following nerves pass through the jugular foramen?
(1) Vagus nerve
(2) Spinal accessory nerve
(3) Glossopharyngeal nerve
(4) Hypoglossal nerve

71. Which of the following statements concerning the orbicularis oculi muscle are true?
(1) It is supplied by temporal and zygomatic branches of the facial nerve.
(2) Its function is to close the eyelids.
(3) Its paralysis results in spilling of tears.
(4) It is branchiomeric in origin.

72. Which of the following nerves provide innervation for the dura of the middle cranial fossa?
(1) Vagus nerve
(2) Facial nerve
(3) Hypoglossal nerve
(4) Trigeminal nerve

73. Which of the following statements concerning the carotid sinus are true?
(1) It is located at the origin of the external carotid artery.
(2) It is innervated by the facial nerve.
(3) It functions as a chemoreceptor.
(4) It is stimulated by changes in blood pressure.

74. Which structures below pass through the optic canal?
(1) Optic nerve
(2) Ophthalmic vein
(3) Ophthalmic artery
(4) Ophthalmic nerve

75. Which of the following conditions result from severance of the greater petrosal nerve?
(1) Decreased lacrimal gland secretion
(2) Decreased diameter of the pupil
(3) Dryness in the nose and palate
(4) Decreased parotid gland secretion

76. Which of the following fibers are contained in the nerve of the pterygoid canal?
(1) Preganglionic sympathetic fibers
(2) Taste fibers from the palate
(3) Postganglionic parasympathetic fibers
(4) Postganglionic sympathetic fibers

77. The digastric muscle is innervated by the
(1) spinal accessory nerve.
(2) trigeminal nerve.
(3) ansa cervicalis.
(4) facial nerve.

78. Contraction of the tensor tympani muscle and the stapedius muscle prevents damage to the eardrum and middle ear ossicles. These muscles are innervated by the
(1) chorda tympani.
(2) trigeminal nerve.
(3) auditory nerve.
(4) facial nerve.

79. The pupillary light reflex can be eliminated by cutting the
(1) short ciliary nerves.
(2) oculomotor nerve.
(3) optic nerve.
(4) ciliary ganglion.

80. A patient's pupil remains small when room lighting is subdued. This condition may indicate damage to the
(1) trochlear nerve.
(2) superior cervical ganglion.
(3) oculomotor nerve.
(4) cervical sympathetic trunk.

81. Which of the following nerves can be damaged if the corneal surface is dry due to a lack of moistening fluid?
(1) Terminal portion of the lacrimal nerve
(2) Zygomatic branch of the maxillary nerve
(3) Greater petrosal nerve
(4) Lesser petrosal nerve

82. Which statements below concerning the frontal sinus are true?
(1) It extends into the parietal bone.
(2) It communicates with the superior nasal meatus.
(3) It receives sensory innervation from the maxillary nerve.
(4) It is supplied with blood by branches of the ophthalmic artery.

83. Which of the following statements concerning the tongue are true?
(1) It has intrinsic and extrinsic muscle attachments.
(2) It contains lymphoid tissue.
(3) It receives fibers from cranial nerves V, VII, IX, and XII.
(4) It receives blood from a branch of the internal carotid artery.

84. A patient can move his eyeball normally and see distant objects clearly but cannot focus on near objects. This condition may indicate damage to the
(1) ciliary ganglion.
(2) oculomotor nerve.
(3) short ciliary nerves.
(4) superior cervical ganglion.

85. Which of the following structures enter the orbit through the superior orbital fissure and the common tendinous ring?
(1) Frontal nerve
(2) Lacrimal nerve
(3) Trochlear nerve
(4) Abducens nerve

86. Which of the following nerves innervate the muscles attached to the styloid process?
(1) Facial nerve
(2) Hypoglossal nerve
(3) Glossopharyngeal nerve
(4) Vagus nerve

87. A functional impediment to the act of swallowing probably can result from damage to the
(1) vagus nerve.
(2) hypoglossal nerve.
(3) mandibular division of the trigeminal nerve.
(4) buccal branch of the facial nerve.

Answers and Explanations

1–B. When the dorsal scapular artery arises from the third part of the subclavian artery, it replaces a deep branch of the transverse cervical artery.

2–C. The costocervical trunk is a short trunk from the subclavian artery that arches to the neck of the first rib, where it divides into the deep cervical and supreme intercostal arteries. Consequently, it is not closely associated with the isthmus of the thyroid gland. A low tracheotomy is a surgical incision of the trachea through the neck, below the isthmus of the thyroid gland.

3–C. The external laryngeal nerve supplies the cricothyroid muscle. Damage to this nerve results in inability to tense the vocal cords. The rima glottidis is widened by the posterior cricoarytenoid muscle.

4–A. The phrenic nerve contains motor and sensory fibers but no cutaneous nerve fibers.

5–C. The sphenoid sinus opens into the sphenoethmoidal recess in the nasal cavity.

6–B. The internal laryngeal nerve accompanies the superior laryngeal artery.

7–D. The glossopharyngeal nerve supplies sensory innervation of the mucosa of the pharynx.

8–C. The supraclavicular nerve is a cutaneous branch of the cervical plexus and supplies skin over the pectoralis major and deltoid muscles.

9–D. The deep cervical artery is a branch of the costocervical trunk. The inferior thyroid, transverse cervical, and subscapular arteries are branches of the thyrocervical trunk. The ascending cervical artery may arise from the transverse cervical artery or the inferior thyroid artery.

10–E. The phrenic nerve descends on the superficial surface of the anterior scalene muscle and enters the thorax by passing between the subclavian artery and vein.

11–B. The glossopharyngeal nerve innervates the stylopharyngeus muscle. This muscle is a landmark for locating the glossopharyngeal nerve because the nerve curves posteriorly around the lateral margin of the stylopharyngeus muscle as the nerve enters the pharyngeal wall.

12–A. The external laryngeal branch of the superior laryngeal nerve supplies the cricothyroid and inferior pharyngeal constrictor muscles.

13–E. The afferent nerve fibers from the larynx are carried by the internal laryngeal branch of the superior laryngeal nerve and the recurrent laryngeal nerve.

14–E. The optic nerve and the ophthalmic artery pass through the optic canal.

15–D. Both the anterior and posterior ethmoidal nerves are branches of the nasociliary nerve and both send branches to the ethmoidal air cells.

16–C. Postganglionic parasympathetic fibers originating in the pterygopalatine ganglion (sphenopalatine) innervate glands in the palate.

17–B. The duct of the submandibular gland passes between the mylohyoid and hyoglossus muscles, where it is crossed laterally by the lingual nerve.

18–E. The vomer bone forms part of the nasal septum.

19–C. Bell's palsy (facial paralysis) can involve inflammation of the cornea leading to corneal ulceration, which probably is attributable to an absence of the blinking reflex due to paralysis of the muscles that close the eyelid.

20–B. The inferior sagittal sinus and the greater cerebral vein are direct tributaries to the straight sinus in the cranial cavity.

21–E. The abducens nerve (CN VI) innervates the lateral rectus muscle, which abducts the eyeball.

22–C. Bilateral severance of the vagus nerve (CN X) may result in death.

23–C. Rupture of the middle meningeal artery in the cranial cavity causes an epidural hemorrhage.

24–E. Preganglionic neurons of the parasympathetic nervous system are located in the brainstem (cranial outflow) and sacral spinal cord segments S2–S4 (sacral outflow).

25–A. The anterior two-thirds of the tongue is innervated by the lingual nerve of the mandibular division of the trigeminal nerve (CN V).

26–E. The pituitary gland lies in the hypophyseal fossa of the sella turcica of the sphenoid bone, which lies immediately posterior and superior to the sphenoid sinus and medial to the cavernous sinus. Its roof is formed by the diaphragma sellae.

27–E. The carotid sheath contains the vagus nerve, the common carotid artery, and the internal jugular vein. The sympathetic trunk lies posterior to the carotid sheath.

28–E. The sella turcica is part of the sphenoid bone.

29–B. When parasympathetic fibers are stimulated, the ciliary muscle contracts, making the lens of the eye thicker. Accommodation for objects close to the eye is mediated by parasympathetic nerve action; thus, the lens becomes thicker during this type of accommodation.

30–B. A horizontal cut through the neck will sever the inferior arteries and the common carotid arteries, but not the external carotid arteries.

31–B. The dural venous sinus nearest the pituitary gland is the cavernous sinus.

32–C. The pterygopalatine fossa lies between the pterygoid plates of the sphenoid and palatine bone below the apex of the orbit. It contains the pterygopalatine ganglion and communicates with the middle cranial cavity through the foramen rotundum and with the infratemporal fossa through the pterygomaxillary fissure.

33–B. The tensor tympani is innervated by the mandibular branch of the trigeminal nerve.

34–D. The posterior cricoarytenoid muscle is the only muscle that abducts the vocal cords during quiet breathing.

35–D. The oculomotor nerve carries parasympathetic fibers to the constrictor pupillae and the ciliary muscle. The secretomotor fibers for lacrimal secretion come through the pterygopalatine ganglion.

36–C. Nerves that supply the muscles of the eyeball and tongue are not of branchiomeric origin. Special visceral efferent (SVE) nerve fibers in the trigeminal, facial, glossopharyngeal, vagus, and accessory nerves innervate a variety of skeletal muscles in the head and neck, and are branchiomeric (non-somitic) in origin because they originate from the branchial arches.

37–B. The oblique arytenoid and aryepiglottic muscles can prevent food from entering the larynx and trachea during the process of swallowing. The cricopharyngeus fibers of the inferior pharyngeal constrictors act as a sphincter that prevents air from entering the esophagus.

38–D. The carotid sheath contains the vagus nerve, the common and internal carotid arteries, and the internal jugular vein.

39–D. The tensor tympani is innervated by the mandibular division of the trigeminal nerve.

40–C. The posterior ethmoidal air cells drain into the superior nasal meatus.

41–D. The internal laryngeal nerve is a branch of the superior laryngeal nerve and an indirect branch of the vagus nerve, and it is sensory to the mucosa of the larynx above the vocal cord. The external laryngeal nerve supplies the cricothyroid and inferior pharyngeal constrictor muscles.

42–D. Veins of the brain are direct tributaries of the dural venous sinuses. The emissary veins connect the dural venous sinuses with the veins of the scalp; whereas, the diploic veins lie in channels in the diploë of the skull and communicate with the dural sinuses, the veins of the scalp, and the meningeal veins. The pterygoid venous plexus communicates with the cavernous sinus through an emissary vein.

43–D. When the parasympathetic fibers to the eyeball are stimulated, the pupil constricts and the ciliary muscle contracts, resulting in a thicker lens and enhanced vision for near objects (accommodation). Contraction of capillaries in the iris and enhanced ability to see distant objects (flattening of the lens) result from stimulation of sympathetic nerves.

44–D. Oculomotor, abducens, trochlear, and ophthalmic nerves all lie in the wall of the cavernous sinus. The mandibular division of the trigeminal nerve does not lie in the wall of the cavernous sinus.

45–D. Preganglionic parasympathetic fibers run in the tympanic nerve, or Jacobson's nerve (a branch of the glossopharyngeal nerve), and then in the lesser petrosal nerve to reach the otic ganglion, where they synapse with postganglionic neurons.

46–C.

47–B.

48–E. The straight sinus runs along the line where the falx cerebri attaches to the tentorium cerebelli.

49–D. The transverse sinus runs laterally and forward in the convex outer border of the tentorium cerebelli.

50–C. The superior sagittal sinus lies in the superior convex border of the falx cerebri.

51–A. The cavernous sinus communicates directly with the ophthalmic veins.

52–B. The sigmoid sinus becomes continuous with the internal jugular vein at the jugular foramen.

53–B. The vagus nerve provides motor innervation to the intrinsic muscles of the larynx through the recurrent and external laryngeal nerves.

54–D. The lingual nerve carries general sensation from the anterior two-thirds of the tongue.

55–E. The glossopharyngeal nerve carries the preganglionic parasympathetic fibers that run in the tympanic and lesser petrosal nerves and synapse in the otic ganglion with cell bodies of postganglionic parasympathetic fibers, which run in the auriculotemporal nerve and innervate the parotid gland.

56–A. The hypoglossal nerve innervates all intrinsic muscles of the tongue and all extrinsic muscles of the tongue except for the palatoglossus muscle, which is innervated by the vagus nerve.

57–E. The glossopharyngeal nerve carries general somatic sensory information from the tympanic cavity, the authority tube, and the mastoid air cells.

58–E. The superior orbital fissure transmits the ophthalmic nerve.

59–D. The foramen lacerum transmits nothing; however, its superior portion is occupied by the internal carotid artery.

60–B. The foramen spinosum transmits the middle meningeal artery.

61–C. The foramen rotundum transmits the maxillary division of the trigeminal nerve. Injury to this nerve causes a loss of general sensation of the maxillary teeth.

62–A. The foramen ovale transmits the mandibular division of the trigeminal nerve. Injury to this nerve causes a loss of sensation of the temporomandibular joint.

63–A. The stylomastoid foramen transmits the facial nerve, which innervates the muscles of facial expression.

64–E. The jugular foramen transmits the glossopharyngeal nerve, which carries sensory fibers from the posterior one-third of the tongue.

65–D. The sphenopalatine foramen transmits the sphenopalatine nerve, which carries sensory fibers from the mucosa of the nasal septum, the posterior lateral nasal wall, and the anterior portion of the hard palate.

66–C. The pterygomaxillary fissure is the passageway between the infratemporal fossa and the pterygopalatine fossa for the terminal part of the maxillary artery.

67–B. The pterygoid canal transmits the nerve of the pterygoid canal, which contains preganglionic parasympathetic fibers to the lacrimal gland.

68–C. The middle ear cavity and the sphenoid sinus are separated from the middle cranial fossa by a thin layer of bone.

69–E. The palatine tonsil is located on each side of the oropharynx, between the palatoglossal and palatopharyngeal folds. It is innervated by the glossopharyngeal nerve and receives blood from branches of the facial, maxillary, and ascending pharyngeal arteries.

70–A. The jugular foramen transmits the glossopharyngeal, accessory, and vagus nerves and the internal jugular vein.

71–E. The orbicularis oculi muscle, derived from the second pharyngeal (or hyoid) arch, is innervated by the facial nerve and is active in eye closure. Its paralysis results in drooping of the lower eyelid and spilling of tears.

72–D. The cranial dura is innervated by the ophthalmic division of the trigeminal nerve in the anterior cranial fossa, the maxillary and mandibular divisions of the trigeminal nerve in the middle cranial fossa, and the vagus and hypoglossal nerves in the posterior cranial fossa.

73–D. The carotid sinus is a spindle-shaped dilatation of the origin of the internal carotid artery. It is a pressoreceptor that is stimulated by changes in blood pressure. The carotid sinus is innervated by the carotid sinus branch of the glossopharyngeal nerve and by a branch of the pharyngeal branch of the vagus nerve.

74–B. The optic canal transmits the optic nerve and ophthalmic artery. The ophthalmic nerve and vein pass through the superior orbital fissure.

75–B. The greater petrosal nerve transmits parasympathetic (preganglionic) fibers, which are secretomotor fibers, to the lacrimal glands and mucous glands in the nasal cavity and palate.

76–C. The nerve of the pterygoid canal (vidian nerve) contains taste fibers from the palate, postganglionic sympathetic fibers, and preganglionic parasympathetic fibers.

77–C. The digastric anterior belly is innervated by the trigeminal nerve; the digastric posterior belly is innervated by the facial nerve.

78–C. The tensor tympani muscle is innervated by the trigeminal nerve and the stapedius muscle is innervated by the facial nerve.

79–E. The efferent limbs of the reflex arc concerned in the pupillary light reflex (e.g., constriction of the pupil in response of illumination of the retina) are composed of parasympathetic preganglionic fibers in the oculomotor nerve and parasympathetic postganglionic fibers in the short ciliary nerves. The afferent limbs of this reflex are optic nerve fibers.

80–C. When the pupil remains small in a dimly lit room, it is an indication that sympathetic fibers that originate from the cervical sympathetic trunk and innervate the dilator pupillae (radial muscles of the iris) are damaged.

81–A. The secretomotor fibers to the lacrimal gland are parasympathetic fibers that run in the facial, greater petrosal, vidian, maxillary, zygomatic, and lacrimal nerves.

82–D. The frontal sinus lies in the frontal bone, communicates with the middle nasal sinus, and is innervated by the supraorbital branch of the ophthalmic nerve.

83–A. The dorsum of the posterior two-thirds of the tongue has nodular masses of lymphoid follicles, which are collectively called the lingual tonsil. The anterior two-thirds of the tongue receives general sensory innervation from the trigeminal nerve (CN V) through the lingual nerve and taste sensation from the facial nerve (CN VII) through the chorda tympani. The posterior one-third of the tongue receives general and taste sensation from the glossopharyngeal nerve (CN IX). Its root near the epiglottis receives taste sensation from the vagus nerve through the internal laryngeal nerve. Its intrinsic and extrinsic muscles receive motor innervation from the hypoglossal nerve (CN XII).

84–B. The patient cannot focus on close objects (accommodation) due to damage to the parasympathetic fibers in the ciliary ganglion or in the short ciliary nerves. The patient can see distant objects clearly because the long ciliary nerve carries sympathetic fibers to the dilator pupillae. The ability to move the eyeball normally indicates that the oculomotor, trochlear, and abducens nerves are intact.

85–D. The trochlear, lacrimal, and frontal nerves and the ophthalmic vein enter the orbit through the superior orbital fissure and outside the common tendinous ring. The abducens nerve enters the orbit through the superior orbital fissure and the common tendinous ring.

86–A. The stylohyoid muscle is innervated by the facial nerve; the styloglossus muscle by the hypoglossal nerve; and the stylopharyngeus muscle by the glossopharyngeal nerve.

87–B. Swallowing involves movements of the tongue to push the food into the oropharynx, elevation of the soft palate to close the entrance of the nasopharynx, elevation of the hyoid bone and the larynx to close the opening into the larynx, and contraction of the pharyngeal constrictors to move the food through the pharynx. The mandibular division of the trigeminal nerve supplies the suprahyoid muscles (e.g., the digastric anterior belly and mylohyoid muscles). The vagus nerve innervates the muscles of the palate, the larynx, and the pharynx. The hypoglossal nerve supplies all of the tongue muscles except the palatoglossus muscle, which is innervated by the vagus nerve.

Comprehensive Examination

DIRECTIONS: Each of the numbered items or incomplete statements in this section is followed by answers or by completions of the statement. Select the **one** lettered answer or completion that is **best** in each case.

1. Which of the following nerves accompanies the posterior humeral circumflex artery through the quadrangular space?
(A) Radial nerve
(B) Axillary nerve
(C) Thoracodorsal nerve
(D) Suprascapular nerve
(E) Accessory nerve

2. Which two structures of the brachial plexus would have to be cut to destroy the function of all abductors of the upper extremity?
(A) Middle trunk and posterior cord
(B) Middle trunk and lateral cord
(C) Lower trunk and lateral cord
(D) Upper trunk and posterior cord
(E) Lower trunk and medial cord

3. Which muscle below acts to flex the elbow but is not innervated by the musculocutaneous, median, or ulnar nerves?
(A) Biceps brachii muscle
(B) Brachioradialis muscle
(C) Brachialis muscle
(D) Flexor digitorum longus muscle
(E) Extensor digitorum longus muscle

4. Each of the following statements concerning the medial epicondyle of the humerus is correct EXCEPT
(A) it provides an attachment for many of the wrist flexors.
(B) it is more prominent than the lateral epicondyle.
(C) it is closer to the basilic vein than to the cephalic vein.
(D) it is grooved posteriorly by the ulnar nerve.
(E) it is the point at which the brachial artery divides into the ulnar and radial branches.

5. Which of the following conditions is caused by damage to the radial nerve in the spiral groove?
(A) Numbness over the medial side of the forearm
(B) Inability to oppose the thumb
(C) Weakness in pronating the forearm
(D) Weakness in abducting the arm
(E) Inability to extend the hand

6. Which of the following nerves innervates the muscle that inserts into the pisiform bone?
(A) Axillary nerve
(B) Radial nerve
(C) Musculocutaneous nerve
(D) Median nerve
(E) Ulnar nerve

7. Abduction of the fingers would be impaired most by paralysis of the
(A) ulnar nerve.
(B) median nerve.
(C) radial nerve.
(D) musculocutaneous nerve.
(E) ancillary nerve.

8. Each of the following statements characterizes the medial meniscus of the knee joint EXCEPT
(A) it is nearly circular.
(B) it is attached to the tibial collateral ligament.
(C) it is larger than the lateral meniscus.
(D) it lies outside the synovial cavity.
(E) it is more frequently torn in injuries than the lateral meniscus.

9. Which of the following actions is most seriously affected by paralysis of the deep peroneal nerve?
(A) Plantar flexion of the foot
(B) Dorsiflexion of the foot
(C) Abduction of the toes
(D) Eversion of the foot
(E) Adduction of the toes

10. The first vascular channel likely to be obstructed or occluded by an embolus from the deep veins of a lower limb would be a
(A) branch of a renal vein.
(B) branch of a coronary artery.
(C) sinusoid of the liver.
(D) branch of one of the pulmonary veins.
(E) branch of one of the pulmonary arteries.

11. A patient presents with flat feet. The foot is displaced laterally and everted, and the head of the talus is no longer supported. Which of the following ligaments probably is stretched?

(A) Plantar calcaneonavicular
(B) Calcaneofibular
(C) Anterior tibiofibular
(D) Lateral talocalcaneal
(E) Anterior tibiotalar

12. Which of the following ligaments is important in preventing forward displacement of the femur on the tibia when the weight-bearing knee is flexed?

(A) Anterior meniscofemoral ligament
(B) Fibular collateral ligament
(C) Oblique popliteal ligament
(D) Posterior cruciate ligament
(E) Anterior cruciate ligament

13. The artery that runs with the great cardiac vein in the anterior interventricular sulcus of the heart branches directly from the

(A) circumflex branch of the left coronary artery.
(B) marginal branch of the right coronary artery.
(C) left coronary artery.
(D) right coronary artery.
(E) marginal branch of the left coronary artery.

14. The coronary arteries, which supply blood to the heart, branch directly from the

(A) pulmonary trunk.
(B) ascending aorta.
(C) right ventricle.
(D) descending aorta.
(E) right atrium.

15. Which set of conditions below describes the respective positions of the pulmonary valve, the aortic valve, and both atrioventricular valves during ventricular systole?

(A) Open, open, closed
(B) Closed, closed, open
(C) Closed, open, closed
(D) Open, closed, open
(E) Open, closed, closed

16. The mediastinum contains each of the following structures EXCEPT the

(A) thymus gland (or remnant).
(B) esophagus.
(C) trachea.
(D) lungs.
(E) heart.

17. All of the following statements concerning the ribs are correct EXCEPT

(A) the intercostal nerves, arteries, and veins are associated with the costal grooves.
(B) the tubercles articulate with the spinous processes of the vertebrae.
(C) the costal cartilages attach the upper 10 pairs of ribs to the sternum.
(D) the upper 7 pairs of ribs are called true ribs.
(E) the lower 2 pairs of ribs are called floating ribs.

18. Each statement below concerning the respiratory system is correct EXCEPT

(A) the left lung has a lingula.
(B) cartilaginous rings are found in the main bronchi.
(C) the left lung has a smaller volume than the right lung.
(D) the main bronchus branches into the lobar bronchi at the carina.
(E) the right lung usually receives a single bronchial artery.

19. Which of the following structures receives its blood supply from both the celiac and superior mesenteric arteries?

(A) Liver
(B) Spleen
(C) Pancreas
(D) Ileum
(E) Descending colon

20. All of the following veins are part of the portal venous system EXCEPT the

(A) left colic vein.
(B) splenic vein.
(C) superior rectal vein.
(D) appendicular vein.
(E) hepatic vein.

21. An indirect inguinal hernia occurs

(A) lateral to the inferior epigastric artery.
(B) between the inferior epigastric artery and the obliterated umbilical artery.
(C) medial to the obliterated umbilical artery.
(D) between the median umbilical fold and the obliterated umbilical artery.
(E) between the median umbilical fold and the inferior epigastric artery.

22. Which of the following statements concerning the internal abdominal oblique muscle is correct?

(A) It forms the floor of the inguinal canal.
(B) Its aponeurosis contributes to the posterior wall of the inguinal canal.
(C) Its aponeurosis aids in the formation of the conjoint tendon.
(D) Its aponeurosis contributes to the formation of the posterior layer of the rectus sheath below the arcuate line.
(E) Its muscle fibers travel in the same general direction as those of the external abdominal oblique muscle.

23. Parasympathetic fibers in the distal portion of the inferior mesenteric plexus are branches of the

(A) vagus nerve.
(B) pelvic splanchnic nerves.
(C) sacral splanchnic nerves.
(D) lesser splanchnic nerves.
(E) greater splanchnic nerves.

24. Carcinoma of the uterus can spread directly to the labia majora through lymphatics that follow the

(A) ovarian ligament.
(B) suspensory ligament of the ovary.
(C) round ligament of the uterus.
(D) uterosacral ligaments.
(E) pubocervical ligaments.

25. The pelvic outlet is formed by all of the following structures EXCEPT the

(A) sacrotuberous ligament.
(B) inferior pubic ramus.
(C) pubic tubercle.
(D) ischial tuberosity.
(E) coccyx.

26. Each of the following statements contrasting male and female pelvic structures is true EXCEPT

(A) in females, the pelvic inlet is oval; in males, it is more heart-shaped.
(B) the depth of the entire pelvis is generally greater in females than in males.
(C) the angle formed by the inferior pubic rami is greater in females than in males.
(D) the sacrum is wider in females than in males.
(E) the lips of the pubis are more everted in males than in females.

27. Which of the following statements concerning the deep perineal space is correct?

(A) It is formed superiorly by the perineal membrane.
(B) It contains the superficial transverse perineal muscles.
(C) In males, it contains a segment of the dorsal nerve of the penis.
(D) It is formed inferiorly by Colles' fascia.
(E) In females, it contains the greater vestibular glands.

28. If the urethra tears distal to the urogenital diaphragm, urine could collect in the

(A) retropubic space.
(B) medial aspect of the thigh.
(C) ischiorectal fossa.
(D) superficial perineal space.
(E) paravesical fossa.

29. Each of the following statements concerning the scrotum is true EXCEPT

(A) it is homologous to the labia majora in females.
(B) most of its lymphatic drainage enters superficial inguinal lymph nodes.
(C) it is supplied with blood by the testicular artery.
(D) it has a dartos layer of fascia and muscle that is continuous with Colles' layer of superficial fascia in the perineum.
(E) it is innervated anteriorly by the ilioinguinal nerve.

30. Which of the following muscles attaches to the disk of the temporomandibular joint?

(A) Masseter muscle
(B) Temporalis muscle
(C) Medial pterygoid muscle
(D) Lateral pterygoid muscle
(E) Buccinator muscle

31. A patient with ptosis of the left eyelid probably has a damaged left

(A) trochlear nerve.
(B) abducens nerve.
(C) oculomotor nerve.
(D) ophthalmic nerve.
(E) facial nerve.

32. The superior petrosal sinus lies in the margin of the

(A) tentorium cerebelli.
(B) falx cerebri.
(C) falx cerebelli.
(D) diaphragma sellae.
(E) straight sinus.

33. Which of the following statements best explains the importance of arachnoid granulations?

(A) They allow cerebrospinal fluid to pass from the subarachnoid space into venous sinuses of the dura mater.
(B) They are a storage area for cerebrospinal fluid.
(C) They increase the surface area available for the production of cerebrospinal fluid.
(D) They provide a shunt that allows cerebrospinal fluid to return to the ventricles of the brain.
(E) They receive blood from the diploë of the skull.

34. The great cerebral vein drains into the

(A) superior sagittal sinus.
(B) inferior sagittal sinus.
(C) cavernous sinus.
(D) transverse sinus.
(E) straight sinus.

35. The palatine tonsil receives a branch from each of the following arteries EXCEPT the

(A) lesser palatine artery.
(B) facial artery.
(C) lingual artery.
(D) superior thyroid artery.
(E) ascending pharyngeal artery.

36. Which of the following muscles has a tendon that loops around the pterygoid hamulus?

(A) Tensor tympani
(B) Tensor veli palatini
(C) Levator veli palatini
(D) Superior pharyngeal constrictor
(E) Stylohyoid

37. The common carotid artery usually bifurcates at the level of the

(A) thyroid isthmus.
(B) cricoid cartilage.
(C) angle of the mandible.
(D) superior border of the thyroid cartilage.
(E) jugular notch.

38. Which of the following nerves is the main sensory supply for skin over the anterior triangle of the neck?

(A) Great auricular nerve
(B) Transverse cervical nerve
(C) Superior ramus of the ansa cervicalis
(D) Inferior ramus of the ansa cervicalis
(E) Superior laryngeal nerve

39. The pterygomandibular raphe is a major common origin for the

(A) superior and middle pharyngeal constrictor muscles.
(B) middle and inferior pharyngeal constrictor muscles.
(C) superior pharyngeal constrictor and buccinator muscles.
(D) medial and lateral pterygoid muscles.
(E) tensor veli palatini and levator veli palatini muscles.

40. The communication between the infratemporal fossa and the pterygopalatine fossa is the

(A) pharyngeal canal.
(B) pterygopalatine canal.
(C) sphenopalatine foramen.
(D) pterygomaxillary fissure.
(E) petrotympanic fissure.

41. Which group of locations below best describes the normal position of the dens of the axis in relation to the anterior arch of the atlas, the cruciform ligament, and the body of the axis?

(A) Anterior, posterior, superior
(B) Anterior, anterior, inferior
(C) Posterior, anterior, superior
(D) Anterior, posterior, inferior
(E) Posterior, posterior, superior

42. Which of the following muscles indents the submandibular gland and divides it into superficial and deep parts?

(A) Hyoglossus muscle
(B) Genioglossus muscle
(C) Styloglossus muscle
(D) Superior constrictor muscle
(E) Mylohyoid muscle

43. Which of the following structures is connected to the nasopharynx by the auditory tube?

(A) Vestibule of the inner ear
(B) Middle ear
(C) Semicircular canals
(D) External ear
(E) Inner ear

DIRECTIONS: Each group of items in this section consists of lettered options followed by a set of numbered items. For each item, select the **one** lettered option that is most closely associated with it. Each lettered option may be selected once, more than once, or not at all.

Questions 44–48

Match each statement below with the muscle it describes.
(A) Pectoralis major muscle
(B) Latissimus dorsi muscle
(C) Anterior serratus muscle
(D) Infraspinatus muscle
(E) Long head of triceps muscle

44. Forms the anterior axillary fold; functions to flex and adduct the arm

45. Arises from the scapula; innervated by branches from the radial nerve

46. Arises from the thoracodorsal fascia; this muscle and the teres major muscle form the posterior axillary fold

47. Forms the anterior wall of the axilla; innervated by the lateral and medial cords of the brachial plexus

48. Helps stabilize the glenohumeral joint; innervated by a branch from the suprascapular nerve

Questions 49–53

Match each statement below with the muscle it describes.
(A) Interossei muscle
(B) Lumbrical muscles
(C) Flexor digitorum profundus muscle
(D) Flexor digitorum superficialis muscle
(E) Extensor digitorum communis muscle

49. Flexes interphalangeal joints; innervated by the median and ulnar nerves

50. Originates from the radial side of tendons of another muscle

51. Flexes interphalangeal joints; innervated solely by the median nerve

52. Extends interphalangeal joints when metacarpophalangeal joints are flexed

53. Inserts into extensor expansion; adducts and abducts the fingers

Questions 54–58

Match each statement below with the nerve it describes.
(A) Femoral nerve
(B) Obturator nerve
(C) Pudendal nerve
(D) Superior gluteal nerve
(E) Sciatic nerve

54. Innervates a muscle that also receives innervation from the sciatic nerve

55. Enters the gluteal region through the greater sciatic foramen and exits this region at the inferior border of the gluteus maximus muscle

56. Enters the gluteal region through the greater sciatic foramen and exits this region through the lesser sciatic foramen in close proximity to the ischial spine

57. Innervates the tensor facia lata muscle

58. Innervates the gracilis muscle

Questions 59–63

Match each statement below with the artery it describes.
(A) Right coronary artery
(B) Superior intercostal artery
(C) Left coronary artery
(D) Internal thoracic artery
(E) Pulmonary artery

59. A branch of the costocervical trunk

60. The superior epigastric artery is a branch of this artery

61. Carries the major blood supply for the anterior portion of the interventricular septum

62. Anterior intercostal arteries are branches of this artery

63. Usually carries the major blood supply for the posterior portion of the interventricular septum

Questions 64–68

Match each statement below with the ligament it describes.

(A) Broad ligament
(B) Round ligament of the uterus
(C) Ovarian ligament
(D) Suspensory ligament of the ovary
(E) Cardinal ligament

64. Enters the deep inguinal ring

65. A double layer of mesentery that attaches to the lateral surface of the uterus

66. Homologous to the most superior portion of the gubernaculum in males

67. Extends from the ovary to the dorsolateral body wall

68. A uterine support that extends from the cervix and the lateral fornices of the vagina to the pelvic wall

Questions 69–73

Match each statement below with the nerve it describes.

(A) Hypoglossal nerve
(B) Recurrent laryngeal nerve
(C) Chorda tympani nerve
(D) Lingual nerve
(E) Glossopharyngeal nerve

69. Provides motor innervation to the intrinsic muscles of the larynx

70. Carries general sensation from the anterior two-thirds of the tongue

71. Carries special visceral sensation from the anterior two-thirds of the tongue

72. Provides motor innervation to the intrinsic muscles of the tongue

73. Carries sensation from pressure receptors in the carotid sinus

Questions 74–77

Match each description below with the appropriate lettered site or structure in this radiograph of the bones of the hand.

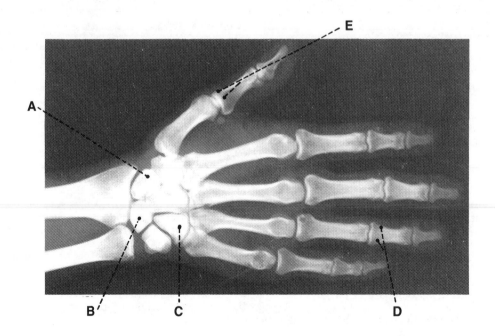

74. Lunate bone

75. Hamate bone

76. Site of attachment of the muscles that form the thenar eminence

77. Site of tendinous attachment of the flexor digitorum superficialis muscle

Questions 78–81

Match each description below with the appropriate lettered structure in this computed tomogram showing a sectional view of the abdomen.

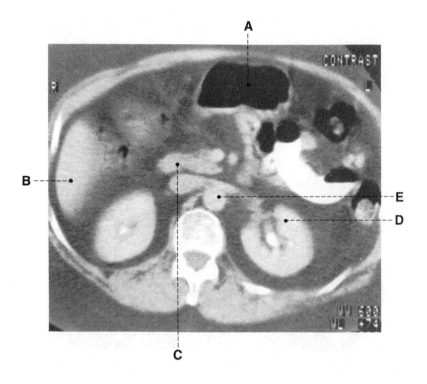

78. Pyloric portion of the stomach

79. Left kidney

80. Portal vein

81. Right lobe of the liver

Questions 82 and 83

Match each description below with the appropriate lettered structure in this radiograph showing a frontal view of the head.

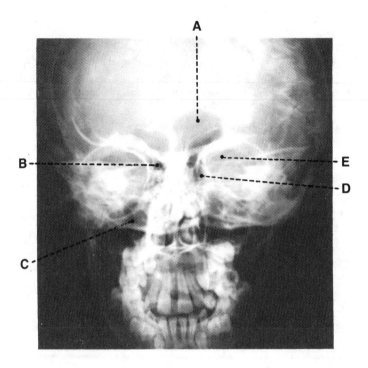

82. Frontal sinus

83. Superior orbital fissure

DIRECTIONS: Each question below contains four suggested answers of which **one or more** is correct. Choose answer

> **A** if **1, 2, and 3** are correct
> **B** if **1 and 3** are correct
> **C** if **2 and 4** are correct
> **D** if **4** is correct
> **E** if **1, 2, 3, and 4** are correct

84. When the hand is in a resting supine position, the radius is in articulation with which of the following bones at the radiocarpal joint?

(1) Triquetrum bone
(2) Lunate bone
(3) Trapezium bone
(4) Scaphoid bone

85. An injury to the thoracodorsal nerve probably would affect the strength of which of the following movements?

(1) Abduction of the arm
(2) Lateral rotation of the arm
(3) Adduction of the scapula
(4) Extension of the arm

86. In Duchenne-Erb palsy, the nerve fibers in the roots of C5 and C6 of the brachial plexus usually are damaged. Which of the following muscles are most likely to be paralyzed?

(1) Biceps brachii muscle
(2) Flexor carpi ulnaris muscle
(3) Brachioradialis muscle
(4) Adductor pollicis muscle

87. The cell bodies of nerve fibers in the lateral antebrachial cutaneous nerve are located in the

(1) collateral (prevertebral) ganglia.
(2) sympathetic chain (paravertebral) ganglia.
(3) lateral horn of the spinal cord.
(4) dorsal root ganglia.

88. Which of the following structures cross the wrist and enter the palm deep to the flexor retinaculum?

(1) Flexor pollicis longus tendon
(2) Ulnar nerve
(3) Median nerve
(4) Superficial palmar branch of the radial artery

89. Which of the following statements concerning the function of interosseous muscles are true?

(1) Dorsal interossei muscles adduct; palmar interossei muscles abduct.
(2) Dorsal interossei muscles abduct; palmar interossei muscles adduct.
(3) Palmar interossei muscles are associated with digits one, two, and four.
(4) Dorsal interossei muscles are associated with digits two, three, and four.

90. Which of the following statements concerning the anterior cruciate ligament of the knee joint are true?

(1) It becomes taut during extension of the leg.
(2) It helps prevent hyperflexion of the knee joint.
(3) It resists posterior displacement of the femur on the tibia.
(4) It resists anterior displacement of the femur on the tibia.

91. Which of the following bones are associated with the medial longitudinal arch of the foot?

(1) Talus bone
(2) Medial three metatarsal bones
(3) Navicular bone
(4) Calcaneus bone

92. Which of the following arteries usually supply blood to the hip joint?

(1) Lateral femoral circumflex artery
(2) Obturator artery
(3) Medial femoral circumflex artery
(4) Inferior gluteal artery

93. Cell bodies of nerve fibers in the ventral root of a thoracic spinal nerve are located in the

(1) dorsal root ganglia.
(2) lateral horn of the gray matter of the spinal cord.
(3) sympathetic chain ganglia.
(4) ventral horn of the gray matter of the spinal cord.

94. Which of the following parts of the digestive tract receive parasympathetic innervation from the vagus nerve?

(1) Duodenum
(2) Sigmoid colon
(3) Transverse colon
(4) Descending colon

95. A slowly growing tumor in the head of the pancreas can compress the

(1) duodenojejunal junction.
(2) bile duct.
(3) inferior mesenteric artery.
(4) descending part of the duodenum.

96. Which of the following structures form part of the anterior wall of the inguinal canal?

(1) Transversalis fascia
(2) Aponeurosis of the internal abdominal oblique muscle
(3) Conjoint tendon
(4) Aponeurosis of the external abdominal oblique muscle

97. Which of the following statements concerning the porta hepatis of the liver are true?

(1) It is a transverse fissure where the hepatic portal vein enters the liver.
(2) It lies between the caudate and quadrate lobes.
(3) It is the area where the right and left hepatic ducts exit the liver.
(4) It is the separation point between the fissure for the ligamentum teres hepatis and the fissure for the ligamentum venosum.

98. The inferior hypogastric plexus contains nerve fibers from the

(1) lumbar sympathetic ganglia.
(2) pelvic splanchnic nerves.
(3) sacral sympathetic ganglia.
(4) vagus nerve.

99. Which of the following muscles provide the action involved in closing the jaws?

(1) Medial pterygoid muscle
(2) Masseter muscle
(3) Temporalis muscle
(4) Lateral pterygoid muscle

100. Which of the following nerve structures may be damaged if the nasal cavity is chronically dry due to a lack of glandular secretions?

(1) Pterygopalatine ganglion
(2) Nerve of the pterygoid canal
(3) Facial nerve
(4) Greater petrosal nerve

101. Which of the following structures are found in the right atrium of the heart?

(1) Fossa ovalis
(2) Septomarginal trabecula
(3) Pectinate muscles
(4) Papillary muscles

102. Which of the following structures articulate with the manubrium of the sternum?

(1) The body of the sternum
(2) The first rib
(3) The second rib
(4) The third rib

103. Which of the following arteries branch directly from the left coronary artery?

(1) Nodal artery
(2) Circumflex artery
(3) Marginal artery
(4) Anterior interventricular artery

104. Which of the following structures contain general visceral afferent (GVA) fibers?

(1) Sympathetic chain
(2) Dorsal root
(3) Greater splanchnic nerve
(4) Gray rami communicantes

Answers and Explanations

1–B. The axillary nerve runs posteriorly, accompanying the posterior humeral circumflex artery through the quadrangular space, and supplies the teres minor and deltoid muscles.

2–D. The abductors of the arm are the deltoid and supraspinatus muscles. The deltoid is supplied by the axillary nerve, which arises from the posterior cord of the brachial plexus; the supraspinatus is innervated by the suprascapular nerve, which arises from the upper trunk of the brachial plexus.

3–B. The brachioradialis is innervated by the radial nerve and functions to flex the elbow.

4–E. The brachial artery divides into the radial and ulnar arteries at the level of the radial neck, about 1 cm before the bend of the elbow in the cubital fossa.

5–E. The radial nerve innervates the extensor muscles of the hand. The skin on the medial side of the forearm is innervated by the medial antebrachial nerve. The opponens pollicis and the pronator teres and quadratus muscles are innervated by the median nerve. The abductor of the arm (the deltoid muscle) and the teres minor muscle are innervated by the axillary nerve.

6–E. The flexor carpi ulnaris inserts on the pisiform and is innervated by the ulnar nerve.

7–A. The ulnar nerve innervates the dorsal interossei, which are the abductors of the fingers.

8–A. The medial meniscus is C-shaped or forms a semicircle.

9–B. The deep peroneal nerve innervates the dorsiflexors of the foot, which include the tibialis anterior, extensor hallucis longus, extensor digitorum longus, and peroneus tertius muscles.

10–E. An embolus from the deep veins of the lower limb would travel through the femoral vein, the external and common iliac veins, the inferior vena cava, the right atrium, the right ventricle, the pulmonary trunk, and into the pulmonary arteries, where it could obstruct and occlude these vessels.

11–A. Flat foot is characterized by disappearance of the medial portion of the longitudinal arch, which appears completely flattened. The plantar calcaneonavicular (spring) ligament supports the head of the talus and the medial side of the longitudinal arch.

12–D. The posterior cruciate ligament prevents forward displacement of the femur on the tibia when the knee is flexed; the anterior cruciate ligament prevents backward dislocation of the femur on the tibia when the knee is extended.

13–C. The great cardiac vein is accompanied by the anterior interventricular artery, which is a branch of the left coronary artery.

14–B. The right and left coronary arteries branch from the ascending aorta.

15–A. During ventricular systole (contraction of both ventricles), the pulmonary valve is open, the aortic valve is open, and both atrioventricular valves are closed.

16–D. The mediastinum contains the heart, trachea, esophagus, and the thymus gland; it does not contain the lungs.

17–B. The tubercles of the ribs articulate with the transverse processes of the vertebrae.

18–D. The carina is the point where the trachea divides into the right and left main bronchi. The main bronchi contain cartilaginous rings. The right lung usually receives one bronchial artery, and the left lung receives two bronchial arteries.

19–C. The pancreas receives blood from the superior pancreaticoduodenal branch of the gastroduodenal artery and from the dorsal pancreatic and pancreatic branches of the splenic artery, which arise from the celiac artery. The pancreas also receives blood from the inferior pancreaticoduodenal artery, which branches from the superior mesenteric artery.

20–E. The hepatic veins are systemic veins that drain hepatic blood into the inferior vena cava.

21–A. An indirect inguinal hernia occurs lateral to the inferior epigastric vessels. A direct inguinal hernia arises medial to these vessels.

22–C. The conjoint tendon is formed by the aponeuroses of the internal abdominal oblique and transverse abdominal muscles. The internal abdominal oblique muscle contributes to the formation of the anterior wall and the roof of the inguinal canal and to the cremaster fascia. The posterior wall of the inguinal canal is formed by the aponeurosis of the transverse abdominal muscle and the transversalis fascia, and the floor is formed by the inguinal and lacunar ligaments. The external abdominal oblique muscle travels inferiorly and medially. The internal abdominal oblique muscle runs superiorly and medially.

23–B. The pelvic splanchnic nerves contain parasympathetic preganglionic fibers, which join the inferior mesenteric plexus to supply the descending and sigmoid colons. The vagus nerve provides parasympathetic fibers up to the transverse colon. The greater, lesser, lowest, lumbar, and sacral splanchnic nerves contain sympathetic preganglionic fibers.

24–C. The round ligament of the uterus extends from the uterus, enters the inguinal canal at the deep inguinal ring, emerges from the superficial inguinal ring, and merges with the subcutaneous tissue of the labium majus. Carcinoma of the uterus can spread directly to the labium majus through the lymphatics that follow the ligament.

25–C. The pelvic outlet (lower pelvic aperture) is bounded posteriorly by the sacrum and coccyx; laterally by the ischial tuberosities and sacrotuberous ligaments; and anteriorly by the pubic symphysis, the arcuate ligament, and the rami of the pubis and ischium.

26–E. Due to its everted ischial tuberosities, the female pelvic outlet is larger than the male pelvic outlet.

27–C. The deep perineal space is the space between the superior and inferior fasciae of the urogenital diaphragm. The superficial transverse perineal muscles and the greater vestibular glands are found in the superficial perineal space.

28–D. Extravasated urine can pass into the superficial perineal space. The urine could spread inferiorly into the scrotum, anteriorly around the penis, and superiorly into the abdominal wall, but it could not spread into the thigh because the superficial fascia of the perineum is firmly attached laterally to the ischiopubic rami and connected with the deep fascia of the thigh (the fascia lata).

29–C. The scrotum receives blood from the posterior scrotal branch of the internal pudendal artery and the external pudendal artery.

30–D. The lateral pterygoid muscle is inserted on the articular disk and capsule of the temporomandibular joint.

31–C. Damage to the oculomotor nerve results in ptosis (drooping) of the eyelid because the levator palpebrae superioris is innervated by the oculomotor nerve. The facial nerve innervates the orbicularis oculi, which functions to close the eyelids.

32–A. The superior petrosal sinus lies in the margin of the tentorium cerebelli.

33–A. Arachnoid granulations are tuft-like collections of highly folded arachnoid that project into the superior sagittal sinus and other dural sinuses. They absorb the cerebrospinal fluid into dural sinuses and often produce erosion or pitting of the inner surface of the calvaria.

34–E. The great cerebral vein (vein of Galen) and the inferior sagittal sinus form the straight sinus.

35–D. The palatine tonsil receives blood from the descending palatine branch of the maxillary artery, a palatine branch of the ascending pharyngeal artery, and the dorsal lingual branches of the lingual artery.

36–B. The tendon of the tensor veli palatini muscle curves around the pterygoid hamulus.

37–D. The common carotid artery usually bifurcates into the external and internal carotid arteries at the level of the superior border of the thyroid cartilage.

38–B. The transverse cervical nerve supplies the skin over the anterior cervical triangle; the great auricular nerve supplies the skin behind the auricle and over the parotid gland. The ansa cervicalis supplies the infrahyoid muscles, including the sternohyoid, sternothyroid, and omohyoid muscles.

39–C. The pterygomandibular raphe serves as a common origin for the superior pharyngeal constrictor and buccinator muscles.

40–D. The pterygopalatine fossa communicates laterally with the infratemporal fossa by way of the pterygomaxillary fissure, medially with the nasal cavity through the sphenopalatine foramen, posteriorly with the foramen lacerum through the pterygoid canal, superiorly with the skull through the foramen rotundum, and anteriorly with the orbit through the inferior orbital fissure. The petrotympanic fissure transmits the chorda tympani.

41–C. The dens (odontoid process) of the axis is located posterior to the anterior arch of the atlas, anterior to the cruciform ligament, and superior to the body of the axis.

42–E. The submandibular gland is indented by and divided into superficial and deep parts by the mylohyoid muscle.

43–B. The auditory (eustachian) tube connects the nasopharynx with the middle ear cavity.

44–A. The pectoralis major muscle adducts and medially rotates the arm. The clavicular part rotates the arm medially and flexes it; the sternocostal part depresses the arm and shoulder. The lateral border of the pectoralis major muscle forms the anterior axillary fold.

45–E. The long head of the triceps brachii muscle originates from the infraglenoid tubercle of the scapula and is innervated by branches from the radial nerve.

46–B. The latissimus dorsi muscle arises from the lumbodorsal fascia and, with the teres major muscle, forms the posterior axillary fold.

47–A. The pectoralis major muscle is innervated by the lateral and medial pectoral nerves and forms the anterior wall of the axilla.

48–D. The tendon of the infraspinatus muscle forms the rotator (musculotendinous) cuff and, thus, helps to stabilize the glenohumeral joint. It is innervated by a branch from the suprascapular nerve.

49–C. The flexor digitorum profundus can flex the distal interphalangeal joints. It is innervated by the median and ulnar nerves.

50–B. The lumbricals arise from the radial side of the tendon of the flexor digitorum profundus. They are innervated by the median and ulnar nerves.

51–D. The flexor digitorum superficialis flexes the proximal interphalangeal joints. It is innervated solely by the median nerve.

52–E. The extensor digitorum communis extends the proximal and distal interphalangeal joints when the metacarpophalangeal joints are flexed by the interossei and the lumbricals.

53–A. The dorsal and palmar interossei muscles insert into the extensor expansion and are innervated by the ulnar nerve. The dorsal interossei abduct the finger; the palmar interossei adduct the finger.

54–B. The adductor magnus muscle is innervated by the obturator and sciatic nerves.

55–E. The sciatic nerve enters the gluteal region through the greater sciatic foramen, has no branches in the gluteal region, and exits this region at the inferior border of the gluteus maximus muscle.

56–C. The pudendal nerve enters the gluteal region through the greater sciatic foramen and exits this region through the lesser sciatic foramen in close proximity to the ischial spine.

57–D. The superior gluteal nerve innervates the gluteus medius, gluteus minimus, and tensor fascia lata muscles.

58–B. The obturator nerve innervates the medial muscles of the thigh.

59–B. The superior intercostal and deep cervical arteries branch from the costocervical trunk.

60–D. The internal thoracic artery gives off the anterior intercostal arteries and then terminates at the sixth intercostal space by dividing into the superior epigastric and musculophrenic arteries.

61–C. The anterior interventricular artery branches from the left coronary artery and supplies the anterior portion of the interventricular septum.

62–D. The anterior intercostal arteries are branches of the internal thoracic artery.

63–A. The right coronary artery gives off the posterior interventricular artery, which provides the major blood supply of the posterior portion of the interventricular septum.

64–B. The round ligament of the uterus enters the deep inguinal ring, runs through the inguinal canal, emerges from the superficial inguinal ring, and becomes lost in the labium majus.

65–A. The broad ligament is a double layer of mesentery that attaches to the lateral surface of the uterus.

66–C. The ovarian (proper) ligament is homologous to the most superior portion of the gubernaculum in males.

67–D. The suspensory ligament of the ovary extends from the ovary to the dorsolateral body wall and is composed of the connective tissue around the ovarian vessels.

68–E. The cardinal (lateral cervical) ligament is an important uterine support. It is composed of fibromuscular condensations of pelvic fascia from the cervix and the lateral fornices of the vagina that extend to the pelvic wall.

69–B. The recurrent laryngeal nerve provides motor innervation to the intrinsic muscles of the larynx.

70–D. The lingual nerve carries general sensation from the anterior two-thirds of the tongue.

71–C. The chorda tympani nerve carries special visceral sensation from the anterior two-thirds of the tongue.

72–A. The hypoglossal nerve provides motor innervation to the intrinsic muscles of the tongue.

73–E. The glossopharyngeal nerve carries sensation from pressure receptors in the carotid sinus.

74–B. The lunate bone.

75–C. The hamate bone.

76–E. The site of attachment for the adductor pollicis brevis and the flexor pollicis brevis, which, along with the opponens pollicis, form the thenar eminence

77–D. The site of attachment for the flexor digitorum superficialis muscle.

78–A. The pyloric portion of the stomach.

79–D. The left kidney.

80–C. The portal vein.

81–B. The right lobe of the liver.

82–A. The frontal sinus.

83–E. The superior orbital fissure.

84–C. The radius and the articular disk articulate with the scaphoid, lunate, and triquetral bones at the radiocarpal (wrist) joint; however, the triquetral bone does not articulate with the radius, but instead articulates with the articular disk on the head of the ulna.

85–D. The thoracodorsal nerve innervates the latissimus dorsi muscle, which adducts, extends, and rotates the arm medially.

86–B. In Duchenne-Erb palsy (or upper trunk injury), the nerve fibers in the roots of C5 and C6 of the brachial plexus (roots of anterior primary rami of cervical nerves 5 and 6) are damaged. The biceps brachii, which is innervated by the musculocutaneous nerve (C5–C7), and the brachioradialis, which is innervated by the radial nerve (C5–T1), usually are paralyzed. The flexor carpi ulnaris and adductor pollicis muscles are not paralyzed because they are innervated by the ulnar nerve, which is formed by the roots of C8 and T1.

87–C. The lateral antebrachial cutaneous nerve contains general somatic afferent fibers, which have cell bodies located in the dorsal root ganglia, and sympathetic postganglionic (general visceral efferent) fibers, which have cell bodies located in the sympathetic chain ganglia.

88–B. Structures entering the palm deep to the flexor retinaculum include the median nerve and the tendons of the flexor pollicis longus, flexor digitorum profundus, and flexor digitorum superficialis muscles.

89–C. The dorsal interossei muscles abduct the fingers, and they are attached to the second, third, and fourth digits. The palmar interossei muscles are attached to the second, fourth and fifth digits, and they adduct these digits. The thumb is the first digit.

90–B. The anterior cruciate ligament of the knee joint prevents posterior displacement of the femur on the tibia and limits hyperextension of the knee joint. This ligament is lax when the knee is flexed and becomes taut when the knee is extended.

91–E. The medial longitudinal arch of the foot is formed by the talus, calcaneus, navicular, cuneiforms, and medial three metatarsal bones; whereas, the lateral longitudinal arch is formed by the calcaneus, cuboid, and lateral two metatarsal bones.

92–E. The hip joint receives blood from branches of the medial and lateral femoral circumflex, superior and inferior gluteal, and obturator arteries.

93–C. The ventral root of a thoracic spinal nerve contains sympathetic preganglionic fibers, which have cell bodies located in the lateral horn of the gray matter of the spinal cord, and general somatic efferent fibers, which have cell bodies located in the ventral horn of the gray matter of the spinal cord.

94–B. The vagus nerve supplies parasympathetic fibers to the thoracic and abdominal viscera, including the duodenum and the transverse colon. The descending and sigmoid colons and other pelvic viscera are innervated by the pelvic splanchnic nerves.

95–C. The duodenum surrounds the head of the pancreas, and the common bile duct traverses the head of the pancreas. The inferior mesenteric vessels arise from the aorta behind the neck of the pancreas and descend across the uncinate process of the pancreas. The duodenojejunal junction is in contact with the inferior portion of the body of the pancreas.

96–C. The anterior wall of the inguinal canal is formed by the aponeuroses of the external and internal abdominal oblique muscles. The conjoint tendon is formed by the aponeuroses of the internal abdominal oblique and transverse abdominal muscles.

97–E. The porta hepatis of the liver is a transverse fissure where the hepatic portal vein enters the liver and the hepatic ducts leave the liver. It lies between the caudate and quadrate lobes and marks the separation point between the fissure for the ligamentum teres hepatis (round ligament of the liver), which lies to the left of the quadrate lobe, and the fissure for the ligamentum venosum, which lies to the left of the caudate lobe.

98–A. The inferior hypogastric (pelvic) plexus contains parasympathetic fibers from the lumbar and sacral sympathetic ganglia and pelvic splanchnic nerves.

99–A. The medial pterygoid, masseter, and temporalis muscles are involved in closing the jaws. The action of the lateral pterygoid muscles opens the jaws.

100–E. The parasympathetic secretomotor fibers for mucous glands in the nasal cavity run in the facial nerve, the greater petrosal nerve, the nerve of the pterygoid canal, and the pterygopalatine ganglion.

101–B. The right atrium contains the fossa ovalis and pectinate muscles. The right ventricle contains the septomarginal trabecula (moderate band) and papillary muscles.

102–A. The third rib articulates with the body of the sternum.

103–C. The nodal and marginal arteries are branches of the right coronary artery.

104–A. The gray rami communicantes contain sympathetic postganglionic (GVE) fibers, which have their cell bodies located in the chain ganglia.

Suggested Readings

Agur AMR. (1991) *Grant's Atlas of Anatomy*, Ninth Edition. Williams & Wilkins Company, Baltimore.

Hollinshead WH, Rosse C. (1985) *Textbook of Anatomy*, Fourth Edition. Harper & Row, Philadelphia.

Lachman E, Faulkner KK. (1981) *Case Studies in Anatomy*, Third Edition. Oxford University Press, New York.

Moore KL. (1985) *Clinically Oriented Anatomy*, Second Edition, Williams & Wilkins, Baltimore.

O'Rahilly R. (1986) *Gardner-Gray-O'Rahilly's Anatomy, A Regional Study of Human Stucture*, Fifth Edition. WB Saunders, Philadelphia.

Woodburne RT, Burkel WE. (1988) *Essentials of Human Anatomy*, Eighth Edition. Oxford University Press, New York.

Index

Note: Page numbers in italics denote illustrations, those followed by *t* denote tables, those followed by Q denote questions, and those followed by E denote explanations.